The Cosmic Spacetime

The Cosmic Spacetime

Fulvio Melia

CRC Press
Taylor & Francis Group
Boca Raton London New York

CRC Press is an imprint of the
Taylor & Francis Group, an **informa** business

First edition published 2021
by CRC Press
6000 Broken Sound Parkway NW, Suite 300, Boca Raton, FL 33487-2742

and by CRC Press
2 Park Square, Milton Park, Abingdon, Oxon, OX14 4RN

© 2021 Taylor & Francis Group, LLC

CRC Press is an imprint of Taylor & Francis Group, LLC

ISBN: 9780367532192 (hbk)
ISBN: 9781003081029 (ebk)

Typeset in LMRoman
by Nova Techset Private Limited, Bengaluru & Chennai, India

To my Sweiteh.

Contents

Preface

BIG or small, young or old, we all marvel at the splendor of our Universe. We are struck by its unfathomable extent and agonize over the mystery of its birth. In a sense, our ancestors have been cosmologists from the beginning of written history, probably even earlier. Each succeeding civilization forges its own framework for understanding our place in the cosmos, whether it be immanent, transcendent, or physicalist. At its very foundation, our communion with nature meanders back and forth as new ideas and discoveries rescind broken concepts and sculpt their replacements. Without doubt, however, the greatest revolution in this subject was inspired by the development of the general theory of relativity. Indeed, many trace the foundation of modern cosmology to the year 1917, when Einstein himself applied his description of space interlaced with time to the cosmic setting, shortly after completing his landmark work with the notable mathematicians David Hilbert and Emmy Noether. As it turned out, Einstein's initial ideas on this topic were incorrect, but his proposal established an important foundation for later thinking.

The steps that have brought us from that inauspicious beginning to where we are today tell one of the most remarkable stories in all of science. And many books have been written in the past century to chronicle various stages in this development, especially in recent decades when a cascade of new observational discoveries seemingly threatens to overwhelm our lagging theoretical understanding. But rarely has any of these works focused on the very backbone of modern cosmology—the cosmic spacetime. As we shall document in this book, the evidence we have today compels us to cease merely *using* the Friedmann-Lemaître-Robertson-Walker (FLRW) metric—one of the most famous solutions to Einstein's equations—and instead initiate a deeper probe into the nature of spacetime on the grandest scale.

The raison d'être for this book is to take what we have learned from decades of yeoman's work measuring, stretching, dissecting and collating the myriad pieces of data informing our view of the cosmos, much of this under the aegis of FLRW, and seek a more fundamental understanding of FLRW itself.

This excursion into the largest possible scales will delight our senses with the surprising insights it provides on local physics, such as the critical link we shall uncover between the Local Flatness Theorem in general relativity—a mathematically formal expression of the equivalence principle—and the symmetries embedded within the metric coefficients. We shall discover that the largest causally-connected region in existence gives physical meaning to the origin of rest-mass energy for some of our tiniest particles. And this is only the beginning of a deeper appreciation we hope to develop for FLRW and its ultimate relevance to the description of cosmic origin and evolution.

Today, observational cosmology is simply too large to fit within a compact monograph. This book discusses only a sample of the most intriguing, groundbreaking measurements and discoveries, notably those most directly relevant to FLRW. As its title suggests, it really is a book about the cosmic spacetime, which will be our focus from beginning to end.

The material covered within its pages is perhaps too technical for it to be viable as a popular science book, but not so technical as to preclude the possibility of introducing the principal players who have made the most enduring contributions to this field. Its level is such that anyone with a solid foundation in graduate physics and astronomy should be able to easily follow its contents. And young professionals, such as postdocs and early career individuals working in general relativity and cosmology, will find it a useful introductory resource.

This work would not have been possible without the career-long support I have received from the National Science Foundation, the National Aeronautics and Space Administration, and the Alfred P. Sloan Foundation, all of whom have generously sponsored my research. It is through the work they have funded that the subject of this book has evolved into one of my abiding interests.

I am also deeply grateful to the numerous students and colleagues who have participated in this research, without whom I could not possibly have completed even a small fraction of our published body of work. Finally, to Patricia, Marcus, Eliana, Adrian, and Ella and to my parents, whose guidance has been priceless, I extend my enduring love and gratitude.

Fulvio Melia
Tucson, Arizona
November, 2019

I

Space and Time

CHAPTER 1

Introduction

T O the extent permitted by current observational testing, we shall be exploring the cosmic spacetime framed in the language of Albert Einstein's (1879–1955) general theory of relativity. Our view of the Universe has undergone substantial revision over history's meandering timeline, often trammeled by false starts and disappointing endings, sometimes relying too heavily on recycled ideas for its spatial extent and temporal beginnings. What is unique about the present paradigm, however, is the strict mathematical formalism implied by the weaving of space and time into a single entity, subject to the constraining postulates and principles of Einstein's theory. As we chart a course through the landscape of ever improving measurement precision, general relativity leaves very little room for the type of wild speculation and unquantifiable musings featured in cosmological models prior to the twentieth century.

Modern cosmology is a comprehensive study of—dare we say it—everything on scales larger than our solar system. It concerns itself with the contents of the Universe, their physical attributes and dynamical implications. Moreover, using general relativity as its contextual basis, we are also compelled to consider cosmic history, thanks to the Birkhoff-Jebsen theorem (Chapter 7), which precludes any possibility of a steady-state existence. The Universe evolves and the need to address its beginnings is unavoidable. Though already implied by the writings of Aleksandr Friedmann (1888–1925), Georges Lemaître (1894–1966), Edwin Hubble (1889–1953) and other contemporaries at the dawn of modern cosmology (Chapter 4), we now know for sure that we must uncover how and when the cosmos began in order to fully understand it today. This issue was formalized as one of the most fundamental questions in science following the work of Penrose, Steven Hawking (1942–2018) and others in the latter part of the twentieth century, when they proved that a Universe strictly governed by the general theory of relativity cannot avoid a spacetime singularity [1, 2]. With the introduction of quantum mechanics, the physics of the very early Universe may avoid such incomprehensible anomalies, though one must still contend with inconceivably small timescales, $t \sim 10^{-43}$ sec, known as the Planck time. One of the most profound questions we face today is "What happened at $t < 10^{-43}$ sec?" Our exploration of the cosmic spacetime must therefore be fully engaged with both space *and* time.

Yet the concept of a 'beginning' was not completely avoided by our ancestors. Even some of the earliest attempts at understanding the heavens were already cosmogonies, not merely cosmologies, because several cultures in antiquity sought to make sense of how the Universe had begun, as well as what it actually is. According to thinkers in the great river cultures of ancient Egypt and Mesopotamia, existence emerged from a dynamic world, encompassing humans and gods alike [3]. In fact, they scripted the origin and evolution of the Universe in mythological terms, not scientifically, so their stories were also theogonies as much as they were cosmogonies.

This begs the question of how one should even view the meaning of time, for the obvious reason that a misinterpretation may lead to contradictory conclusions.[1] If $t = 0$ heralded the beginning of the Universe, then what was $t < 0$? This conundrum has occupied the minds of scientists and philosophers throughout history. Take St. Augustine (354–430), for example, who in the fifth century C.E. declared that God's creation of the world must have included the origin of time itself—a theme not unlike some of the explanations offered today, though the recent versions have more rigor and scientific clout. He argued that time did not exist before the Universe was created, writing "The world was made, not in time, but simultaneously with time" [4].

Actually, the debate concerning the nature of time and the beginning of the Universe goes back much farther in history, at least to the age of the Roman poet Titus Lucretius Carus (c. 99 B.C.E.–55 B.C.E.) and the Greek philosopher Aristotle (384 B.C.E–322 B.C.E.)—some of it also in the arguments of the Stoic philosopher Zeno of Citium (c. 336 B.C.E.–265 B.C.E.). In his well-studied text *De Rerum Natura* (On the Nature of Things), Lucretius pointed out that the shortness of human history is proof that the Universe is of finite age [5]. He asked: "If there was no origin of the heavens and Earth from generation, and if they existed from all eternity, how is it that other poets, before the time of the Theban War, and the destruction of Troy, have not also sung of other exploits of the inhabitants of Earth?" Lucretius believed that the whole Universe is quite modern and not very old. Several hundred years earlier, Zeno had made similar arguments against the eternity of the world based on the observed erosion on Earth's surface [6].

An opposing view, which is no longer tenable with the application of general relativity to cosmology, was Aristotle's position, which he expressed in his *De Caelo* (On the Heavens), that the cosmos is in steady state and eternal, while local changes are merely restricted to the sublunary world [7]. Aristotle's argument falls in line with the "why not sooner?" proponents in cosmology, who wonder what could have changed in an otherwise eternal timeline to create the Universe just once at $t = 0$ [8].

In the modern era, the steady state theory of the Universe was championed by Fred Hoyle (1915–2001), Hermann Bondi (1919–2005) and Thomas Gold (1920–2004) [9, 10], who proposed in the middle of the twentieth century that the Universe is eternal, both backwards and forwards, with a constant average density of matter in spite of an observationally confirmed continual expansion. Whether a persistent matter creation could sustain such a balance was the subject of heated debate. As we shall see

---

[1]We shall examine this issue more substantively in Section 2.2 and in the introduction to the general theory of relativity.

throughout this book, cosmological observations now completely obviate any possibility of such a steady-state Universe, but Hoyle, Bondi and Gold sustained the debate for many years. Quite famously, on 28 March 1949, during a BBC broadcast, Hoyle coined—what he meant pejoratively—the term, "Big Bang" for any cosmological theory that assumed its origin in an explosive event. He often expressed the viewpoint that a temporal beginning of the Universe could be equated to divine intervention and was therefore unscientific. In one of his most influential review articles, he wrote "The passionate frenzy with which the big-bang cosmology is clutched to the corporate scientific bosom evidently arises from a deep-rooted attachment to the first page of Genesis, religious fundamentalism at its strongest" [11].

The enthusiasm with which Hoyle, Bondi and Gold promoted their steady-state cosmology may be better appreciated given the historical precedence they derived from Einstein himself. The foundations of modern cosmology were laid in 1917, when Einstein proposed the first model of the Universe based on his newly minted general theory of relativity [12]. Though his world view was no doubt revolutionary, today we recognize many ideas and assumptions that he borrowed from earlier epochs. His Universe was uniform and spatially closed (i.e., it had positive spatial curvature; see Section 4.1). As such, it was finite, though without a boundary, consistent with the prevailing view that there was only a limited number of stars in existence. The key point, however, is that his Universe was static, which led to his famous introduction of a cosmological constant, Λ, into his field equations in order to prevent a cosmic collapse or expansion (Section 2.7). Einstein's Universe lacked a temporal dimension, so he did not concern himself with the question of origin. By the time Hoyle, Bondi and Gold entered the fray, the expansion of the Universe was well established, so their version of a steady-state cosmos had to include this new feature, which they justified on the basis of an eternal creation of matter in lieu of a single Big Bang event.

But as incomprehensible as the beginning of the Universe may seem from a scientific perspective, we have no choice but to address it. The observational and theoretical arguments in favour of a "Big Bang" are completely overwhelming, as we shall chronicle in the following chapters. Take the second law of thermodynamics, for example. Discovered in the 1850's, this unassailable pillar of physical theory implies that all natural processes should evolve towards equilibrium with uniform temperature. Extrapolating to the future, this law predicts that the Universe should attain a state of "heat death," in which all physical processes will eventually terminate. Reversing the extrapolation backwards, it is impossible to avoid the conclusion that the present state of the Universe requires an initiation of our physical laws at a finite time in the past. Some may recognize this argument more readily in terms of Rudolf Clausius' (1822–1888) definition of entropy—the incessant pursuit of a closed system to reach its most probable configuration. Viewing the Universe as such a self-contained environment, a physicist would argue that if it existed forever, it should have reached its maximum entropy long ago [13]. We still see order and structure, however, which could not exist in a maximum entropy state, so the Universe must have evolved to its current configuration over a finite time. It must have had a beginning, echoing the arguments of Lucretius and Zeno several thousand years ago.

1.1 HOT BIG BANG COSMOLOGY

We enter this story not long after Einstein proposed his now defunct steady-state model. In the 1920's, the Russian physicist Friedmann discovered that Einstein's field equations do not merely describe an eternal Universe, but also an entirely different class of dynamical models with a time-dependent size. Some of these cosmologies contract, others expand and assault our senses with their implied beginning at $t = 0$. Friedmann reported it this way [14]: "The time since the creation of the world is the time which has passed from the moment at which space was a point ($R = 0$) to the present state ($R = R_0$)." (He used R as a measure of distance.)

For the first time in history, the idea that the Universe had an origin emerged not from philosophical or religious doctrine but from fundamental physics. In Chapters 4 and 5 we shall trace the inspirational steps that eventually led to the acquisition of a 'cosmic spacetime,' with landmark contributions from Einstein and Friedmann, of course, but also from Willem de Sitter (1872–1934), Lemaître, Hubble, and George Gamow (1904–1968) (Figure 1.1).

Lemaître's seminal proposal in 1931 that the Universe began with an explosive event [15] was independently transformed into a quantitative theory of the early Universe, based primarily on nuclear physics. Lemaître's challenge, "une cosmogonie vraiment complète devrait expliquer les atomes comme les soleils" (meaning that a truly complete cosmogonie should be able to account for the observed solar abundances) [16] was first accepted by the future Nobel laureate Subrahmanyan Chandrasekhar (1910–1995) and his student Louis Henrich (1908–2004) [17], and later by the Russian-American physicist Gamow [18]. Supposing that the expanding cosmos began on a much smaller spatial scale, one could reasonably assume that matter densities and temperature increased inexorably towards the beginning, forming a hot cauldron of nuclear reactions and the creation of the elements. But as often happens with such exploratory ideas, the initial assumptions were too simple, and the earliest calculations assumed equilibrium processes at a fixed temperature of roughly 10^{10} K and matter densities of about 10^9 kg m^{-3}, resulting in several gross inconsistencies with the observed abundances. Chandrasekhar and Henrich concluded that some non-equilibrium process was required.

With his student Ralph Alpher (1921–2007), and Robert Herman (1914–1997) and Hans Bethe (1906–2005), Gamow later published several remarkable papers [19] in 1948 establishing the essential features of what we now call "Hot Big Bang Cosmology." The frantic pace at which these individuals published that year, in various collaborative combinations, somewhat clouds how the credit should be apportioned [20]. All would agree, however, that the inspiration for this work, and the eventual prescient prediction that the Universe today ought to be bathed in the glow of cooled microwave radiation left over from the initial hot inferno [21], began with Gamow.

The subsequent discovery of the cosmic microwave background radiation (CMB) elevated the Hot Big Bang Model to the paradigmatic stature it has enjoyed ever since. But ironically, the cosmic microwave radiation was not detected in a systematic search, motivated by sound theoretical argument. It would indeed have happened this way, by Robert Dicke's (1916–1997) group at Princeton, which was building an

Figure 1.1 George Gamow in the late 1960's, shortly before he died. Together with his student Ralph Alpher, Gamow wrote a seminal paper on 'Big Bang' nucleosynthesis, outlining how hydrogen and helium were produced during the Universe's earliest moments. In 1948, he and his collaborators published a series of papers that eventually led to the prediction of a cosmic microwave background with a blackbody temperature of a few degrees Kelvin, very close to its measured value. (Courtesy AIP Emilio Segre Visual Archives)

experiment to detect the cooled remnant of a hot, early Universe, had it not been for the accidental discovery [22] of an excess noise temperature of about 3.5 ± 1 K by Arno Penzias and Robert Wilson during their commissioning of a 20-foot horn antenna built for telecommunications (Figure 1.2). The Princeton group confirmed the measurement just a few months later [23], reporting a blackbody spectrum with a temperature of 3.0 ± 0.5 K at a wavelength of 3.2 cm. In fact, it was news of the Princeton experiment that led Penzias and Wilson to interpret their excess noise temperature as being extraterrestrial.

Sadly, neither of these two historic publications even mentioned the papers by Gamow, Alpher and Herman, who missed out on the immediate celebrity (and possibly a fraction of the Nobel Prize) they deserved. This was fully rectified by 1971, after Gamow's death [24], but of this oversight at the time of the discovery, Dicke wrote [25] "There is one unfortunate and embarrassing aspect of our work on the fire-ball radiation. We failed to make an adequate literature search and missed the more important papers of Gamow, Alpher and Herman. I must take the major blame for this, for the others in our group were too young to know these old papers. In ancient times I had heard Gamow talk at Princeton but I had remembered his model universe as cold and initially filled only with neutrons."

What we know today about the cosmic microwave background (CMB) constitutes some of the most compelling evidence ever assembled in support of a prevailing paradigm—in this case, the Hot Big Bang model of the Universe. Measurements by three independent satellites, COBE [26, 27], WMAP [28] and *Planck* [29], and ground-based facilities, such as the Atacama Cosmology Telescope [30] and the South Pole Telescope [31], have produced incredibly detailed and informative maps of the microwave radiation and a blueprint of cosmic evolution. The CMB informs our theoretical modeling in unparalleled ways, as we shall document later in this book, particularly Section 6.4 and Chapter 11.

A recent snapshot of the entire sky produced by *Planck* is shown in Figure 1.3. As it turns out, the CMB temperature is almost uniform everywhere (with an updated value of 2.72548 ± 0.00057 K), though not exactly. With the sensitivity of the latest instruments, tiny fluctuations at the level of approximately 1 part in 100,000 have emerged, on spatial scales ranging from a fraction of $1°$ to tens of degrees. Mining this fossilized archive has become indispensable for modeling the formation of structure, starting in the quantum domain (Sections 6.7, 11.2 and 13.1), all the way to the condensations of matter that formed stars, galaxies and clusters (Chapter 13). In the standard model of cosmology (Chapter 6), the CMB anisotropies would have imprinted themselves onto this image about 380,000 years after the Big Bang, well past the point where quantum fluctuations classicalized into self-gravitating, over-dense regions, but well before the appearance of the earliest known (Population III) stars at $t \sim 200$–300 Myr. Figure 1.3 thus records the seeds of all the structure we see today.

Figure 1.2 Robert Woodrow Wilson (left) and Arno A. Penzias in front of the microwave horn antenna they developed for telecommunications and subsequently used to scan the skies for sources of radiation. In 1964, they serendipitously discovered the cosmic microwave background radiation which, unknown to them, had earlier been predicted through the work of Gamow, Alpher and Herman in 1948–1949. (Courtesy AIP Emilio Segre Visual Archives)

1.2 LARGE-SCALE STRUCTURE

But the conversion of this information into a predicted three-dimensional distribution of galaxies has taken considerable effort and the development of powerful, modern computers. More importantly, cosmologists have had to hypothesize the influence of two dark components in the cosmic fluid, about which we still know virtually nothing, even though their existence is inferred with overwhelming confidence from the gravitational impact they have on the expansion dynamics and the growth of density fluctuations.

Figure 1.4 shows a series of snapshots tracing the simulated evolution of the matter content in the Universe, starting 24 Myr after the Big Bang on the left and ending with the filamentary structure we see today on the right. This is one of the more recently completed calculations, known as the 'Bolshoi' model [32], after the Russian word for 'big' or 'grand.' Bolshoi is certainly not the first attempt to mimic the evolution of the Universe in this way, but it is currently the best. The '2005 Millenium' simulation is considered to be the pioneering effort in this field [33], but Bolshoi is now superior because of the improved accuracy and precision of the input data from the most recent cosmological measurements, and the more advanced computational technology and analysis algorithms.

The clumpy filaments in this figure represent mostly the aggregation of so-called 'dark matter'—one of these two mysterious and elusive unseen components that nonetheless appear to dominate most of the energy density in the Universe. The first hint of its existence was inferred in the 1930's by Zwicky from the dynamics of galaxy clusters [34], but it would take another half century before its average density could be estimated reliably for it to be recognized as an integral agent in the formation of structure. The evidence for its existence today appears to be unassailable. It includes the flattened rotation curves in galaxies, the unusually high internal kinetic energy in groups and clusters of galaxies, large-scale cosmic flows and gravitational lensing (see, e.g., Section 12.5). The latter phenomenon, in particular, which was actually predicted by Zwicky himself [35], produces a readily measurable general relativistic effect due to the presence of a gravitating source that is not luminous or baryonic. The distorted images of background sources due to the passage of lightrays near intervening mass concentrations show that dark matter is present in the outskirts of the halos [36], in the intra-cluster medium [37] and in the general mass field [38].

The fact that it behaves like ordinary matter gravitationally, but does not couple to the electromagnetic field and is thus not visible to us, is the reason we call it 'dark' *matter*. The filaments in Figure 1.4 give the impression that we are looking at luminous matter because galaxies are associated with dark matter halos using a method known as 'halo abundance matching,' which allocates more luminous galaxies to halos whose dark-matter particles are moving faster. This is consistent with observations that show both spiral and elliptical galaxies having greater mass and luminosity when their constituent stars have a higher velocity.

In Chapter 6 we shall see that the most precise measurements available to us today, principally from a detailed analysis of the fluctuation distribution in the CMB temperature maps [29], suggest that—in the context of the standard model

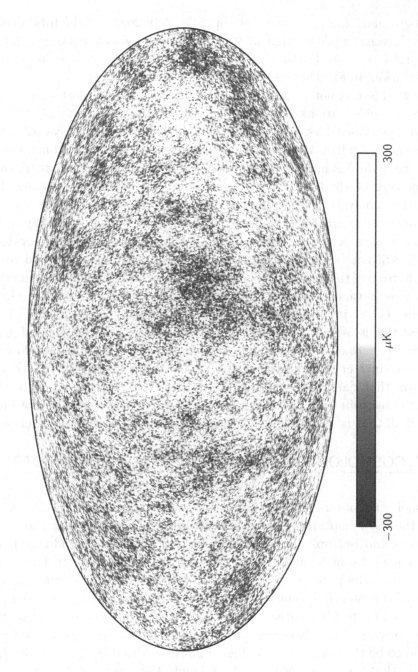

300

μK

−300

Figure 1.3 Anisotropies observed in the cosmic microwave background radiation by the *Planck* satellite. In these Galactic coordinates, our line-of-sight in the middle is towards the the Galactic center; longitude increases from 0 to 180 degrees towards the bottom, then restarts at 180 degrees at the top and continues to 360 degrees back to the middle. The latitude ranges from −90 degrees on the right edge to +90 degrees to the left. The tiny fluctuations represent variations in temperature of roughly 1 part in 100,000, indicated by the scale on the right in units of μK. (Courtesy European Space Agency and the Planck Collaboration)

of cosmology—dark matter represents approximately 26.6% of the total energy density in the Universe, while baryonic matter constitutes a much less significant fraction of only $\sim 4.9\%$. We, our Earth, the Sun and the rest of the solar system are merely a 'contamination' in an otherwise dark, non-baryonic Universe.

And yet, this was not even the biggest surprise from the last three decades of ever improving observations. That mantle rests with the discovery of 'dark energy,' our second mysterious dark component, and the largest fractional contributor to the Universe's energy budget. We now have very strong evidence of its influence on cosmic dynamics from various lines of evidence, especially the fluctuation distribution and power spectrum of the CMB. But its existence was confirmed first and foremost via the earlier analysis of Type Ia supernovae [39, 40, 41], for which the principal investigators were awarded the Nobel Prize in Physics. Dark energy may turn out to be Einstein's cosmological constant Λ after all, though there is growing evidence that it probably will not (see, e.g., Chapter 12). In addition, a cosmological constant is problematic from a theoretical physics point of view [42]. The basic scenario would interpret Λ as a vacuum energy due to quantum loop corrections at the Planck energy scale (i.e., $\sim 10^{19}$ GeV), where all the fundamental forces in principle become unified. But this is over 120 orders of magnitude larger than the value of Λ required to represent the measured dark energy. Such an incredible disparity would be viewed as a profound mystery, should the reality of Λ ever be confirmed. In either case, it now appears that dark energy represents no less than approximately 68.5% of all the energy density in the Universe, meaning that we know virtually nothing about $\sim 95.1\%$ of all that is—a perplexing situation, at once alarming and exciting.

1.3   THE COSMOLOGICAL PRINCIPLE: HOMOGENEITY & ISOTROPY

The Bolshoi simulations, and actually the entire contents of this book, rest on arguably the most impactful assumption made in cosmology—that the Universe is homogeneous and isotropic on large scales, two symmetries individually hewn from an all-embracing Cosmological Principle. Viewed as 'Occam's razor' for the simplest possible model, this principle has been the starting point for cosmology since the beginning of the twentieth century, though various naive forms of it emerged much earlier. According to his treatise on the world's cosmological principles [43], Konrad Rudnicki writes that "According to the oldest Indian traditions, the Universe is understood to be the body of the highest, infinite spiritual being, and thus has some of his properties. If we attempt to render this into the language of contemporary science, we arrive at the following formulation: The Universe is infinite in space and time and is infinitely heterogeneous." The ancient Greeks believed that Earth is the natural center of the Universe, which would argue for isotropy, though not homogeneity, while the Copernican cosmological principle (see below) maintains that the Universe as observed from any planet looks the same as it does from Earth, suggesting that no place in the Universe is more special than any other.

In the framework of general relativity, one begins by assuming that the large-scale evolution of spacetime is described by a uniform application of Einstein's field equations everywhere (Section 2.7), and that the ensuing description is based on an isotropic perspective for all the fundamental observers, i.e., those associated with the average motion of matter in the Universe. As we shall see in Chapter 4, this ensures that our cosmic spacetime may be anchored by a single time coordinate, known as the *cosmic* time, t, which advances at an equal rate along every individual worldline starting at the Big Bang. The Cosmological Principle thus endows the metric with a very high degree of symmetry, simplifying it considerably (Section 4.2).

Nevertheless, this assumption becomes meaningful only when related to actual astronomical observations which, however, are totally dependent on telescope and detector technology. Based on what we can see today, cosmological observations are statistically isotropic about us, when averaged over large scales, meaning the size of galaxy clusters and above. For example, there is no preferred direction pointing to an apparent 'center' of the Universe. And the CMB is highly isotropic, with a temperature that varies across the sky by less than 1 part in 10,000, after we subtract Earth's motion relative to the cosmos (roughly 250 km s^{-1}), which creates a dipole with a strength of 1 part in a thousand (see Equation 6.68) [29].

Anisotropies in the matter distribution between us and the surface of last scattering create fluctuations in the CMB temperature map of less than 1 part in 100,000, as seen by COBE, WMAP and *Planck* (Section 1.1). Together with other kinds of observation, the CMB thus completely supports the conclusion that our past lightcone— i.e., that portion of the Universe with which we are causally connected from our past— is statistically isotropic. Other data include the National Radio Astronomy Observatory (NRAO) Very Large Array (VLA) Sky 1.4 GHz continuum survey (NVSS), covering the entire sky north of $-40°$ declination, which has been exploited for the statistical analysis of galaxy number counts across the sky [44, 45]. The inferred radio galaxy distribution is also fully consistent with statistical isotropy within our past lightcone.

But this is really as far as we can go in confirming the Cosmological Principle with actual measurements. To suggest that the entire Universe is isotropic about us requires some inductive justification, specifically that the high degree of isotropy within our past lightcone is indicative of what we would see everywhere, should our location within the cosmic spacetime be arbitrary. Establishing observational evidence for spatial homogeneity is even harder.

Various arguments have been used to justify the second symmetry in the Cosmological Principle. One of them stands out above the rest. Not surprisingly, it overlaps with the reason we believe the Universe is isotropic even beyond our past lightcone. The first claim made in support of homogeneity is a philosophical one: when we assume this symmetry as an initial condition, it directly affects the evolving geometry later on. None of our observations thus far have required anything more complex for a reasonable interpretation, so the simplest assumption appears to be correct.

A more scientific justification invokes the idea that all of the physical processes used to describe cosmic evolution thus far assume homogeneity. For example, we shall see in Chapter 6 that an inflationary phase is required early in the expansion of the standard-model Universe to fix a very awkward temperature-horizon problem. Such a rapid acceleration naturally leads to homogeneous regions in the cosmos and, more

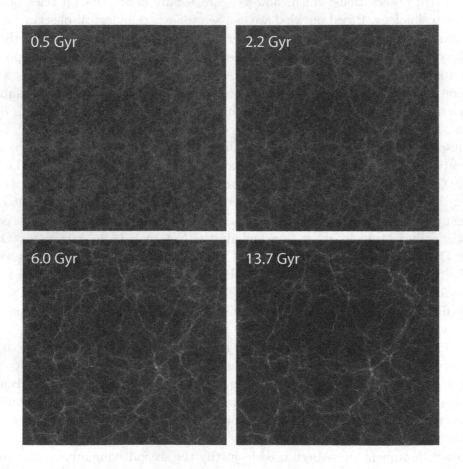

Figure 1.4   Four snapshots of the Bolshoi simulation, showing the evolution in time of a cubic volume of the Universe about 1 billion light-years across, containing 8.6 billion particles of dark matter, each with a mass of approximately 2×10^8 $M_\odot$. The clock in the simulation started when the Universe was 24 million years old and continued as the large-scale structure evolved towards the present time. (Courtesy Anatoly Klypin, New Mexico State University; Joel R. Primack, University of California, Santa Cruz; and Stefan Gottlöber, AIP, Germany)

to the point, one of these would include our entire causally-connected spacetime. Of course, this would not ensure that the entire Universe is strictly homogeneous, though homogeneity could then describe the distribution of such causally-connected pockets. This argument, however, is not very convincing because very little here is based on actual observation. Inflation may or may not have happened. Indeed, we shall demonstrate in Chapter 11 that the observational evidence seems to argue more and more against the most basic slow-roll inflationary paradigm, leaving us with a considerable amount of effort to make inflation work self-consistently—if at all.

Yet another theoretical justification is based on the concept of uniform thermal histories. The uniformity of objects we see in the sky, e.g., the types of galaxy we detect at large distances compared to what we see locally, suggests they all had a similar evolution, and deduce from this that the Universe must be spatially homogeneous. A clear counterargument would be a measurement showing that the ages of distant objects were incompatible with local estimates, but no such indication has yet been found.

The most generally accepted reason in favor of homogeneity is related to the comment we made above concerning the isotropy of the Universe even on scales beyond our past lightcone. It is not difficult to prove rigorously, and is intuitively rather obvious, that if all observers see an isotropic Universe, then it must also be homogeneous [46]. Indeed, the theorem is even simpler than this: the Universe must be homogeneous if only three spatially separated observers see isotropy. Anchored to the Earth, however, we cannot realistically base our argument on actual measurements. The best we can do is to replace the required observations with a supposition—that we are not privileged observers, often called the *Copernican* Principle, after Nicolaus Copernicus (1473–1543), the Polish Catholic priest who proposed that the center of the Universe was the Sun, not the Earth, thereby greatly simplifying the motion of the planets. At the same time, this sweeping shift in paradigm downgraded Earth from its exalted erstwhile status as the center of all things.

The Copernican Principle seems plausible on a cosmic scale because we have never found major changes to the physical conditions anywhere. Together with the isotropy we see in our past lightcone, this uniformity would suggest an isotropic Universe, and thus a homogeneous one as well.

The stage is now set. We have a clear view of the problems we must contend with. Our study of the cosmic spacetime will be greatly informed by the many exquisite observational clues gleaned from the data, some of which we have introduced in this chapter. As far as we can tell, the Universe is expanding with a very high degree of symmetry, tightly constraining the physics in its earliest moments of existence, promising to stretch our communion with nature well beyond our comfort zone in the sublunary world. Much of the cosmic substance is dark and unknown, but if the general theory of relativity is correct, its beautiful and elegant structure will guide our interpretation of key physical concepts, such as dark energy's equation-of-state. Only then can we seriously start to think about how to extend the standard model of particle physics to possibly accommodate this mysterious new substance.

CHAPTER 2

Theoretical Background

2.1 BASIC CONSIDERATIONS

THE human endeavor to understand the natural world has meandered through various lines of inquiry, some evolving in parallel with scientific achievements, others veering off into featureless superstition and soulless magical realms. Needless to say, our appreciation for the wondrous workings of nature on the largest imaginable scales has grown in concert with the timely application of scientific breakthroughs in all disciplines of Physics. Our development of theoretical cosmology has benefited greatly from the principles and methods we have learned in the laboratory, on Earth, our local corner of the Galaxy and, more recently, via measurements made of the largest structures and distances using a wide assortment of instruments.

Though its meaning may have been much less expansive than what we think today, cosmology existed before the advent of relativity. When all that humans could see and understand was their local environment, simple measures of distance and time as the Greeks understood them, and then Galileo Galilei (1564–1642) and Sir Isaac Newton (1642–1727), were more than sufficient. Indeed, the earliest considerations of the relevance of general relativity to cosmology, some of them by Albert Einstein (1879–1955) himself [12], didn't really make use of spacetime's most telling and exotic features.

Einstein's motivation for proposing his cosmological theory was a desire to encapsulate Mach's principle in a closed homogeneous world model. We now know that several of his ideas were wrong to begin with, but such was the inauspicious marriage of relativity and cosmology. Einstein rejected the notion, held seriously by many at the time, that the Universe was essentially a concentrated island of stars surrounded by an otherwise asymptotically featureless realm. Ernst Mach's (1838–1916) principle—that a particle's inertia is generated by matter in the Universe—could not hold under such conditions, he argued, because a particle escaping this island and moving into flat space would still retain its inertial properties according to relativity theory, yet violate Mach's mechanism for creating that inertia. Of course, we now know that Mach's principle is actually not fully consistent with general relativity anyway, because a free-falling observer is completely oblivious to the gravitational field generated by the rest of the Universe.

With his closed world model, Einstein was likewise rejecting the idea of an infinite material Newtonian universe, which he argued would produce an arbitrarily large local potential and stellar kinetic energy. Several years later, the Harvard mathematician George David Birkhoff (1884–1944) published a remarkable book describing what we now refer to as the *Birkhoff-Jebsen theorem*, that elegantly and simply explains why the local dynamics in the Universe is completely independent of its contents at arbitrarily large distances—so long as the Universe is isotropic, not even necessarily homogeneous.[1]

But over the course of a century since then, we have learned that the Universe is expanding, implying—among other things—that it began in a cauldron of exceedingly high concentration and energy. Though we still marvel at the fact that our locally developed physical theories can even be valid under such extreme conditions, there is no longer any doubt that to correctly explain what we see, general relativity and its implied constraints on the cosmic spacetime are absolutely indispensable elements of any comprehensive cosmological theory. And so we begin by establishing the foundational tenets of relativity that will gradually inform our observational and theoretical engagement with the principal topic of this book.

2.2 SPECIAL RELATIVITY

The nature of space and the meaning of time have been scrutinized by ancient thinkers to the modern era, though what they are still remains somewhat of a mystery. Without question, however, the greatest contribution to our view of the Universe was Einstein's general theory of relativity, which takes space and time and folds them into a single interwoven unit.

In retrospect, we might wonder why it took so long to address the apparent inconsistencies in the classic interpretation of space and time head-on. But even today it is difficult for us to provide a compelling answer to some of the peculiarities uncovered by the early thinkers. Take, for example, one of Zeno of Elea's arguments, known as 'The Arrow Paradox.' Zeno (not to be confused with the later Zeno of Citium mentioned in Chapter 1) was born around 495 BC in the Greek colony of Elea in southern Italy, but very little is known about him [47]. Only eight of his paradoxes have survived, thanks to the efforts of subsequent writers, such as Aristotle. Zeno's goal was to demonstrate that motion, change, time, and plurality are mere illusions—that in reality the Universe is singular and immutable.

The Arrow Paradox seems illogical, perhaps even confusing, but it cannot be discarded very easily. Aristotle himself abandoned Zeno's paradoxes without really proving their (assumed) fallacy, but they were revived many centuries later by mathematicians such as Bertrand Russell (1872–1970) [48] and Lewis Carroll (1832–1898).

Hypothesizing that absolute space and absolute time exist separately of each other, Zeno imagined an arrow launched along a straight line. But he could not resolve the paradox that if the arrow exists distinctly at a sequence of discrete instants in

---

[1]We shall study this very important theorem and its corollary in Sections 4.3, 5.4.3 and, especially, 7.2.

time, and if no motion is discernible in any given instant, there cannot be any motion from one instant to the next.

In today's parlance, we would describe this situation in terms of a photograph we might take of the arrow in flight; it would be very difficult for us to tell from the image whether the arrow were moving or not. With an increasing shutter speed, any blurring would disappear and eventually there would be no difference at all in the appearance of a moving arrow and its stationary counterpart. Zeno correctly inferred an inconsistency with the concept of a time independent of space from the fact that the arrow cannot possibly 'know' from one instant to the next whether it is moving. How can the causality be transmitted forward in time through a sequence of such instants, he argued, when motion does not exist in any of them?

So what *is* different between the moving and non-moving arrows? We no longer accept Zeno's denial of any real physical motion, but an answer to his question may only be found through the physical principles founded by Isaac Newton and developed later by Albert Einstein and his contemporaries during the twentieth century. Modern physics has concluded, along with Zeno, that the classical view of space and time was simply wrong to begin with—that in fact motion could not be possible in a universe constructed according to the old ideas.

In his highly influential book—the *Philosophiae Naturalis Principia Mathematica*—Newton provided the first significant break from the ancient view of the Universe, though still describing time as a one-dimensional continuum existing on its own, and space as an absolute 3-dimensional substrate. But the novelty of his concept was the idea that space does not have a center, and that it is not always necessary to describe motion relative to a special, fixed frame.

According to Newton, an observer can always tell when he or she is accelerating, which he ascribed to the influence of an absolute space, against which acceleration is manifested. But as long as the observer is moving at constant velocity within that space, thereby sitting in what we now refer to as an *inertial* frame, the laws of nature are not affected by the motion.

It would take another 150 years before physics would confront a major crisis. By the middle of the nineteenth century, James Clarke Maxwell (1831–1879) had achieved a surprising synthesis of all electric and magnetic phenomena. The most astonishing result of all was his demonstration that light itself is an electromagnetic occurrence, propagating at a fixed speed completely independent of anyone's perspective. Of course, the most compelling argument in support of this was Michelson and Morley's demonstration in 1887, that light travels at the same speed along Earth's orbital motion and transverse to it, proving that the speed is identical in every case. This was a problem because according to Newton, an individual could certainly measure someone else's speed relative to himself—even if no constant motion could be discerned relative to absolute space—so two observers surely could not infer the *same* speed for light.

By now, the resolution to this paradox is quite well known. Space and time do not exist separately of each other. Had The Arrow been taken more seriously, some aspects of special relativity might have been anticipated hundreds of years before

Albert Einstein finally grasped the true meaning of light's aberrant behavior early in the twentieth century.

Since two people moving past each other claim that light has the same speed, the distances they measure in ascertaining how far it travels in a given time cannot be the same. Any empirical evidence that defies our intuitive concept of speed, such as the constancy of the speed of light, hints at the fact that the measurement of distance and time depends on who is making the measurement.

Newton was correct in declaring that space has no center; but the constancy of the speed of light means that absolute space and absolute time do not exist either. Zeno was right after all, though not exactly for the reason he imagined.

A merger of space and time was accomplished with Einstein's special theory of relativity, which begins with two postulates: (1) that the laws of physics are the same for all non-accelerated individuals (though even this restriction would eventually be overcome with the development of the general theory), and (2) that the speed of light in vacuum is independent of the motion of all observers and sources, and is observed to have the same value c. The first of these implies that we cannot infer an absolute velocity relative to space. The second postulate is a much more dramatic departure from classical thought, for it requires not only a lack of universality in the determination of physical quantities but, more importantly, it also demands an *intertwining* of the measured time and space.

An important consequence of this result is that simultaneity cannot be universal. Since time flows differently for different individuals, one observer may decide that two events seen some distance away occur at exactly the same time, while a second individual moving relative to the first, will see one event happening before the other. Time exists in a certain region of space only because the constituents at that location are changing. We infer that time has passed because the inner workings of the clock have caused its hand to move across its face. Time and movement are inseparable concepts. If all physical processes within a given region of space were to slow down to half their normal speed, time would correspondingly 'flow' at half the rate with which it flows elsewhere.

With special relativity, everything that affects space bears on the change within it and hence also affects time. Two and a half thousand years ago, Zeno correctly anticipated the fundamental problem with absolute space and an independent time. In the classical mindset, there really was no way to distinguish a moving arrow from a stationary one during the briefest possible instant.

Einstein's theory of special relativity provides an elegant resolution of Zeno's paradox. Space and time are not independent, it turns out, and the two are coupled by the constant speed of light. Any motion alters the conditions for simultaneity, and therefore events observed by someone moving with the arrow do not coincide with those seen from the ground. This distinction alone is sufficient to render the moving arrow distinguishable from its stationary counterpart.

2.3 FOUR-DIMENSIONAL SPACETIME

The fact that the speed of light c is constant for all observers means that there must exist certain relations between the coordinates to preserve the invariance of c, even though time intervals Δt and lengths Δr may separately change. The 'spacetime interval'

$$(\Delta s)^2 \equiv (c\,\Delta t)^2 - |\Delta \mathbf{x}|^2 \tag{2.1}$$

is the most useful of these formulations. It is trivially zero along any light path, and every observer measures the same value. However, the applicability of Δs is far more extensive than this simple mathematical restatement of the second postulate of special relativity, since it is constant for any linked pair of events, whether or not they are associated with massless particles.

The transformation of coordinates must relate t and $\mathbf{x} \equiv (x, y, z)$ in one frame (say, the lab frame) to t' and $\mathbf{x}' \equiv (x', y', z')$ in the frame moving with velocity $\mathbf{v}$ relative to the first, in such a way that the invariance of $(\Delta s)^2$ is preserved. With some guesswork, one can easily show that the two sets of coordinates must be related via the so-called Lorentz transformation,

$$
\begin{aligned}
x' &= x \,, \\
y' &= y \,, \\
z' &= \gamma \left(z - \frac{v}{c}\, ct \right) , \\
ct' &= \gamma \left(ct - \frac{v}{c}\, z \right) ,
\end{aligned}
\tag{2.2}
$$

in which the Lorentz factor

$$\gamma \equiv \left[1 - (v/c)^2 \right]^{-1/2} \tag{2.3}$$

has a minimum value of 1, but can increase indefinitely as $v \to c$. There exit other transformations that also preserve the speed of light—for instance, a rigid rotation. Together, all these possible solutions constitute the homogeneous Lorentz group. The first postulate of special relativity requires that there be perfect symmetry between the two observers, so the reciprocal transformation to (2.2) may differ from it only in the sign of $\mathbf{v}$, as seen from the perspective of the second frame.

The coordinates in Equation (2.1) are but one example in which a grouping of four physical quantities transmute into each other under a Lorentz transformation from one frame to another. Since the founding of special relativity early in the twentieth century, we have come to recognize that such 4-groupings are the manifestation of the four-dimensional structure of spacetime. Quantities such as $x^\mu \equiv (x^0, x^1, x^2, x^3)$ or, equivalently, $x^\mu \equiv (x^0, x, y, z)$, where $x^0 \equiv ct$, involve both a magnitude and a direction in 3-space. The constancy of c forces a constraint between the 3-space coordinates $\mathbf{x}$ and time t, so it is clear why four components must be involved in any transformation of frames.

In cases where a measurable element is independent of direction, however, one would expect the number of transmutable quantities in a grouping to be less than four. For example, the number of marbles in a box is independent of the latter's orientation. Some physical quantities require only denumerability, and for them, a single function of the spacetime coordinates is sufficient. Being solitary members of their respective groupings, these physical quantities should therefore be invariant under a transformation from one frame to the next.

One such quantity is the spacetime interval Δs that we have already introduced. Experiment shows that it must be invariant for light, but it is not difficult to demonstrate from the transformation in Equation (2.2) that Δs must be invariant for any pair of coupled events, regardless of whether they are linked by light or slower moving particles. The only difference is that, whereas Δs is always zero for photons, it is non-zero for massive particles. We can see this trivially for a pair of events, one at $x^\mu = (0,0,0,0)$ and the other at $x^\mu = (x^0, x^1, x^2, x^3)$, for which

$$
\begin{aligned}
(\Delta s)^2 &= (x^0)^2 - (x^1)^2 - (x^2)^2 - (x^3)^2 \\
&= \gamma^2 \left(x'^0 + \beta\, x'^3 \right)^2 - (x'^1)^2 - (x'^2)^2 - \gamma^2 \left(x'^3 + \beta\, x'^0 \right)^2 \\
&= (x'^0)^2 - (x'^1)^2 - (x'^2)^2 - (x'^3)^2 \\
&= (\Delta s')^2
\end{aligned}
\tag{2.4}
$$

where, following convention, we define $\beta \equiv v/c$.

The interval Δs is a single function of the coordinates and is invariant under a Lorentz transformation, so it is a *scalar* in four-dimensional spacetime. It may therefore be used directly to compare the intervals of space (and/or time) in one frame with those measured in another. The rest frame is particularly important because $\Delta\mathbf{x} = 0$. The interval Δs is then entirely due to the passage of time, $\Delta\tau$ which, for obvious reasons, is called the *proper* time:

$$
\Delta s = c\,\Delta\tau .
\tag{2.5}
$$

In any other frame,

$$
\begin{aligned}
\Delta s &= \Delta x^0 \left[1 - \left(\Delta\mathbf{x}/\Delta x^0 \right)^2 \right]^{1/2} \\
&= c\,\Delta t\,(1 - \beta^2)^{1/2} \\
&= c\,\Delta t/\gamma ,
\end{aligned}
\tag{2.6}
$$

resulting in an effect known as time dilatation (or stretching) from one frame to the next, since $\Delta t = \gamma\,\Delta\tau$.

In four-dimensional spacetime, we must also describe interactions for which four components are insufficient. For example, the transport of a momentum vector in a direction independent of its pointing, requires at least nine different components. For physical quantities such as these, the minimal groupings involve four-vectors of four-vectors, constituting 16 functions of the coordinates that transmute into each other under a Lorentz transformation.

Physical variables such as these are generically called *tensors*, whose *rank* indicates how many functions of the coordinates belong to each set. The interval Δs is a scalar, or a tensor of rank 0, because it is a solitary function whose value does not change under a transformation. The grouping of four quantities x^μ is a four-vector, or a tensor of rank 1. As we have just noted, rank 2 tensors are four-vectors of four-vectors, constituting 16 intermixing functions. Together, tensors form a language that permits us to write invariant physical laws, in full compliance with the first postulate of special relativity.

The properties of a tensor depend on its rank k, determined by how it transforms under the translation

$$x^\mu \rightarrow x'^\mu \ . \tag{2.7}$$

A set of four ordered numbers $V^\alpha = (V^0, V^1, V^2, V^3)$ that transform according to the rule

$$V'^\alpha = \frac{\partial x'^\alpha}{\partial x^\beta} V^\beta \ , \tag{2.8}$$

is a *contravariant* vector. It is also useful to introduce a *covariant* vector, which transforms according to the prescription

$$V'_\alpha = \frac{\partial x^\beta}{\partial x'^\alpha} V_\beta \ . \tag{2.9}$$

In each of these, α and β take on the four possible values $(0, 1, 2, 3)$, corresponding to the four (space and time) coordinates of four-dimensional spacetime. Note that in this expression, and elsewhere in this book unless otherwise noted, a repeated index means that the term in which the index appears is to be summed over all its possible values. So, for example, $x^\alpha x_\alpha \equiv x^0 x_0 + x^1 x_1 + x^2 x_2 + x^3 x_3$.

The coefficients in the transformation of Equations (2.8) and (2.9) are derived from the chain rule of differentiation for dx^α, i.e., $dx'^\alpha = (\partial x'^\alpha / \partial x^\beta) \, dx^\beta$. We know that x^μ is a four-vector in this spacetime, and we've determined its transformation properties empirically, so we might as well use this knowledge to execute a transformation of all such four-vectors. Of course, this procedure is based on the assumption that all four-vectors transform identically to x^μ.

Contravariant and covariant tensor indices represent the same physics; they are introduced as separate entities only for mathematical convenience. For example, they allow us to easily handle a differentiation with respect to a contravariant coordinate x^μ, which evidently produces a covariant four-vector:

$$\frac{\partial}{\partial x'^\alpha} = \frac{\partial x^\beta}{\partial x'^\alpha} \frac{\partial}{\partial x^\beta} \ . \tag{2.10}$$

A tensor $T^{\alpha\beta}$ of rank $k = 2$ is a four-vector of four-vectors, and is therefore a grouping of 16 functions of x that transform according to an obvious generalization of the rules for scalars and vectors:

$$T'^{\alpha\beta} = \frac{\partial x'^\alpha}{\partial x^\gamma} \frac{\partial x'^\beta}{\partial x^\delta} T^{\gamma\delta} \tag{2.11}$$

and, correspondingly,

$$T'_{\alpha\beta} = \frac{\partial x^\gamma}{\partial x'^\alpha}\frac{\partial x^\delta}{\partial x'^\beta} T_{\gamma\delta} . \tag{2.12}$$

The *contraction* of a covariant four-vector with a contravariant four-vector produces an invariant quantity, as in

$$V'^\alpha W'_\alpha = \frac{\partial x'^\alpha}{\partial x^\beta} V^\beta \frac{\partial x^\gamma}{\partial x'^\alpha} W_\gamma$$

$$= \frac{\partial x^\gamma}{\partial x^\beta} V^\beta W_\gamma$$

$$= \delta^\gamma_\beta V^\beta W_\gamma$$

$$= V^\beta W_\beta . \tag{2.13}$$

In contrast, the product $S_{\alpha\beta} \equiv V_\alpha W_\beta$ is not a scalar, but rather a tensor of rank 2, since

$$S'_{\alpha\beta} = V'_\alpha W'_\beta$$

$$= \frac{\partial x^\gamma}{\partial x'^\alpha} V_\gamma \frac{\partial x^\delta}{\partial x'^\beta} W_\delta$$

$$= \frac{\partial x^\gamma}{\partial x'^\alpha}\frac{\partial x^\delta}{\partial x'^\beta} S_{\gamma\delta} . \tag{2.14}$$

And in an evident generalization to tensors of higher rank, $T_\alpha{}^\alpha$ is a scalar, whereas $W_\alpha \equiv T_\alpha{}^\beta V_\beta$ is a covariant four-vector.

In differential form, the square of the interval $(\Delta s)^2 \to (ds)^2$ in Equation (2.4) may be written

$$(ds)^2 = \eta_{\alpha\beta}\, dx^\alpha\, dx^\beta , \tag{2.15}$$

where

$$[\eta_{\alpha\beta}] \equiv \begin{pmatrix} 1 & 0 & 0 & 0 \\ 0 & -1 & 0 & 0 \\ 0 & 0 & -1 & 0 \\ 0 & 0 & 0 & -1 \end{pmatrix} \tag{2.16}$$

is called the *metric tensor*. In special relativity, its definition is one of convenience, simplifying the notation, e.g., for ds. For a transformation between non-inertial observers (see Section 2.4 below), however, the generalized form of $\eta^{\alpha\beta} = \eta_{\beta\alpha}$ represents the inertial forces arising from a frame's acceleration.

From Equation (2.15), it is also clear that

$$dx_\beta = \eta_{\beta\alpha}\, dx^\alpha, \tag{2.17}$$

where $dx_\beta = (x_0, x_1, x_2, x_3) = (x^0, -x^1, -x^2, -x^3)$, demonstrating that a contraction of a four-vector with the metric tensor is the general procedure for converting a contravariant index into a covariant one. This also works in reverse, converting a covariant index into a contravariant one. To see this, we note that the coordinates x^α are linearly independent, so the condition

$$x^\alpha = \eta^{\alpha\beta}(\eta_{\beta\gamma}x^\gamma) \tag{2.18}$$

is satisfied as long as

$$\eta^{\alpha\beta}\eta_{\beta\gamma} = \delta^\alpha_\gamma. \tag{2.19}$$

Thus, in special relativity, the coefficients $\eta^{\alpha\beta}$ and $\eta_{\beta\gamma}$ are identical when written in terms of Cartesian coordinates, and so

$$dx^\beta = \eta^{\beta\alpha}dx_\alpha. \tag{2.20}$$

The constancy of the interval ds highlights the principal difference between special relativity and its predecessor—the framework of transformations attributed to Galileo. In Galilean relativity, the magnitude of a 3-vector is invariant under a rotation or translation, though its components (i.e., its direction cosines) may individually vary; but it contains no mixing of spatial coordinates with time, because the classical view holds that time is a 'universal fluid' independent of space or its contents.

Special relativity adopts from the very beginning the empirical fact that time should instead be viewed as a measure of the rate of change, which must then necessarily depend on the specific frame of reference. Nature tells us that it is the contraction $x^\mu x_\mu$ that must be preserved, not just $x^i x_i$ (with $i = 1, 2, 3$), which then becomes the four-dimensional analog of the 3-dimensional magnitude $|\mathbf{x}|$ in classical mechanics.

In order to use the language of four-dimensional spacetime to write physical laws that remain invariant under a coordinate transformation, we must first understand how to interpret a particle's 'inertia' when a force is applied to it. The fact that inertia is not itself an invariant quantity presents an obvious difficulty in using forces to quantify the effects on a particle due to the influence of others. Not knowing *a priori* how to calculate the force on a moving object, we have no direct means of translating a physical law known in terms of frame-dependent language into its invariant form mandated by special relativity. But we generally do know how to write down a description of the interaction in at least one frame—the one in which the particle being acted on is instantaneously at rest. The *rest mass*, m, representing the particle's inertia in its own frame, is a quantity upon which all observers can agree. We shall therefore always mean this value when we refer to the 'mass.' Since distances and time vary from observer to observer, however, we do not expect the measured inertia to have the same value in every frame.

An observer in the particle's rest frame would infer its inertia m based on the acceleration he or she measures when a Newtonian force $\mathbf{F} \equiv (F^1, F^2, F^3) = m\mathbf{a}$ is applied. To turn this law into a form upon which all observers can agree, we must recast it using a combination of four-dimensional tensors. A reasonable guess is to write the four-dimensional version f^α of $\mathbf{F}$ in the form

$$f^\alpha \equiv m \frac{d^2 x^\alpha}{d\tau^2} . \tag{2.21}$$

Clearly, f^α reduces to $\mathbf{F} = m\mathbf{a}$ in the particle's rest frame, so we know that it is correct there. But note that because the rest-frame inertia m and proper time τ are scalar quantities, the transformation properties of $m\, d^2 x^\alpha/d\tau^2$ are fully specified by those of x^α alone. The right-hand-side of Equation (2.21) is therefore a four-vector, and this expression must thus retain the same form in all frames. As written, this generalization of Newton's equation of motion must therefore be an invariant physical law, valid for all observers.

Further, since the time rate of change of the linear momentum $\mathbf{p}$ is considered to be an equivalent representation of the applied force $\mathbf{F}$, a reasonable invariant representation of this statement is the following:

$$\frac{dp^\alpha}{d\tau} = f^\alpha , \tag{2.22}$$

where evidently

$$p^\alpha \equiv m \frac{dx^\alpha}{d\tau} . \tag{2.23}$$

This is a four-vector whose components in the particle's rest frame reduce to $p^0 = mc$ and $\mathbf{p} = \mathbf{0}$. In any other frame, where $dt = \gamma d\tau$,

$$p^0 = \gamma mc ,$$
$$\mathbf{p} = \gamma m\mathbf{v} . \tag{2.24}$$

Note, however, that the contraction of p^α with itself always produces an invariant quantity proportional to the particle's rest mass:

$$p^\alpha p_\alpha = \gamma^2 m^2 c^2 - \gamma^2 m^2 |\mathbf{v}|^2 = m^2 c^2 = \text{constant} . \tag{2.25}$$

We can see now how inertia is modified by special relativity. An observer moving relative to the particle infers a mass γm, which varies with speed v, since $\gamma = [1 - (v/c)^2]^{-1/2}$. In addition,

$$\gamma mc = \frac{mc}{\sqrt{1 - (v/c)^2}}$$
$$\approx mc \left[1 + \frac{1}{2} \left(\frac{v}{c} \right)^2 + \frac{3}{8} \left(\frac{v}{c} \right)^4 + \ldots \right]$$
$$\approx mc + \frac{1}{2c} m v^2 \tag{2.26}$$

for small values of v/c. So $p^0 c$ is the sum of the classical kinetic energy $(mv^2/2)$ and another form of energy that does not vanish even when $v = 0$. For obvious reasons, we refer to the latter as the particle's *rest energy*, and to the sum

$$E \equiv \gamma mc^2 \tag{2.27}$$

as its *relativistic energy*. Thus,

$$p^0 = E/c \,, \tag{2.28}$$

and from Equation (2.25), we therefore obtain the general expression relating the particle's inertia and momentum to its energy:

$$E^2 = |\mathbf{p}|^2 c^2 + m^2 c^4 \,. \tag{2.29}$$

Why inertia m is associated with a rest energy is still not completely understood, even a century after Einstein developed this interpretation via Equation (2.29). As we shall see in Chapter 10, however, the role played by our gravitational horizon in cosmology now provides a likely explanation for this remarkable property. There we shall learn that the rest energy $E = mc^2$ appears to represent a particle's gravitational binding energy to that portion of the Universe with which we (the observers) are causally connected.

For the moment, the equations we have derived in this section are all we shall need from special relativity. The next step is to introduce a transformation between accelerated frames, which will lead us directly into the general theory and its use as a description of gravity.

2.4 ACCELERATED FRAMES

In order to develop a more general relativity theory, we must be able to express the interval ds in terms of arbitrary coordinate systems, which may not be Cartesian or inertial. Using the notation $X^\alpha \equiv (X^0, X^1, X^2, X^3)$ to represent the generalized coordinates, we can transform Equation (2.15) with the chain rule of differentiation:

$$\begin{aligned}
(ds)^2 &= \eta_{\gamma\delta} \, dx^\gamma \, dx^\delta \\
&= \eta_{\gamma\delta} \frac{\partial x^\gamma}{\partial X^\alpha} dX^\alpha \frac{\partial x^\delta}{\partial X^\beta} dX^\beta \\
&= \left(\eta_{\gamma\delta} \frac{\partial x^\gamma}{\partial X^\alpha} \frac{\partial x^\delta}{\partial X^\beta} \right) dX^\alpha \, dX^\beta \,.
\end{aligned} \tag{2.30}$$

We write this as

$$(ds)^2 = g_{\alpha\beta} \, dX^\alpha \, dX^\beta \,, \tag{2.31}$$

where the metric tensor $\eta_{\alpha\beta}$ has been supplanted by its more general form

$$g_{\alpha\beta} \equiv \eta_{\mu\nu} \frac{\partial x^\mu}{\partial X^\alpha} \frac{\partial x^\nu}{\partial X^\beta} \,, \tag{2.32}$$

in terms of the inertial Cartesian coordinates x^α.

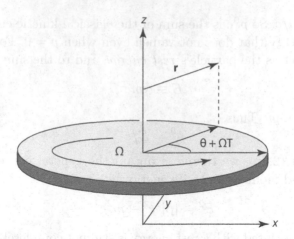

Figure 2.1 The accelerated frame (cT, r, θ, Z) is rotating about the z-axis in the Cartesian frame (ct, x, y, z) with an angular velocity Ω. The angle $\theta + \Omega T$ is measured relative to the x-axis. (Adapted from Melia 2007)

For example, consider the transformation from Cartesian coordinates x^α into a rotating system $X^\alpha \equiv (cT, r, \theta, Z)$ (shown in Figure 2.1). Ignoring for now any effects due to time dilation or length contraction (i.e., assuming that the relative motion has a speed much smaller than c), we immediately see that

$$t = T$$
$$x = r\,\cos(\theta + \Omega T)$$
$$y = r\,\sin(\theta + \Omega T)$$
$$z = Z\ , \tag{2.33}$$

and leaving the evaluation of the individual coefficients $g_{\alpha\beta}$ (using Equation 2.32) as a straightforward exercise for the reader, we write

$$(ds)^2 = (1 - \Omega^2 r^2/c^2)\,c^2\,(dT)^2 - (dr)^2 - r^2\,(d\theta)^2 - (dZ)^2 - 2\Omega\,r^2\,d\theta\,dT\ . \tag{2.34}$$

In this case, the metric tensor is

$$[g_{\alpha\beta}] = \begin{pmatrix} (1 - \Omega^2\,r^2/c^2) & 0 & -\Omega\,r^2 & 0 \\ 0 & -1 & 0 & 0 \\ -\Omega\,r^2 & 0 & -r^2 & 0 \\ 0 & 0 & 0 & -1 \end{pmatrix} \tag{2.35}$$

(remembering that $g_{\alpha\beta}$ is symmetric, so that $g_{0\theta} = g_{\theta 0}$). We note that, whereas $\eta_{\alpha\beta}$ is diagonal, this is no longer true for a general transformation, as illustrated here by the appearance of the term proportional to $d\theta\,dT$. Following a particle moving radially in the accelerated frame, we cannot ignore the contribution to ds from an azimuthal motion arising from the angular acceleration. In fact, writing the last term in Equation (2.34) as $2\Omega\, r^2\, \dot{\theta}\, (dT)^2$, we can merge it with the first term, thereby showing that modifications to this metric arise solely from the nonzero value of $\dot{\theta}$.

Another important difference between the metric coefficients $\eta_{\alpha\beta}$ and their more generalized form $g_{\alpha\beta}$, is that the latter are no longer necessarily independent of the coordinates themselves. Thus, differentiation of the $g_{\alpha\beta}$ can lead to the emergence of additional terms in the equations of motion, representing inertial forces associated with the frame's acceleration. As we have seen, the equation of motion for a particle in an inertial frame x^α (see Equation 2.21) is

$$m\frac{du^\alpha}{d\tau} = f^\alpha \,, \tag{2.36}$$

where

$$u^\alpha \equiv \frac{dx^\alpha}{d\tau} \tag{2.37}$$

is the four-velocity, and f^α is the four-force. A transformation of Newton's law into another inertial frame obviously leaves it unchanged because that is how special relativity is constructed. But let us now transform this equation into a non-inertial frame, and see how the properties of the $g_{\alpha\beta}(x)$ coefficients bring out the effects of an acceleration.

According to the transformation law for a four-vector, the force (which we here label with an overbar) measured in the non-inertial frame is

$$\bar{f}^\alpha = \frac{\partial X^\alpha}{\partial x^\delta}\, f^\delta$$

$$= m\frac{\partial X^\alpha}{\partial x^\delta}\left(\frac{\partial x^\delta}{\partial X^\beta}\frac{d^2 X^\beta}{d\tau^2} + \frac{\partial^2 x^\delta}{\partial X^\beta\,\partial X^\gamma}\frac{dX^\beta}{d\tau}\frac{dX^\gamma}{d\tau}\right)$$

$$= m\frac{d^2 X^\alpha}{d\tau^2} + m\frac{\partial^2 x^\delta}{\partial X^\beta\,\partial X^\gamma}\frac{\partial X^\alpha}{\partial x^\delta}\frac{dX^\beta}{d\tau}\frac{dX^\gamma}{d\tau} \,. \tag{2.38}$$

The generalized form of Newton's law in the accelerated frame is therefore

$$m\frac{d^2 X^\alpha}{d\tau^2} = \bar{f}^\alpha - m\Gamma^\alpha{}_{\beta\gamma}\frac{dX^\beta}{d\tau}\frac{dX^\gamma}{d\tau} \,, \tag{2.39}$$

where the coefficients

$$\Gamma^\alpha{}_{\beta\gamma} \equiv \frac{\partial^2 x^\delta}{\partial X^\beta\,\partial X^\gamma}\frac{\partial X^\alpha}{\partial x^\delta} \tag{2.40}$$

are known as the Christoffel symbols.

Equation (2.39) is useful in discerning the geometry of spacetime by permitting us to follow the paths taken by particles moving solely under the influence of inertial forces. These paths, known as *geodesics*, are described by the equations

$$\frac{d^2 X^\alpha}{d\tau^2} + \Gamma^\alpha{}_{\beta\gamma}\frac{dX^\beta}{d\tau}\frac{dX^\gamma}{d\tau} = 0 \,, \tag{2.41}$$

which follow from the equation of motion (2.39) in the absence of any other force $\bar{f}^\alpha$. Flat spacetime has no curvature because $g_{\alpha\beta} = \eta_{\alpha\beta}$ and $\Gamma^\alpha{}_{\beta\gamma} = 0$. In that case, the particle paths are all straight lines. We shall return to the curvature arising in other spacetimes, particularly those associated with gravity, later in this chapter.

In preparation for that step, we should point out that it is often unnecessary to calculate the Christoffel symbols directly from Equation (2.40), since the curvature of spacetime is already embedded within the metric coefficients $g_{\alpha\beta}$. To see how we can express $\Gamma^\alpha{}_{\beta\gamma}$ directly in terms of these coefficients, consider the derivative

$$\bar{g}_{\alpha\beta,\gamma} \equiv \frac{\partial}{\partial X^\gamma} \left(\frac{\partial x^\lambda}{\partial X^\alpha} \frac{\partial x^\mu}{\partial X^\beta} g_{\lambda\mu} \right) \tag{2.42}$$

where, following our temporary convention, an overbar denotes a tensor being evaluated in the non-inertial frame in terms of the coordinates X^α. That is,

$$\bar{g}_{\alpha\beta,\gamma} = \frac{\partial x^\lambda}{\partial X^\alpha} \frac{\partial x^\mu}{\partial X^\beta} \frac{\partial x^\nu}{\partial X^\gamma} g_{\lambda\mu,\nu} + g_{\lambda\mu} \frac{\partial^2 x^\lambda}{\partial X^\alpha \partial x^\nu} \frac{\partial x^\mu}{\partial X^\beta} \frac{\partial x^\nu}{\partial X^\gamma}$$
$$+ g_{\lambda\mu} \frac{\partial^2 x^\mu}{\partial X^\beta \partial x^\nu} \frac{\partial x^\lambda}{\partial X^\alpha} \frac{\partial x^\nu}{\partial X^\gamma} . \tag{2.43}$$

If we now permute the indices $\alpha \to \beta \to \gamma$ and $\lambda \to \mu \to \nu$, and take the symmetry of $g_{\mu\nu}$ into account, we find that

$$\bar{g}_{\beta\gamma,\alpha} + \bar{g}_{\gamma\alpha,\beta} - \bar{g}_{\alpha\beta,\gamma} = \frac{\partial x^\lambda}{\partial X^\alpha} \frac{\partial x^\mu}{\partial X^\beta} \frac{\partial x^\nu}{\partial X^\gamma} (g_{\mu\nu,\lambda} + g_{\nu\lambda,\mu} - g_{\lambda\mu,\nu})$$
$$+ 2g_{\rho\sigma} \frac{\partial^2 x^\rho}{\partial X^\alpha \partial X^\beta} \frac{\partial x^\sigma}{\partial X^\gamma} . \tag{2.44}$$

Already, the second term on the right-hand side of this expression bears some resemblance to the right-hand side of Equation (2.40). In fact, it is not difficult to show that

$$\Gamma^\delta{}_{\alpha\beta} = \frac{1}{2} g^{\delta\gamma} [g_{\beta\gamma,\alpha} + g_{\gamma\alpha,\beta} - g_{\alpha\beta,\gamma}] \tag{2.45}$$

for, in the non-inertial frame, using the result in Equation (2.44),

$$\bar{\Gamma}^\delta{}_{\alpha\beta} = \frac{1}{2} \frac{\partial X^\delta}{\partial x^\epsilon} \frac{\partial X^\gamma}{\partial x^\eta} g^{\epsilon\eta} \left[\frac{\partial x^\lambda}{\partial X^\alpha} \frac{\partial x^\mu}{\partial X^\beta} \frac{\partial x^\nu}{\partial X^\gamma} (g_{\mu\nu,\lambda} + g_{\nu\lambda,\mu} - g_{\lambda\mu,\nu}) \right.$$
$$\left. + 2g_{\rho\sigma} \frac{\partial^2 x^\rho}{\partial X^\alpha \partial X^\beta} \frac{\partial x^\sigma}{\partial X^\gamma} \right] . \tag{2.46}$$

That is,

$$\bar{\Gamma}^\delta{}_{\alpha\beta} = \frac{\partial X^\delta}{\partial x^\epsilon} \frac{\partial x^\lambda}{\partial X^\alpha} \frac{\partial x^\mu}{\partial X^\beta} \Gamma^\epsilon{}_{\lambda\mu} + \frac{\partial^2 x^\epsilon}{\partial X^\alpha \partial X^\beta} \frac{\partial X^\delta}{\partial x^\epsilon} . \tag{2.47}$$

But if x^α are the coordinates in a Cartesian inertial frame, then all of the $\Gamma^\epsilon_{\lambda\mu}$ are zero (trivially, since then $g_{\alpha\beta} = \eta_{\alpha\beta}$ in Equation 2.45), and therefore

$$\bar{\Gamma}^\delta_{\alpha\beta} = \frac{\partial^2 x^\epsilon}{\partial X^\alpha \partial X^\beta} \frac{\partial X^\delta}{\partial x^\epsilon} . \tag{2.48}$$

In Equation (2.45), we have introduced the matrix of coefficients $g^{\delta\gamma}$, the inverse of $g_{\delta\gamma}$, that is,

$$g^{\delta\gamma} g_{\lambda\delta} = \delta^\gamma_\lambda . \tag{2.49}$$

The definition (Equation 2.32) for $g_{\alpha\beta}$ ensures that the metric tensor does have an inverse,

$$g^{\delta\gamma} \equiv \eta^{\alpha\beta} \frac{\partial X^\delta}{\partial x^\alpha} \frac{\partial X^\gamma}{\partial x^\beta} \tag{2.50}$$

for, using the product rule

$$\frac{\partial X^\delta}{\partial x^\alpha} \frac{\partial x^\beta}{\partial X^\delta} = \delta^\beta_\alpha , \tag{2.51}$$

we find that

$$\begin{aligned}
g^{\delta\gamma} g_{\lambda\delta} &= \eta^{\alpha\beta} \frac{\partial X^\delta}{\partial x^\alpha} \frac{\partial X^\gamma}{\partial x^\beta} \eta_{\mu\nu} \frac{\partial x^\mu}{\partial X^\lambda} \frac{\partial x^\nu}{\partial X^\delta} \\
&= \eta^{\nu\beta} \frac{\partial X^\gamma}{\partial x^\beta} \eta_{\mu\nu} \frac{\partial x^\mu}{\partial X^\lambda} \\
&= \frac{\partial X^\gamma}{\partial x^\beta} \frac{\partial x^\beta}{\partial X^\lambda} \\
&= \delta^\gamma_\lambda ,
\end{aligned} \tag{2.52}$$

as required.

Thus, in general, whether or not the Christoffel symbols are calculated in a non-inertial frame, Equation (2.48) agrees with the definition in Equation (2.40). It is evident that Equation (2.45) is therefore a convenient method for calculating $\Gamma^\delta_{\alpha\beta}$ (or $\bar{\Gamma}^\delta_{\alpha\beta}$ if we want to explicitly emphasize that the Christoffel symbols are being calculated in a non-inertial frame) once the metric tensor $g_{\alpha\beta}$ is known.

As one might imagine, the Christoffel symbols are far more useful in general relativity than the single application we have seen thus far in Equation (2.39). For example, let us examine how the derivative of a covariant four-vector A_α transforms between a Cartesian-inertial and non-inertial frame:

$$\bar{A}_{\alpha,\beta} \equiv \frac{\partial \bar{A}_\alpha}{\partial X^\beta} = \frac{\partial}{\partial X^\beta} \left(\frac{\partial x^\lambda}{\partial X^\alpha} A_\lambda \right) . \tag{2.53}$$

Carrying through all the partial derivatives, we find that

$$\bar{A}_{\alpha,\beta} = \frac{\partial x^\lambda}{\partial X^\alpha} \frac{\partial x^\nu}{\partial X^\beta} A_{\lambda,\nu} + \frac{\partial^2 x^\lambda}{\partial X^\alpha \partial X^\beta} \frac{\partial X^\gamma}{\partial x^\lambda} \bar{A}_\gamma . \tag{2.54}$$

Correspondingly, for a contravariant vector,

$$\bar{A}^\alpha_{,\beta} = \frac{\partial X^\alpha}{\partial x^\lambda} \frac{\partial x^\nu}{\partial X^\beta} A^\lambda_{,\nu} - \frac{\partial^2 x^\lambda}{\partial X^\beta \partial X^\gamma} \frac{\partial X^\alpha}{\partial x^\lambda} \bar{A}^\gamma . \tag{2.55}$$

Clearly, neither $\bar{A}_{\alpha,\beta}$, nor $\bar{A}^{\alpha}{}_{,\beta}$, transforms as a tensor, which greatly mitigates their usefulness in general relativity since we cannot incorporate them into any invariant expression of the physical laws.

But notice that

$$\bar{A}_{\alpha,\beta} - \bar{\Gamma}^{\gamma}{}_{\alpha\beta}\,\bar{A}_{\gamma} = \frac{\partial x^{\lambda}}{\partial X^{\alpha}}\frac{\partial x^{\nu}}{\partial X^{\beta}}\,A_{\lambda,\nu}\,, \tag{2.56}$$

and since $\Gamma^{\epsilon}{}_{\lambda\nu} = 0$ in a Cartesian inertial frame, we can modify this expression slightly to read:

$$\bar{A}_{\alpha,\beta} - \bar{\Gamma}^{\gamma}{}_{\alpha\beta}\,\bar{A}_{\gamma} = \frac{\partial x^{\lambda}}{\partial X^{\alpha}}\frac{\partial x^{\nu}}{\partial X^{\beta}}\left(A_{\lambda,\nu} - \Gamma^{\epsilon}{}_{\lambda\nu}\,A_{\epsilon}\right)\,. \tag{2.57}$$

So the quantity $A_{\lambda,\nu} - \Gamma^{\epsilon}{}_{\lambda\nu}\,A_{\epsilon}$ transforms as a covariant second rank tensor, and we identify it as the covariant derivative of A_{α}, calling it

$$A_{\alpha;\beta} \equiv A_{\alpha,\beta} - \Gamma^{\delta}{}_{\alpha\beta}\,A_{\delta}\,. \tag{2.58}$$

Analogously,

$$A^{\alpha}{}_{;\beta} \equiv A^{\alpha}{}_{,\beta} + \Gamma^{\alpha}{}_{\gamma\beta}\,A^{\gamma}\,. \tag{2.59}$$

This process generalizes in an intuitively obvious way for tensors of higher rank. For example,

$$A^{\alpha\beta}{}_{\gamma;\delta} = A^{\alpha\beta}{}_{\gamma,\delta} + \Gamma^{\alpha}{}_{\epsilon\delta}A^{\epsilon\beta}{}_{\gamma} + \Gamma^{\beta}{}_{\epsilon\delta}A^{\alpha\epsilon}{}_{\gamma} - \Gamma^{\epsilon}{}_{\gamma\delta}A^{\alpha\beta}{}_{\epsilon}\,. \tag{2.60}$$

For future reference, we list here several additional properties of the covariant and contravariant derivatives that the reader may easily verify from the definitions:

$$\text{Product Rule: } \left(A^{\alpha}B^{\beta}\right)_{;\gamma} = A^{\alpha}{}_{;\gamma}B^{\beta} + A^{\alpha}B^{\beta}{}_{;\gamma} \tag{2.61}$$

$$\text{Contraction: } (A^{\alpha}B_{\alpha})_{;\gamma} = (A^{\alpha}B_{\alpha})_{,\gamma} \tag{2.62}$$

$$\text{Gradient: } A_{;\alpha} = A_{,\alpha} \tag{2.63}$$

$$\text{Curl: } A_{\alpha;\beta} - A_{\beta;\alpha} = A_{\alpha,\beta} - A_{\beta,\alpha} \tag{2.64}$$

$$\text{Contravariant: } A^{\alpha;\beta} \equiv g^{\beta\gamma}A^{\alpha}{}_{;\gamma}\,. \tag{2.65}$$

Equations (2.62) and (2.63) follow from the fact that A and $A^{\alpha}B_{\alpha}$ are scalars, and their derivatives therefore do not incur terms analogous to the second one on the right-hand side of Equation (2.54). Equation (2.64) results from the symmetry of $\Gamma^{\gamma}{}_{\alpha\beta}$, i.e., from the fact that $\Gamma^{\gamma}{}_{\alpha\beta} = \Gamma^{\gamma}{}_{\beta\alpha}$ (see Equation 2.48).

The covariant divergence of a four-vector appears often in physical laws (e.g., Maxwell's equations) and is most elegantly expressed with the introduction of a new quantity—the determinant

$$g \equiv \det[g_{\alpha\beta}] \tag{2.66}$$

of the metric tensor. In a non-inertial frame,

$$\bar{g}_{\alpha\beta} = \frac{\partial x^{\gamma}}{\partial X^{\alpha}}\,g_{\gamma\delta}\,\frac{\partial x^{\delta}}{\partial X^{\beta}}\,, \tag{2.67}$$

so that

$$\bar{g} \equiv \det[\bar{g}_{\alpha\beta}] = \left\|\frac{\partial x}{\partial X}\right\|^2 g , \tag{2.68}$$

where the factor multiplying g is the square of the Jacobian for this particular transformation. In a Cartesian inertial frame, $g = -1$ (see Equation 2.16), so g must always be negative—in every frame—since the Jacobian in Equation (2.68) is squared.

Now, the covariant divergence of a four-vector may be written

$$A^\alpha{}_{;\alpha} = A^\alpha{}_{,\alpha} + \Gamma^\alpha{}_{\alpha\beta} A^\beta \tag{2.69}$$

(remembering that $\Gamma^\alpha{}_{\alpha\beta}$ is symmetric). But

$$\Gamma^\alpha{}_{\alpha\beta} = \frac{1}{2} g^{\alpha\gamma}[g_{\beta\gamma,\alpha} + g_{\gamma\alpha,\beta} - g_{\alpha\beta,\gamma}]$$
$$= \frac{1}{2} g^{\alpha\gamma} g_{\gamma\alpha,\beta} , \tag{2.70}$$

since $g^{\alpha\gamma}g_{\beta\gamma,\alpha} = g^{\gamma\alpha}g_{\beta\gamma,\alpha} = g^{\alpha\gamma}g_{\beta\alpha,\gamma} = g^{\alpha\gamma}g_{\alpha\beta,\gamma}$. If we let $[g_{\gamma\alpha}]$ denote the matrix of coefficients $g_{\gamma\alpha}$, then this equation may also be written

$$\Gamma^\alpha{}_{\alpha\beta} = \frac{1}{2} \mathrm{Tr} \left\{ [g_{\gamma\alpha}]^{-1} \frac{\partial}{\partial x^\beta} [g_{\gamma\alpha}] \right\} , \tag{2.71}$$

where Tr is the trace of the matrix enclosed within the curly brackets, i.e., the sum of its diagonal elements. Thus,

$$\Gamma^\alpha{}_{\alpha\beta} = \frac{1}{2} \frac{\partial}{\partial x^\beta} \ln(-g) . \tag{2.72}$$

This can be shown by considering the variation in $\ln(-\det A)$ due to a variation δx^β in x^β, where A is any arbitrary matrix with a negative determinant:

$$\delta \ln(-\det A) = \ln(-\det[A + \delta A]) - \ln(-\det A)$$
$$= \ln\left(\frac{\det[A + \delta A]}{\det A}\right)$$
$$= \ln\left(\det\{A^{-1}[A + \delta A]\}\right)$$
$$= \ln\left(\det[1 + A^{-1}\delta A]\right) . \tag{2.73}$$

For small δA, this gives

$$\delta \ln(-\det A) \to \ln\left(1 + \mathrm{Tr}[A^{-1}\delta A]\right)$$
$$\to \mathrm{Tr}\left(A^{-1}\delta A\right) . \tag{2.74}$$

Since $\delta A = (\partial A/\partial x^\beta)\,\delta x^\beta$, and similarly for $\delta \ln(\det A)$, the result in Equation (2.72) then follows by taking the coefficient multiplying δx^β on both sides of Equation (2.74) and substituting $[g_{\gamma\alpha}]$ for A.

Equation (2.72) may also be written in the more convenient form

$$\Gamma^{\alpha}{}_{\alpha\beta} = \frac{1}{\sqrt{-g}} \frac{\partial}{\partial x^{\beta}} \sqrt{-g} \ . \tag{2.75}$$

So returning to Equation (2.69), we see that

$$A^{\alpha}{}_{;\alpha} = \frac{\partial A^{\alpha}}{\partial x^{\alpha}} + \frac{1}{\sqrt{-g}} \frac{\partial}{\partial x^{\beta}} \sqrt{-g} \, A^{\beta} \ , \tag{2.76}$$

and therefore

$$A^{\alpha}{}_{;\alpha} = \frac{1}{\sqrt{-g}} \frac{\partial}{\partial x^{\alpha}} \left(\sqrt{-g} A^{\alpha} \right) \ , \tag{2.77}$$

the non-inertial effects now all subsumed in g.

In constructing differential equations to represent invariant physical laws, it will also be useful to know that individual indices may be raised or lowered using the metric tensor without introducing additional terms. This follows from *Ricci's Lemma*, which states that

$$g_{\alpha\beta;\gamma} = 0 \ . \tag{2.78}$$

We can prove this directly from Equation (2.58), using the definition of the Christoffel symbols in Equation (2.45):

$$
\begin{aligned}
g_{\alpha\beta;\gamma} &= g_{\alpha\beta,\gamma} - \Gamma^{\epsilon}{}_{\alpha\gamma}\, g_{\epsilon\beta} - \Gamma^{\epsilon}{}_{\beta\gamma}\, g_{\alpha\epsilon} \\
&= g_{\alpha\beta,\gamma} - \frac{1}{2} g_{\epsilon\beta}\, g^{\epsilon\delta} [g_{\gamma\delta,\alpha} + g_{\delta\alpha,\gamma} - g_{\alpha\gamma,\delta}] \\
&\quad - \frac{1}{2} g_{\alpha\epsilon}\, g^{\epsilon\delta} [g_{\gamma\delta,\beta} + g_{\delta\beta,\gamma} - g_{\beta\gamma,\delta}] \\
&= g_{\alpha\beta,\gamma} - \frac{1}{2}[g_{\gamma\beta,\alpha} + g_{\beta\alpha,\gamma} - g_{\alpha\gamma,\beta} \\
&\quad + g_{\gamma\alpha,\beta} + g_{\alpha\beta,\gamma} - g_{\beta\gamma,\alpha}] \\
&= 0 \ . \tag{2.79}
\end{aligned}
$$

Similarly,

$$g^{\alpha\beta}{}_{;\gamma} = 0 \ . \tag{2.80}$$

2.5 RELATIVISTIC CONTINUUM MECHANICS

The metric coefficients $g_{\alpha\beta}$ generally vary across a given spacetime domain. Thus, the derivation of the equations of general relativity must be based on conservation laws applied to continuously changing physical conditions. We shall begin by considering the relativistically correct equations for the conservation of mass, momentum, and energy in a Cartesian inertial frame. The transformation to any other arbitrary frame may then be made using the tools we developed in the previous section. To facilitate the notation in cases of possible confusion, we shall use a subscript '0' to denote quantities measured in a locally inertial frame *comoving* with the fluid.

Consider an element of matter with rest mass Δm_0 and density ρ_{m0} moving with velocity $\mathbf{v}$ relative to an observer. As a result of length contraction, its volume element is $\Delta V = \Delta V_0/\gamma$, where $\gamma = [1 - (v/c)^2]^{-1/2}$, and since Δm_0 is invariant, the observer infers a density

$$\rho_m = \gamma \rho_{m0} \ . \tag{2.81}$$

Defining the four-velocity in the usual way (see Equation 2.37), and noting that $dt = \gamma \, d\tau$ (as we pointed out above, we shall first derive the conservation laws in a Cartesian inertial frame, so the coordinates will be represented using lower case letters), we may write

$$u^\alpha = (\gamma c, \gamma \mathbf{v}) \ . \tag{2.82}$$

This then permits us to define the mass-current four-vector in terms of u^α and the rest-frame density ρ_{m0}:

$$\begin{aligned} S^\alpha &\equiv \rho_{m0} u^\alpha \\ &= (\rho_{m0} \gamma c, \rho_{m0} \gamma \mathbf{v}) \\ &= (\rho_m c, \rho_m \mathbf{v}) \ . \end{aligned} \tag{2.83}$$

Let us now follow this volume element ΔV as it moves along its trajectory specified by $\mathbf{v}$. In a time dt, this volume changes by an amount

$$d(\Delta V) = \int_{\Delta \Sigma} d\mathbf{a} \cdot \mathbf{v} \, dt \ , \tag{2.84}$$

where $\Delta \Sigma$ is the surface surrounding ΔV, and $d\mathbf{a}$ is the infinitesimal area. Thus, by Gauss's divergence theorem, we may write

$$d(\Delta V) = dt \int_{\Delta V} d^3x \, \vec{\nabla} \cdot \mathbf{v} \ , \tag{2.85}$$

and in the limit $\Delta V \to 0$,

$$\frac{1}{\Delta V} \frac{d(\Delta V)}{dt} \approx \vec{\nabla} \cdot \mathbf{v} \ . \tag{2.86}$$

The rest mass within ΔV must be conserved, and therefore

$$\frac{d}{dt} (\rho_m \Delta V) = 0 \tag{2.87}$$

which, by the chain rule of differentiation, yields

$$\frac{d\rho_m}{dt} \Delta V + \rho_m \frac{d(\Delta V)}{dt} = 0 \ . \tag{2.88}$$

Thus, using Equation (2.86), we arrive at the expression for the conservation of rest mass

$$\frac{d\rho_m}{dt} + \rho_m \vec{\nabla} \cdot \mathbf{v} = 0 \ . \tag{2.89}$$

However, the variation in ρ_m may come either from an actual evolution in time, or from motion of the fluid through an inhomogeneous region. By the chain rule of differentiation,

$$\frac{d}{dt} = \frac{\partial}{\partial t} + \mathbf{v} \cdot \vec{\nabla} , \qquad (2.90)$$

and therefore

$$\frac{\partial \rho_m}{\partial t} + \vec{\nabla} \cdot (\rho_m \mathbf{v}) = 0 . \qquad (2.91)$$

In the language of four-dimensional spacetime, the continuity of mass equation therefore reads

$$S^\alpha{}_{,\alpha} = 0 . \qquad (2.92)$$

We may derive an equation for the conservation of momentum in a similar fashion. For simplicity, though—and since this is really the only case we need to consider in cosmology—we shall assume that the force is parallel (or anti-parallel) to the velocity. In that case, the relativistic force (Equation 2.21) is

$$f^\alpha = \left(\gamma \frac{Fv}{c}, \gamma \frac{F\mathbf{v}}{v} \right) , \qquad (2.93)$$

in terms of the Newtonian force F.

Since we are dealing with mass densities, it is necessary for us to define an analogous force density

$$\phi^\alpha \equiv \frac{1}{\Delta V_0} f^\alpha \qquad (2.94)$$

or, in component form,

$$\phi^\alpha = \left(\frac{\xi v}{c}, \frac{\xi \mathbf{v}}{v} \right) , \qquad (2.95)$$

where $\xi \equiv F/\Delta V$ is the Newtonian force density. Thus, for a continuously distributed mass, Equation (2.36) may also be written

$$\rho_{m0} \frac{du^\alpha}{d\tau} = \phi^\alpha . \qquad (2.96)$$

Now, with the goal of transforming this expression into a continuity equation like (2.92), we write

$$\rho_{m0} \frac{\partial u^\alpha}{\partial x^\beta} \frac{dx^\beta}{d\tau} = \phi^\alpha , \qquad (2.97)$$

or

$$\rho_{m0} \frac{\partial u^\alpha}{\partial x^\beta} u^\beta = \phi^\alpha . \qquad (2.98)$$

That is,

$$\frac{\partial}{\partial x^\beta} \left(\rho_{m0} u^\alpha u^\beta \right) - u^\alpha \frac{\partial}{\partial x^\beta} \left(\rho_{m0} u^\beta \right) = \phi^\alpha . \qquad (2.99)$$

But $\rho_{m0} u^\beta = S^\beta$, and $S^\beta{}_{,\beta} = 0$, so

$$\theta^{\alpha\beta}{}_{,\beta} = \phi^\alpha , \qquad (2.100)$$

where the matrix of coefficients

$$\theta^{\alpha\beta} \equiv \rho_{m0} u^{\alpha} u^{\beta} \tag{2.101}$$

is known as the kinetic energy-momentum tensor. Equation (2.100) is the kinetic energy-momentum analog of the continuity equation (2.92), in the sense that the components $\theta^{\alpha\beta}$ are conserved when there is no force on the fluid, i.e., when $\phi^{\alpha} = 0$.

In matrix form, Equation (2.101) is written

$$[\theta^{\alpha\beta}] = \begin{pmatrix} \rho_e & c\mathbf{g} \\ c\mathbf{g} & g^i v^j \end{pmatrix} , \tag{2.102}$$

where

$$\rho_e \equiv \gamma \rho_m c^2 \tag{2.103}$$

is the fluid's energy density, and

$$\mathbf{g} \equiv \gamma \rho_m \mathbf{v} \tag{2.104}$$

is its (vector) momentum density. From these components, we also identify the momentum current $g^i v^j$ (of the ith component of momentum in the jth coordinate direction) and the energy current $\rho_e \mathbf{v} = c^2 \mathbf{g}$. Of course, in its own frame, the fluid possesses only a rest-mass energy density:

$$[\theta_0^{\alpha\beta}] = \begin{pmatrix} \rho_{m0} c^2 & 0 \\ 0 & 0 \end{pmatrix} . \tag{2.105}$$

Sometimes it is also convenient to write Equation (2.100) in a form analogous to the mass continuity equation (2.91). From the $\alpha = 0$ component,

$$\theta^{0\beta}{}_{,\beta} = \phi^0 , \tag{2.106}$$

we find that

$$\frac{\partial \rho_e}{\partial t} + \vec{\nabla} \cdot (\rho_e \mathbf{v}) = \xi v , \tag{2.107}$$

which is the continuity equation for energy. The right-hand side is the power density, which can add or remove energy from the fluid (depending on whether the force is parallel or anti-parallel to the velocity). And from the ith component of Equation (2.100),

$$\theta^{i\beta}{}_{,\beta} = \phi^i , \tag{2.108}$$

we obtain the continuity equation for momentum:

$$\frac{\partial g^i}{\partial t} + \vec{\nabla} \cdot (g^i \mathbf{v}) = \phi^i , \tag{2.109}$$

for each of the components $i = (1, 2, 3)$.

The force density ϕ^α has contributions from internal influences arising, e.g., from pressure gradients, and from influences outside the fluid, such as an electromagnetic field (but not gravity because in general relativity, the effects of gravity are manifested through the curvature of spacetime). It is often convenient to show this dichotomy explicitly by writing

$$\phi^\alpha = \phi^\alpha_{\text{elas}} + \phi^\alpha_{\text{ext}} \tag{2.110}$$

and, moreover, to characterize the 'elastic' forces in terms of a *stress* tensor $\mathcal{S}^{\alpha\beta}$, where

$$\phi^\alpha_{\text{elas}} \equiv -\mathcal{S}^{\alpha\beta}{}_{,\beta} \ . \tag{2.111}$$

Then, if we define the energy-momentum tensor for matter,

$$T_{\text{m}}^{\alpha\beta} \equiv \theta^{\alpha\beta} + \mathcal{S}^{\alpha\beta} \ , \tag{2.112}$$

Equation (2.100) becomes

$$T_{\text{m}}^{\alpha\beta}{}_{,\beta} = \phi^\alpha_{\text{ext}} \ , \tag{2.113}$$

demonstrating that only external forces can alter the total energy and momentum densities of the fluid.

For example, when the external influence is due to an electromagnetic field, we may write

$$\phi^\alpha_{\text{ext}} = -T_{\text{em}}^{\alpha\beta}{}_{,\beta} \ , \tag{2.114}$$

where in obvious notation, $T_{\text{em}}^{\alpha\beta}$ is the energy-momentum tensor for this field. For this particular case, we may even consider the matter and fields together as a closed system, defining the total energy-momentum tensor

$$T^{\alpha\beta} \equiv T_{\text{m}}^{\alpha\beta} + T_{\text{em}}^{\alpha\beta} \ , \tag{2.115}$$

with the consequence that

$$T^{\alpha\beta}{}_{,\beta} = 0 \ . \tag{2.116}$$

This is as far as we need to go in our development of relativistic continuum mechanics, but before we proceed to the Principle of Equivalence, there are two additional issues we need to address. First, although the covariant physical laws we just derived contain all of the equations of motion needed to describe the behavior of the fluid, relations such as (2.113) must be generalized when applied to non Cartesian inertial frames. As we saw in Equation (2.57), the derivative of a tensor incurs additional terms. Thus, e.g., in any arbitrary (possibly accelerated) frame,

$$T_{\text{m}}^{\alpha\beta}{}_{;\beta} = \phi^\alpha_{\text{ext}} \ . \tag{2.117}$$

Second, most applications in cosmology adopt the so-called *perfect fluid* approximation, in which the stress tensor (and hence the energy-momentum tensor) simplifies considerably. This happens because in a perfect fluid there are no shear forces that would lead to the transport of momentum components in directions other than those associated with the components themselves. Using language that we developed earlier, we would say that a perfect fluid is one in which the momentum current $g^i v^j$ is zero unless $i = j$.

In a frame instantaneously at rest with the fluid, the stress tensor $\mathcal{S}^{\alpha\beta}$ must therefore be diagonal, and since ϕ^0_{elas} is zero in this frame, we have

$$[\mathcal{S}_0^{\alpha\beta}] = \begin{pmatrix} 0 & 0 & 0 & 0 \\ 0 & P_0 & 0 & 0 \\ 0 & 0 & P_0 & 0 \\ 0 & 0 & 0 & P_0 \end{pmatrix}, \qquad (2.118)$$

where P_0 is the pressure. Thus, with Equation (2.105), we see that

$$[T_{\text{m0}}^{\alpha\beta}] = \begin{pmatrix} \rho_{\text{m0}}c^2 & 0 & 0 & 0 \\ 0 & P_0 & 0 & 0 \\ 0 & 0 & P_0 & 0 \\ 0 & 0 & 0 & P_0 \end{pmatrix}. \qquad (2.119)$$

But since most of these matrix components are zero for a perfect fluid, it is actually more convenient to express the matter energy-momentum tensor in the following way. Evidently,

$$T_{\text{m0}}^{00} = \left(\rho_{\text{m0}} + \frac{P_0}{c^2}\right)c^2 - P_0, \qquad (2.120)$$

or

$$T_{\text{m0}}^{00} = \left(\rho_{\text{m0}} + \frac{P_0}{c^2}\right)u_0^0 u_0^0 - P_0\,\eta^{00}, \qquad (2.121)$$

because $\gamma = 1$ for u^α evaluated in the comoving frame. In addition, since $u_0^i = 0$ (for $i \neq 0$) in this frame, we may also write

$$T_{\text{m0}}^{ij} = \left(\rho_{\text{m0}} + \frac{P_0}{c^2}\right)u_0^i u_0^j - P_0\,\eta^{ij}, \qquad (2.122)$$

and

$$T_{\text{m0}}^{0i} = \left(\rho_{\text{m0}} + \frac{P_0}{c^2}\right)u_0^0 u_0^i - P_0\,\eta^{0i}. \qquad (2.123)$$

We may group these together into a single expression,

$$T_{\text{m0}}^{\alpha\beta} = \left(\rho_{\text{m0}} + \frac{P_0}{c^2}\right)u_0^\alpha u_0^\beta - P_0\,\eta^{\alpha\beta}, \qquad (2.124)$$

which clearly has the form of a rank-2 tensor, and must therefore be valid in any inertial frame:

$$T_{\text{m}}^{\alpha\beta} = \left(\rho_{\text{m0}} + \frac{P_0}{c^2}\right)u^\alpha u^\beta - P_0\,\eta^{\alpha\beta}. \qquad (2.125)$$

Figure 2.2 Today's view of the roof next door as seen from Einstein's patent office window in Bern, Switzerland. (Photo by the author)

The contravariant tensor that reduces to Equation (2.125) in the absence of acceleration is thus

$$T_{\mathrm{m}}^{\mu\nu} = \left(\rho_{\mathrm{m}0} + \frac{P_0}{c^2}\right) u^\mu u^\nu - P_0\, g^{\mu\nu}\,, \qquad (2.126)$$

where u^μ is the local value of $dx^\mu/d\tau$ for a comoving fluid element. Note that in all cases, P_0 and $\rho_{\mathrm{m}0}$ are the pressure and rest mass density, respectively, measured by an observer in a locally inertial frame comoving with the fluid at the instant of measurement, and are therefore Lorentz scalars.

2.6 THE PRINCIPLE OF EQUIVALENCE

Two years after the advent of special relativity, Einstein was writing a review on the new physics, which at the time he called *invariance theory*, when he began to wonder how Newtonian gravitation could be modified to make it consistent with the new relativity theory. The idea that took shape in his mind, described later by him as the 'happiest thought of my life,' was that an observer falling from the roof of a house experiences no gravitational field (see Figure 2.2 for a view of the house next door to the Swiss patent office, as seen from Einstein's window in Bern). The unfortunate homeowner is in *free fall* within Earth's gravity, but every loose object in his vicinity falls with him at exactly the same rate, since the acceleration due to Earth's pull is completely independent of inertial mass, as Galileo had deduced several centuries earlier. Thus, the observer considers his free-falling frame of reference to be *inertial*, since nothing within it accelerates relative to anything else nearby.

Regardless of what actually causes the pull of gravity, its effect is entirely equivalent to a uniform acceleration throughout a given volume of space. Einstein called this the *principle of equivalence*, in the sense that one can produce the same effect whether a uniform gravitational field is accelerating everything in the same direction, or whether the frame as a whole is experiencing a uniform acceleration pointing the opposite way.

The 'strong' version of this principle goes one step further, by asserting that the laws of nature in such a free-falling frame take the same form as in a Cartesian inertial frame *without* gravitation. Einstein hypothesized that since all gravitational fields vanish inside a free-falling frame, special relativity ought to apply to all measurements of distances and times within that frame. And in a leap of faith (or inspiration, or both), he argued that the two postulates of the special theory should apply even in cases where we compare the measurements of an observer in this frame with those of a distant observer, for whom the effects of the gravitational field are negligible. Thus, in an elegant and all-encompassing way, the theory of gravity was merged with the framework of special relativity. Gravity is described by its equivalence to an accelerated frame, and its effects are thereby fully incorporated into the laws of physics via the properties of special relativity. This is the essence of the *general* theory of relativity applied to gravity.

When matter (or energy, since energy and mass are equivalent in special relativity[2]) is present, the gravitational field it produces is described by the metric $g_{\alpha\beta}(x)$, different from the 'flat-space' metric $\eta_{\alpha\beta}(x)$. Because this altered $g_{\alpha\beta}(x)$ has an effect on the particle *trajectories*, it is often said that the gravitational field changes the geometry of spacetime. For example, in an inertial frame, the geodesic Equation (2.41) reduces to the form $d^2x^\alpha/d\tau^2 = 0$, which describes a straight-line trajectory. But $d^2X^\alpha/d\tau^2 \neq 0$ in an accelerated frame chosen (via the equivalence principle) to represent the local effect of gravity.

The procedure for describing a gravitational field thus calls for a transformation into the accelerated (i.e., laboratory) frame X^α from a Cartesian inertial frame x^α. According to the principle of equivalence, one can always choose a frame for which this requirement can be satisfied *locally*. (This concept is sometimes formally referred to as *The Local Flatness Theorem* [49]. In Chapter 10, we shall use this theorem to explore a crucial property of the so-called lapse function, g_{tt}, in the Friedmann-Lemaître-Robertson-Walker metric.) To see this explicitly, let us suppose that the metric tensor $g_{\alpha\beta}$, representing the gravitational field, is known at some spacetime

---

[2]In Chapter 10, we shall discuss a likely reason for this equivalence, based on the interpretation of a particle's rest-mass energy as a gravitational binding energy to that portion of the Universe with which it is causally connected.

point $\hat{X}^\alpha$ in the accelerated frame. Then

$$\hat{g}_{\alpha\beta} = g_{\alpha\beta}(\hat{X}) \,, \tag{2.127}$$

from which we can also calculate the Christoffel symbols using Equation (2.45):

$$\hat{\Gamma}^\alpha{}_{\beta\gamma} = \Gamma^\alpha{}_{\beta\gamma}(\hat{X}) \,. \tag{2.128}$$

Thus, finding the local Cartesian inertial frame amounts to uncovering the set of coordinates $x^\mu = x^\mu(\hat{X})$ for which

$$\hat{\Gamma}^\delta{}_{\alpha\beta} = \frac{\partial^2 x^\epsilon}{\partial X^\alpha \partial X^\beta} \frac{\partial X^\delta}{\partial x^\epsilon} \,. \tag{2.129}$$

We can transform this expression into an equation we can solve by separating the two terms on the right-hand side. So let us contract both sides with $\partial x^\eta / \partial X^\delta$, which gives

$$\begin{aligned}
\frac{\partial x^\eta}{\partial X^\delta} \hat{\Gamma}^\delta{}_{\alpha\beta} &= \frac{\partial^2 x^\epsilon}{\partial X^\alpha \partial X^\beta} \frac{\partial X^\delta}{\partial x^\epsilon} \frac{\partial x^\eta}{\partial X^\delta} \\
&= \frac{\partial^2 x^\epsilon}{\partial X^\alpha \partial X^\beta} \delta^\eta_\epsilon \\
&= \frac{\partial^2 x^\eta}{\partial X^\alpha \partial X^\beta} \,.
\end{aligned} \tag{2.130}$$

It is straightforward to see now that this equation has the following 'polynomial' solution,

$$x^\alpha = a^\alpha + b^\alpha{}_\mu \left(X^\mu - \hat{X}^\mu \right) + \frac{1}{2} b^\alpha{}_\lambda \hat{\Gamma}^\lambda{}_{\mu\nu} \left(X^\mu - \hat{X}^\mu \right) \left(X^\nu - \hat{X}^\nu \right) + \dots \,, \tag{2.131}$$

where a^α and $b^\alpha{}_\lambda$ are constants. For any given spacetime point $\hat{X}^\alpha$, we can therefore always find the Cartesian inertial frame in which x^α is given in terms of the known $\Gamma^\alpha{}_{\beta\gamma}(\hat{X})$.

The idea that the effects of gravity on the motion of particles may be described with the geodesic Equation (2.41) suggests we examine their behavior in the weak-field limit, where this formulation should reduce to Newton's theory. When particles are moving slowly, $\gamma|\mathbf{v}| \ll \gamma c$ in Equation (2.82), so

$$\frac{d^2 X^\alpha}{d\tau^2} \approx -\Gamma^\alpha{}_{00} \frac{dX^0}{d\tau} \frac{dX^0}{d\tau} \tag{2.132}$$

or, more simply,

$$\frac{d^2 X^\alpha}{d\tau^2} \approx -c^2 \Gamma^\alpha{}_{00} \left(\frac{dT}{d\tau} \right)^2 \,. \tag{2.133}$$

Now suppose the field is weak, so that

$$g_{\alpha\beta} = \eta_{\alpha\beta} + h_{\alpha\beta} \,, \tag{2.134}$$

with $|h_{\alpha\beta}| \ll 1$. Then, for a stationary metric, i.e., $g_{\alpha\beta,0} = 0$, we have

$$
\begin{aligned}
\Gamma^{\alpha}{}_{00} &= \frac{1}{2} g^{\alpha\gamma} \left[g_{0\gamma,0} + g_{\gamma 0,0} - g_{00,\gamma} \right] \\
&= -\frac{1}{2} g^{\alpha\gamma} g_{00,\gamma} \\
&= -\frac{1}{2} g_{00}{}^{,\alpha} \\
&= -\frac{1}{2} h_{00}{}^{,\alpha} .
\end{aligned}
\tag{2.135}
$$

For $\alpha = 0$, Equation (2.132) therefore gives

$$
\frac{d^2 T}{d\tau^2} = 0 ,
\tag{2.136}
$$

or

$$
\frac{dT}{d\tau} = 1
\tag{2.137}
$$

(since $dT \to d\tau$ in the non-relativistic limit), whereas for $\alpha = i = (1,2,3)$,

$$
\frac{d^2 X^i}{d\tau^2} = -\frac{1}{2} c^2 h_{00,i} .
\tag{2.138}
$$

That is,

$$
\frac{d^2 \mathbf{X}}{d\tau^2} = -\vec{\nabla}\psi ,
\tag{2.139}
$$

in terms of the (defined) gravitational potential

$$
\psi \equiv \frac{1}{2} c^2 h_{00} .
\tag{2.140}
$$

Evidently, Einstein's theory of gravity, based on the equivalence principle, correctly reduces to Newton's law if

$$
g_{00} = 1 + \frac{2\psi}{c^2} .
\tag{2.141}
$$

For example, at Earth's surface,

$$
\psi = -\frac{GM_E}{R_E} ,
\tag{2.142}
$$

in terms of Earth's mass M_E and radius R_E, which has the numerical value $\psi \approx 6.25 \times 10^7$ m^2 s^{-2}. Therefore, $h_{00} \approx 1.39 \times 10^{-9}$, amply justifying the approximation we made in (2.134). But more importantly, this simple analysis demonstrates that—at least in the presence of a weak field—the principle of equivalence leads to a viable theory of gravity within the context of general relativity.[3]

---

[3]Ref. [50] presents a more detailed history of the events that occurred during the 'golden age' of relativity, spanning the period 1960–1975.

2.7 EINSTEIN'S FIELD EQUATIONS

The key question we are now faced with is how to find the metric corresponding to a given gravitational field. We begin with the hypothesis that gravity may indeed be described by a symmetric tensor $g_{\alpha\beta}$ whose source is the matter and/or energy content of the system, and seek a generalization of Poisson's equation,

$$\nabla^2\psi = 4\pi G\rho_{m0} , \tag{2.143}$$

that correctly produces Newton's law in terms of the gravitational potential ψ and its source—the mass density ρ_{m0}.

But this task is far from trivial, for there are at least two principal differences between Einstein's and Newton's theories. First, mass and energy are interchangeable in relativity, so the self-energy of a mass aggregate is itself a source of gravity, leading to non-linear effects that are completely absent in classical mechanics. Second, whereas Newton's theory accounts for gravity as an action-at-a-distance, no physical effect in relativity can propagate faster than light. Therefore, the generalization of Poisson's equation in relativity must incorporate both the contribution to the field from self-energy and the time-dependent constraints imposed on the ensuing acceleration due to the finite propagation time of the gravitational influence.

Though not obvious at first, a useful approach to follow with this problem is to first find a criterion that tells us whether or not $g_{\alpha\beta}$ is in fact a gravitational field. Returning to the discussion that led to Equations (2.58) and (2.59), we recall that the partial derivative of a tensor, e.g., $A^{\alpha}{}_{,\beta\gamma}$, does not itself transform as a tensor when inertial forces are present, because the transformation coefficients are then also functions of the coordinates and must be differentiated too. This creates a complication because the manner in which the additional terms must be added to $A^{\alpha}{}_{,\beta\gamma}$ to produce a covariant derivative like that shown in Equation (2.58) suggests that the result of differentiation depends on the ordering of the coordinates.

We can demonstrate this formally as follows. Let us write out the tensor $A^{\alpha}{}_{;\beta\gamma}$ in terms of the Christoffel symbols, following the rules established in Equations (2.58) and (2.59):

$$
\begin{aligned}
A^{\alpha}{}_{;\beta\gamma} &= (A^{\alpha}{}_{;\beta})_{;\gamma} \\
&= (A^{\alpha}{}_{;\beta})_{,\gamma} + \Gamma^{\alpha}{}_{\epsilon\gamma} A^{\epsilon}{}_{;\beta} - \Gamma^{\epsilon}{}_{\beta\gamma} A^{\alpha}{}_{;\epsilon} \\
&= A^{\alpha}{}_{,\beta\gamma} + \Gamma^{\alpha}{}_{\delta\beta,\gamma} A^{\delta} + \Gamma^{\alpha}{}_{\delta\beta} A^{\delta}{}_{,\gamma} \\
&\quad + \Gamma^{\alpha}{}_{\epsilon\gamma}(A^{\epsilon}{}_{,\beta} + \Gamma^{\epsilon}{}_{\delta\beta} A^{\delta}) - \Gamma^{\epsilon}{}_{\beta\gamma}(A^{\alpha}{}_{,\epsilon} + \Gamma^{\alpha}{}_{\delta\epsilon} A^{\delta}) .
\end{aligned}
\tag{2.144}
$$

If we change the order of β and γ, however, we get instead

$$
\begin{aligned}
A^{\alpha}{}_{;\gamma\beta} &= A^{\alpha}{}_{,\gamma\beta} + \Gamma^{\alpha}{}_{\delta\gamma,\beta} A^{\delta} + \Gamma^{\alpha}{}_{\delta\gamma} A^{\delta}{}_{,\beta} \\
&\quad + \Gamma^{\alpha}{}_{\epsilon\beta}(A^{\epsilon}{}_{,\gamma} + \Gamma^{\epsilon}{}_{\delta\gamma} A^{\delta}) - \Gamma^{\epsilon}{}_{\gamma\beta}(A^{\alpha}{}_{,\epsilon} + \Gamma^{\alpha}{}_{\delta\epsilon} A^{\delta}) .
\end{aligned}
\tag{2.145}
$$

Since $\Gamma^{\alpha}{}_{\beta\gamma} = \Gamma^{\alpha}{}_{\gamma\beta}$ (see Equation 2.40), many of the terms on the right-hand side of these two equations are the same, but some are not. It is straightforward to show that

$$A^{\alpha}{}_{;\beta\gamma} - A^{\alpha}{}_{;\gamma\beta} = R^{\alpha}{}_{\delta\beta\gamma} A^{\delta} , \qquad (2.146)$$

where

$$R^{\alpha}{}_{\delta\beta\gamma} \equiv \Gamma^{\alpha}{}_{\delta\beta,\gamma} - \Gamma^{\alpha}{}_{\delta\gamma,\beta} - \Gamma^{\alpha}{}_{\epsilon\beta}\Gamma^{\epsilon}{}_{\delta\gamma} + \Gamma^{\alpha}{}_{\epsilon\gamma}\Gamma^{\epsilon}{}_{\delta\beta} \qquad (2.147)$$

is known as the Riemann-Christoffel (or *curvature*) tensor. Similarly,

$$A_{\alpha;\beta\gamma} - A_{\alpha;\gamma\beta} = -R^{\delta}{}_{\alpha\beta\gamma} A_{\delta} . \qquad (2.148)$$

If spacetime is flat, by definition there exists a Cartesian inertial frame x^{α} in which $\Gamma^{\alpha}{}_{\beta\gamma} = 0$ everywhere. Therefore, in this frame, the curvature tensor is also (trivially) zero ($R^{\alpha}{}_{\delta\beta\gamma} = 0$), and this must be true in any reference frame because no combination of zero components can produce a non-zero tensor in another coordinate system. Non-zero values of $R^{\alpha}{}_{\delta\beta\gamma}$ indicate the presence of inertial forces due to an acceleration—meaning that the spacetime is curved—therefore signaling the influence of a gravitational field.

The Riemann-Christoffel tensor appears to have some characteristics in common with the potential ψ in Equation (2.143), insofar as it provides some indication that a source of gravity is present nearby. In addition, one might expect the degree of curvature to correlate with the strength of the source, just as the gradient of ψ increases with mass density ρ_{m0}. It therefore makes sense to incorporate $R^{\alpha}{}_{\delta\beta\gamma}$ into a mathematical generalization of Equation (2.143).

Actually, we already had an inkling from Equation (2.141) that this must be the case, since Poisson's equation apparently involves second-order derivatives of the metric coefficients $g_{\alpha\beta}$, which appear explicitly in the definition of the Riemann-Christoffel tensor (Equation 2.147).

Indeed, we see from Equations (2.141) and (2.143) that

$$\nabla^2 g_{00} = \frac{8\pi}{c^2} G \rho_{m0} . \qquad (2.149)$$

But

$$T^{00}_{m0} = \rho_{m0} c^2 , \qquad (2.150)$$

and since $g^{00} = g_{00}$ (see Equations 2.32 and 2.50), we may write

$$\nabla^2 g^{00} = \frac{8\pi G}{c^4} T^{00}_{m} \qquad (2.151)$$

in the weak-field limit, where $T^{00}_{m0} \approx T^{00}_{m}$. As we anticipated, the tensor generalization to Poisson's equation expresses derivatives of the metric coefficients in terms of the energy-momentum tensor, though we expect that in the final analysis the source on the right-hand side should also include $T^{\alpha\beta}_{em}$.

But how must Equation (2.151) be modified to account for situations in which the field is not weak? This was one of the critical questions faced by Einstein, and later by David Hilbert (1862–1943) (Figure 2.3) and Emmy Noether (1882–1935) (Figure 2.4)

Figure 2.3 David Hilbert, as he here appeared in the 1930's, was one of the most influential mathematicians of the twentieth century. As a professor at Göttingen University in the early 1900's, he was instrumental—together with Albert Einstein and Emmy Noether (Figure 2.4)—in producing the correct field equations of general relativity. (Courtesy of the Deutsches Museum, München, Archiv, PT-01509-02)

Noether's contribution to this effort has not always been fully appreciated. A brief account of how the 'correct' equations of general relativity were obtained may be found in ref. [50]. Upon her passing in 1935, Einstein wrote her a glowing obituary in the New York Times. Her final remains now lie on the campus of Bryn Mawr College, just outside the city of Philadelphia (see Figure 2.5). The collective efforts of Einstein, Hilbert and Noether eventually produced the correct field equations by November, 1915. We can already see that the tensor on the left-hand side must depend on at least the second derivatives of $g^{\alpha\beta}$, and since $T^{\alpha\beta}$ is symmetric, this combination of derivatives must be symmetric as well. Also, conservation of the total energy and momentum dictates that its covariant divergence must be zero since $T^{\alpha\beta}{}_{;\beta} = 0$.

Given these requirements, a viable candidate is a contraction of the Riemann-Christoffel tensor $R^{\alpha}{}_{\beta\gamma\delta}$. After some trial and error, the combination that works best is a contraction on α and γ, producing what is now known as the Ricci tensor:

$$R_{\beta\delta} \equiv g^{\alpha\gamma} R_{\alpha\beta\gamma\delta} = R^{\gamma}{}_{\beta\gamma\delta} , \qquad (2.152)$$

though this by itself is not yet sufficient since one can show that $R^{\alpha\beta}{}_{;\beta} \neq 0$. We shall leave it as an exercise for the reader to demonstrate, from the definition of $R^{\alpha}{}_{\beta\gamma\delta}$ in Equation (2.151), that instead of zero, one gets

$$R^{\alpha\beta}{}_{;\beta} = \frac{1}{2} R^{;\alpha} , \qquad (2.153)$$

where the contracted Ricci tensor

$$R \equiv g^{\mu\nu} R_{\mu\nu} \qquad (2.154)$$

is known as the *curvature scalar*. Thus, in order to satisfy the conservation of energy and momentum, which leads to the condition $T^{\alpha\beta}{}_{;\beta} = 0$, $T^{\alpha\beta}$ must be proportional to a tensor of the form

$$G^{\alpha\beta} \equiv R^{\alpha\beta} - \frac{1}{2} g^{\alpha\beta} R , \qquad (2.155)$$

which is known as the Einstein tensor.

Evidently, the generalized form of Poisson's equation may be written

$$G^{\alpha\beta} = K T^{\alpha\beta} , \qquad (2.156)$$

where the proportionality constant K must be determined by requiring that these *field equations* reduce to their known weak-field limit given in Equation (2.151). In the non-relativistic ($|\mathbf{v}| \ll c$) limit, $|T^{ij}| \ll |T^{00}|$ (see Equation 2.126). Therefore,

$$R^{ij} - \frac{1}{2} g^{ij} R \approx 0 , \qquad (2.157)$$

so

$$R^{ij} \approx \frac{1}{2} g^{ij} R . \qquad (2.158)$$

And for a weak-field, $g^{\alpha\beta} \approx \eta^{\alpha\beta}$, in which case the curvature scalar may be written

$$R = R^{\alpha}{}_{\alpha} = R^{00} - R^{ii} . \qquad (2.159)$$

Figure 2.4 In 1915, Emmy Noether (shown here while still in Europe) was a key participant, along with David Hilbert (Figure 2.3), in helping Einstein arrive at the 'correct' equations of general relativity. She emigrated to the United States in 1933, but passed away just a few years later (see Figure 2.5). (Image courtesy of the P. Roquette archives, Heidelberg and the Göttingen State and University Library, Göttingen)

From Equation (2.158) we infer that $R^{ii} = -(3/2)R$, so putting everything together,

$$R = -2R^{00} . \tag{2.160}$$

To complete the Einstein tensor, we also need to find the weak-field limit of $R^{\alpha}{}_{\delta\beta\gamma}$. By definition,

$$R_{00} = R^{\alpha}{}_{0\alpha0} = R_{0000} - R_{i0i0} . \tag{2.161}$$

But for a static field, we may also put

$$R_{0000} \approx 0 , \tag{2.162}$$

and

$$R_{i0i0} = \frac{1}{2}(g_{ii,00} + g_{00,ii} - g_{i0,i0} - g_{i0,i0})$$

$$= \frac{1}{2}\nabla^2 g_{00} . \tag{2.163}$$

Therefore,

$$G_{00} = R_{00} - \frac{1}{2}g_{00} R$$

$$= 2R_{00}$$

$$= -\nabla^2 g_{00} . \tag{2.164}$$

Thus in the weak-field limit, the generalized Poisson equation reduces to Equation (2.151),

$$-\nabla^2 g_{00} = -\nabla^2 g^{00} = KT^{00} , \tag{2.165}$$

but only so long as

$$K = -\frac{8\pi G}{c^4} . \tag{2.166}$$

The gravitational field equations in general relativity are evidently

$$G^{\alpha\beta} \equiv R^{\alpha\beta} - \frac{1}{2} g^{\alpha\beta} R = -\frac{8\pi G}{c^4} T^{\alpha\beta} . \tag{2.167}$$

The appearance of the curvature scalar on the left-hand side of this equation is sometimes inconvenient. If we contract on α and β, we find that

$$g_{\alpha\beta} R^{\alpha\beta} - \frac{1}{2}g_{\alpha\beta} g^{\alpha\beta} R = -\frac{8\pi G}{c^4} g_{\alpha\beta} T^{\alpha\beta} , \tag{2.168}$$

which trivially reduces to the form

$$R = \frac{8\pi G}{c^4} T^{\gamma}{}_{\gamma} . \tag{2.169}$$

Figure 2.5 Noether's final remains now reside in the Cloisters (at the marked stone to the lower left in this image) of the Thomas library at Bryn Mawr College, just outside Philadelphia. (Photo by the author)

Thus, an alternative representation of the field equations is

$$R^{\alpha\beta} = -\frac{8\pi G}{c^4}\left(T^{\alpha\beta} - \frac{1}{2}\,g^{\alpha\beta}\,T^{\gamma}{}_{\gamma}\right)\;. \tag{2.170}$$

We should mention here that Equation (2.167) contains an additional degree of freedom of fundamental importance to cosmology. As we have already seen, the principal constraints that must be satisfied by the left-hand side of this equation are that its covariant divergence must be zero (to comply with the conservation of energy and momentum), and that it must be symmetric (because $T^{\alpha\beta}$ is symmetric). But we can still retain these properties even if we add a term proportional to $g^{\alpha\beta}$. The metric tensor is clearly symmetric, and the first of these conditions is automatically satisfied because of Ricci's Lemma (Equation 2.78). Thus, introducing the *Cosmological* constant[4] Λ, we write the field equations in their most general form:

$$G^{\alpha\beta} - \Lambda\,g^{\alpha\beta} = -\frac{8\pi G}{c^4}\,T^{\alpha\beta}\;. \tag{2.171}$$

---

[4]As we shall see in Chapter 6, Λ appears to play an indispensable role in the current standard model of cosmology. It is one of the leading candidates for what is commonly referred to as 'dark energy.'

Though at first sight it may appear that 16 equations must be solved for a complete description of gravity, in actuality the irreducible set contains far fewer. First, $G^{\alpha\beta}$ (and $g^{\alpha\beta}$ if we include a Λ) is symmetric, so Equation (2.171) can have at most only 10 different components. But the fact that $T^{\alpha\beta}{}_{;\beta} = 0$ also means that there are four non-ignorable constraints. Thus, in total, one is left with six simultaneous equations to work with for a complete solution.

CHAPTER 3

The Black-hole Spacetime

W ITH Einstein's equations now at our disposal, we are in a position to begin examining several important solutions describing the spacetime surrounding a compact object. The Friedmann-Lemaître-Robertson-Walker metric, the backbone of standard cosmology, will be featured in the following chapter. The Schwarzschild and Kerr metrics discussed here inform several important aspects of the cosmic spacetime, notably the role played by the gravitational horizon, so their inclusion should be viewed as an essential foundation for the more elaborate developments throughout the remainder of this book.

We shall see in our derivations that the approach often used to find a solution involves the initial identification of a symmetry in the system, which provides a reasonable guess for the functional form of the coefficients $g_{\alpha\beta}$ in the metric ds (Equation 2.31). This is followed by an evaluation of the Ricci tensor (Equation 2.152) and its substitution into Einstein's field equations which, when solved, provide us with the structure of spacetime. We know this because the derivatives of $g_{\alpha\beta}$ reveal the inertial forces arising from the influence of gravity, and therefore encoded within them is the information telling us how the intervals of space and time relate to each other based on the local curvature.

3.1 SCHWARZSCHILD METRIC

The first significant step in solving Equation (2.167) was taken within only months of its publication. Karl Schwarzschild (1873–1916) (see Figure 3.1), then a professor at Potsdam, had volunteered for military service in 1914, and had been stationed on the Russian front when he received copies of Einstein's papers. He had the foresight, while avoiding becoming a casualty of war, to invoke the highest degree of symmetry he could imagine in order to eliminate as many of the six independent components as possible in this expression.

As we shall see shortly, Schwarzschild produced a description of gravity surrounding a single compact object, such as Earth or the Sun. Sadly, he contracted an illness soon after his discovery, and died upon returning home. At the time, his work was considered to be purely theoretical, with little application to reality. Today, however,

Figure 3.1 Karl Schwarzschild was a professor of Physics at the University of Potsdam when he produced the first complete solution to Einstein's field equations of general relativity. His metric corresponds to a highly simplified, time-independent spacetime surrounding a spherically symmetric mass. (Image courtesy of the Leibniz-Institut für Astrophysik Potsdam [51])

Schwarzschild's name is rightfully associated with several distinguishing features of black holes.

The most general, static, spherically symmetric form of the metric may be written

$$ds^2 = F(r)\,c^2 dT^2 - 2rE(r)c\,dT\,dr - r^2 D(r)\,dr^2$$
$$-C(r)(dr^2 + r^2\,d\theta^2 + r^2\sin^2\theta\,d\phi^2)\,. \qquad (3.1)$$

(Following the notation we established in the previous chapter, we shall designate the time measured in the accelerated frame with a capital letter.) This metric, written in terms of arbitrary functions F, E, C, and D, permits all possible combinations of the coordinate elements without violating spherical symmetry and, in addition, maintains spatial intervals that themselves do not depend on time. Redefining the time

$$T' = T + f(r)\,, \qquad (3.2)$$

so that

$$dT = dT' - \frac{df}{dr}\,dr\,, \qquad (3.3)$$

we shall choose the function $f(r)$ in order to satisfy the condition

$$\frac{df}{dr} = -\frac{rE}{Fc}\,. \qquad (3.4)$$

This choice makes the cross terms vanish. In addition, we may also re-normalize the radial coordinate to

$$r' = \sqrt{C(r)}r\,, \qquad (3.5)$$

which eliminates the coefficient multiplying the angular term of the metric. Finally, relabeling the time coordinate as T and the radial coordinate as r, we arrive at a useful (standard) representation of this metric, written as

$$ds^2 = B(r)c^2 dT^2 - A(r)\,dr^2 - r^2(d\theta^2 + \sin^2\theta\,d\phi^2)\,, \qquad (3.6)$$

where the functions A and B are also arbitrary.

The metric tensor is therefore

$$[g_{\alpha\beta}] = \begin{pmatrix} B(r) & 0 & 0 & 0 \\ 0 & -A(r) & 0 & 0 \\ 0 & 0 & -r^2 & 0 \\ 0 & 0 & 0 & -r^2\sin^2\theta \end{pmatrix}, \qquad (3.7)$$

with a determinant

$$g = -r^4\sin^2\theta\,A(r)B(r)\,. \qquad (3.8)$$

The corresponding inverse metric tensor (see Equation 2.50) is

$$[g^{\alpha\beta}] = \begin{pmatrix} B(r)^{-1} & 0 & 0 & 0 \\ 0 & -A(r)^{-1} & 0 & 0 \\ 0 & 0 & -r^{-2} & 0 \\ 0 & 0 & 0 & -(r\sin\theta)^{-2} \end{pmatrix} . \tag{3.9}$$

Given this simple diagonal form for $g_{\alpha\beta}$, only 13 Christoffel symbols are non-zero, and these may be evaluated directly from Equation (2.45). They are

$$\Gamma^T{}_{Tr} = \Gamma^T{}_{rT} = \frac{B'}{2B}$$

$$\Gamma^r{}_{TT} = \frac{B'}{2A}$$

$$\Gamma^r{}_{rr} = \frac{A'}{2A}$$

$$\Gamma^r{}_{\theta\theta} = -\frac{r}{A}$$

$$\Gamma^r{}_{\phi\phi} = -\frac{r^2 \sin^2\theta}{A}$$

$$\Gamma^\theta{}_{\theta\theta} = -\sin\theta\cos\theta$$

$$\Gamma^\theta{}_{r\theta} = \Gamma^\theta{}_{\theta r} = \frac{1}{r}$$

$$\Gamma^\phi{}_{r\phi} = \Gamma^\phi{}_{\phi r} = \frac{1}{r}$$

$$\text{and} \quad \Gamma^\phi{}_{\theta\phi} = \Gamma^\phi{}_{\phi\theta} = \cot\theta , \tag{3.10}$$

where prime denotes a derivative with respect to r. From here it is straightforward to calculate the components of the Ricci tensor (Equation 2.152), which are

$$R_{TT} = -\frac{B''}{2A} + \frac{B'}{4A}\left(\frac{A'}{A} + \frac{B'}{B}\right) - \frac{1}{r}\frac{B'}{A}$$

$$R_{rr} = \frac{B''}{2B} - \frac{B'}{4B}\left(\frac{A'}{A} + \frac{B'}{B}\right) - \frac{1}{r}\frac{A'}{A}$$

$$R_{\theta\theta} = -r + \frac{r}{2A}\left(-\frac{A'}{A} + \frac{B'}{B}\right) + \frac{1}{A}$$

$$R_{\phi\phi} = \sin^2\theta\, R_{\theta\theta}$$

$$\text{and} \quad R_{\alpha\beta} = 0 \quad \text{(for } \alpha \neq \beta\text{)} . \tag{3.11}$$

Schwarzschild sought a solution to Einstein's equations in vacuum, for which the $T^{\alpha\beta}$ in Equation (2.167) are all zero. It is therefore convenient to use the alternative form (Equation 2.170) of Einstein's equations, since we only have to deal with one term on the left-hand side. In that case, R_{TT}, R_{rr}, $R_{\theta\theta}$, and $R_{\phi\phi}$ are all zero, and forming the sum $R_{TT}/B + R_{rr}/A$, we therefore get

$$-\frac{1}{rA}\left(\frac{B'}{B}+\frac{A'}{A}\right) = 0 , \tag{3.12}$$

so that

$$\frac{B'}{B} = -\frac{A'}{A} , \tag{3.13}$$

which has the immediate solution

$$A(r)\,B(r) = \text{constant} . \tag{3.14}$$

Equation (3.6) is meant to describe the spacetime surrounding a centralized mass distribution, whose influence is expected to drop off with distance away from the source. Therefore, it is perfectly reasonable for us to assume that $g_{\alpha\beta} \to \eta_{\alpha\beta}$ as $r \to \infty$, and it is clear that the constant in Equation (3.14) must therefore be 1. Hence,

$$A(r) = \frac{1}{B(r)} . \tag{3.15}$$

Substituting for the function $A(r)$ in the equation $R_{rr} = 0$ and solving the simple differential equation produced by this step, we therefore find that

$$B(r) = 1 - \frac{k}{r} , \tag{3.16}$$

where k is a constant. But we already know that in the Newtonian limit (where gravity is relatively weak)

$$g_{00} = 1 - \frac{2GM}{rc^2} \tag{3.17}$$

(see Equations 2.141 and 2.142). Therefore, since $B(r) = g_{00}$, we must have

$$k = \frac{2GM}{c^2} . \tag{3.18}$$

And so we arrive at Schwarzschild's vacuum solution for the spacetime surrounding a time-independent, non-rotating, spherically symmetric distribution of mass M:

$$ds^2 = \left(1 - \frac{2GM}{c^2 r}\right)c^2 dT^2 - \left(1 - \frac{2GM}{c^2 r}\right)^{-1} dr^2 - r^2 d\Omega^2 , \tag{3.19}$$

where

$$d\Omega^2 \equiv d\theta^2 + \sin^2\theta\, d\phi^2 . \tag{3.20}$$

At a fixed radius r (and angles θ and ϕ), the accelerated observer measures an interval of time dT, related to the proper time $d\tau$ (measured in the inertial, or free-falling, frame) via the interval ds in Equation (3.19):

$$ds = c\,d\tau = \left(1 - \frac{2GM}{c^2 r}\right)^{1/2} c\,dT \,. \tag{3.21}$$

Evidently,

$$d\tau = \left(1 - \frac{2GM}{c^2 r}\right)^{1/2} dT \,, \tag{3.22}$$

so clocks run slowly in a gravitational field, an effect that produces both a time dilation and a gravitational redshift in radiation emitted near a gravitating body.

This effect is actually not as mysterious as it sounds; its roots may be traced to the dilation of time associated with a Lorentz transformation in special relativity. The distortions to intervals of time and distance are entirely dependent upon the relative velocity of two different observers (see Equation 2.2). One can therefore understand why a relative acceleration causes clocks to run more slowly by thinking about the meaning of time and its dependence on change.

Time elapses when something changes. To measure an interval of time on the clock, the hand must turn across its face. Regardless of how fast the accelerated frame is moving relative to us, its speed will have increased even further during the process of measuring the time interval. Thus, there is no way of avoiding the fact that the other frame's velocity relative to a Cartesian inertial frame at the end of the time interval is different compared to its value at the beginning. No matter how small we make the time interval, the starting and ending velocities are always different—that's the nature of acceleration. According to special relativity, there should be an additional time dilation associated with this increase in speed. The effect is greatly magnified when the acceleration is so great that even a tiny interval can bring the magnitude of the falling frame's velocity close to the speed of light. The ensuing time dilation can then appear to freeze the action completely. Thus, as long as we accept the fact that time intervals are altered during a transformation from one frame to another (but always in such a way as to preserve the constancy of the speed of light), we must also accept the conclusion that the acceleration of one frame relative to another itself incurs an additional time dilation. Gravitational fields, therefore, slow down the passage of time as viewed from distant vantage points, and the retardation effect is greater the stronger the field.

In concordance with the time dilation implied by Equation (3.22), the frequency of radiation emitted in a gravitational field must also change when viewed by a distant observer, producing what is commonly referred to as a *gravitational* redshift. Let a source at radius r_e emit radiation with frequency ν_e, as measured in the proper frame (one that is free of any gravitational effects). In the laboratory (i.e., accelerated) frame, we instead measure a frequency

$$\bar{\nu} = \nu_e \left. \frac{d\tau}{dT} \right|_{r_e} = \nu_e \left(1 - \frac{2GM}{c^2 r_e}\right)^{1/2} . \tag{3.23}$$

Correspondingly, a proper-frame observer at r_0 will infer a frequency

$$\nu_0 = \bar{\nu} \left. \frac{dT}{d\tau} \right|_{r_0} , \qquad (3.24)$$

and the overall change in frequency between the points of emission and detection is therefore given by

$$\left(\frac{\nu_0}{\nu_e} \right) = \left(1 - \frac{2GM}{c^2 r_e} \right)^{1/2} \left(1 - \frac{2GM}{c^2 r_0} \right)^{-1/2} . \qquad (3.25)$$

Since typically the observer views the source of radiation from a great distance (where $r_0 \gg r_e$),

$$\left(\frac{\nu_\infty}{\nu_e} \right) = \left(1 - \frac{2GM}{c^2 r_e} \right)^{1/2} . \qquad (3.26)$$

Notice that the observed frequency goes to zero for light emitted progressively closer to the radius $2GM/c^2$, which must therefore represent a limit of observability for a mass concentration such as this. The behavior of light at this radius coincides with a limitation on r, known as the *static* limit, that also emerges from Equation (3.19). The question here is how close can one get to the source of gravity and still be able to maintain a fixed position in space (hence the term 'static'). This condition translates into the statement that for the static observer, only the passage of time should contribute to the interval ds. Clearly, since $(ds)^2 > 0$ or, equivalently, since $(d\tau)^2 > 0$, the observer can remain static ($dr = d\theta = d\phi = 0$) only if $r > 2GM/c^2$. For smaller radii, no force can maintain a constant value of r since a positive value of $(ds)^2$ would then force $dr \neq 0$, and the observer would be pulled inexorably toward the center at $r = 0$. This critical radius,

$$r_S \equiv \frac{2GM}{c^2} , \qquad (3.27)$$

is named in honor of Karl Schwarzschild; it specifies the location of the black hole's so-called *event horizon*.

This is but one of several different kinds of horizon one may identify in any given spacetime. In the context of black-hole metrics, one sometimes sees this particular one also referred to as a 'gravitational' horizon, principally because its origin is clearly due to the gravitational influence of the central body. As it turns out, however, the gravitational horizon generally has a different meaning than an event horizon, and the two are not necessarily the same. They happen to be identical in this particular case because the Schwarzschild metric is independent of time. But this is not true in cosmology, as we shall explore in much greater detail in Chapter 7, where we shall define these characteristic radii more precisely and describe how they are related to each other.

3.2   KERR METRIC

It took another half century before Roy Kerr (see Figure 3.2) finally managed to find a description of intervals and time surrounding a *rotating* black hole [50]. The Kerr metric finds greater relevance in nature than Schwarzschild's solution because most objects in the Universe spin at least a little—it is virtually impossible to assemble an aggregate of matter without any angular momentum at all. Indeed, conservation of angular momentum can produce furiously spinning objects when their progenitors collapse into ultracompact volumes.

Schwarzschild's metric cannot handle rapidly rotating objects (and is only an approximation for slowly spinning ones) because a rotating source of gravity impacts the spacetime around it in unexpected and challenging ways. The interval ds, for example, is no longer given by Equation (3.19) for the same reason that a mother watching her child jumping onto a merry-go-round cannot measure the total distance he covers by simply counting his steps. Though she may still be able to monitor the passage of time T by tracking the ticks on her watch, the actual distance traveled by the boy is now augmented by the merry-go-round's lateral motion, which carries him along for the ride.

But making a transition from a static mass distribution to a rotating mass is not at all trivial. We shall not reproduce Kerr's complete derivation here, but at least mention the key ideas that led to his important solution.[1] By the early 1960's, it was already known that Einstein's equations could be simplified considerably if the *shear* in the bundles of light moving along geodesics were to drop off quickly as one recedes from the source of gravity. As we learned in the previous chapter (see, e.g., Equation 2.41), a *geodesic* is the shortest distance between two points in spacetime. For flat space—meaning the absence of any gravitational influence—the geodesic is just a straight line. Generally, light rays emitted anywhere in the spacetime follow geodesics outward from their point of origin, because these are the most 'direct' paths they can follow.

A group of geodesics is said to be *shear*-free when these lightrays follow special paths that prevent distortions to an image. The image may shrink, or get enlarged, but it may not be distorted in any other way. Kerr's approach was similar to that of Schwarzschild, choosing coordinates that incorporated rotational symmetry from the outset, and no dependence on time, though he allowed for the possibility of steady rotation in the source. The important new ingredient, however, was the additional assumption of shear-free conditions *everywhere* in the spacetime. (In retrospect, we now know that Schwarzschild's solution is itself shear-free, but this property had not been fully appreciated back then.) Together, these simplifications reduce the number of essential components in Equation (2.167) to a manageable set of four. From there, it was 'simply' a matter of trial and error to find the functions that solve the remaining differential equations.

The Kerr solution is completely specified by the central object's mass, M, and its angular momentum, $\mathbf{J}$. It is also possible to find solutions for a mass that contains charge, Q. However, in nature it is difficult to maintain non-zero values of Q over distances associated with the force of gravity, which is much weaker than the

---

[1]The interested reader can follow this derivation in Kerr's (1963) seminal paper "Gravitational Field of a Spinning Mass as an Example of Algebraically Special Metrics" [52].

Figure 3.2 Roy Kerr, shown here in his office in 1963 after joining the faculty at the University of Texas, Austin, at the time he discovered the metric that now bears his name. (Image used with permission of the University of Texas Center for Relativity)

electromagnetic force. The spacetime interval for a rotating compact object is given by the expression

$$(ds)^2 = A(r,\theta)\,(c\,dT)^2 + C(r,\theta)\,c\,dT\,d\phi - (\Sigma/\Delta)\,(dr)^2 -$$
$$\Sigma\,(d\theta)^2 - D(r,\theta)\,\sin^2\theta\,(d\phi)^2\,, \tag{3.28}$$

where

$$A(r,\theta) \equiv \left(1 - \frac{2GMr}{c^2\Sigma}\right)\,, \tag{3.29}$$

$$C(r,\theta) \equiv \frac{4aGMr\sin^2\theta}{c^2\Sigma}\,, \tag{3.30}$$

$$D(r,\theta) \equiv \left(r^2 + a^2 + \frac{2GMra^2\sin^2\theta}{c^2\Sigma}\right)\,, \tag{3.31}$$

with

$$a \equiv \frac{J}{cM}\,, \tag{3.32}$$

$$\Delta \equiv r^2 - \frac{2GM}{c^2}r + a^2\,, \tag{3.33}$$

and

$$\Sigma \equiv r^2 + a^2\cos^2\theta\,. \tag{3.34}$$

The metric coefficients are therefore

$$[g_{\alpha\beta}] = \begin{pmatrix} A(r) & 0 & 0 & C(r,\theta)/2 \\ 0 & -\Sigma/\Delta & 0 & 0 \\ 0 & 0 & -\Sigma & 0 \\ C(r,\theta)/2 & 0 & 0 & -D(r,\theta)\sin^2\theta \end{pmatrix}. \tag{3.35}$$

We need to point out here that Equation (3.28) is written in terms of so-called Boyer-Lindquist coordinates [53], which reduce to the spherical coordinates of Equation (3.19) only when the spin parameter a goes to zero. Specifically, Boyer-Lindquist coordinates are those for which the Cartesian coordinates (x, y, z) may be written

$$x = \left(r^2 + a^2\right)^{1/2}\sin\theta\,\cos\phi\,,$$
$$y = \left(r^2 + a^2\right)^{1/2}\sin\theta\,\sin\phi\,,$$
$$z = r\cos\theta\,. \tag{3.36}$$

One can trivially show that when the spin parameter a is zero, the Kerr solution (Equation 3.28) reduces to the spherically-symmetric Schwarzschild metric. When a is nonzero, however, two new effects emerge. First, the metric coefficients $g_{\alpha\beta}$ depend on θ, so rotation has broken the spherical symmetry, though we still have azimuthal

Schwarzschild Black Hole

Axis of Rotation

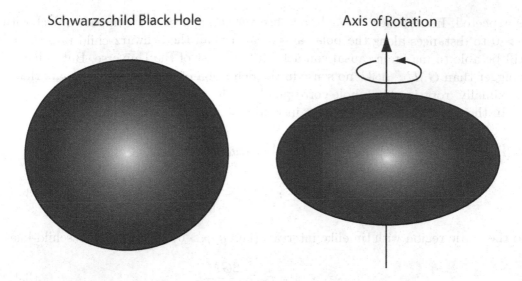

Figure 3.3 For a Schwarzschild (non-spinning) black hole, the event horizon (shown on the left) is strictly spherical. When the black hole is spinning, however, the singularity in the interior is a ring, not a point, and therefore the curvature it induces depends on angle, so the (Kerr 'outer') event horizon (shown on the right) is widest at the equator, tapering off gradually towards the poles. (Image adapted from ref. [50])

symmetry (see Figure 3.3). Second, a transformation into a rotating frame inevitably mixes the angle of rotation (here ϕ) with time. So the metric tensor now has a non-zero off-diagonal term, $g_{T\phi} = g_{\phi T}$.

Let us first consider what happens along the polar axis, $\theta = 0$, where

$$(ds)^2 = \left[1 - \frac{2GMr}{c^2(r^2 + a^2)}\right](c\,dT)^2 - \frac{r^2 + a^2}{\Delta}(dr)^2 - (r^2 + a^2)(d\theta)^2 . \tag{3.37}$$

Along this direction, the corresponding metric tensor is diagonal because the swirling action of the spinning body has no impact on the spacetime. As we discussed earlier for the Schwarzschild solution, *static* observers are permitted only when $g_{TT} > 0$, i.e., only for speeds smaller than c. The limiting radius for this *timelike* region is therefore given by the condition

$$2GMr < c^2(r^2 + a^2) , \tag{3.38}$$

or

$$r > r_+ \equiv \frac{GM}{c^2} + \left(\left[\frac{GM}{c^2}\right]^2 - a^2\right)^{1/2} , \tag{3.39}$$

the negative root corresponding instead to a region inside the limiting radius, which is beyond the derived static limit. The radius r_+ represents the static limit along $\theta = 0$. Note that when $a \to 0$,

$$r_+ \to \frac{2GM}{c^2} = r_S , \tag{3.40}$$

as expected. Interestingly, though, an observer can approach a compact object with $a > 0$ to distances along the polar axis smaller than the Schwarzschild radius, and still be able to maintain causal contact with the rest of the Universe. But values of a bigger than GM/c^2 make no sense in the definition of r_+, and so it appears that a 'maximally' rotating black hole corresponds to the condition $a = r_S/2$.

In the equatorial plane, we have instead $\theta = \pi/2$, and

$$(ds)^2 = \left(1 - \frac{2GM}{c^2 r}\right)(c\,dT)^2 + \frac{4aGM}{c^2 r}\,cdT\,d\phi -$$

$$\frac{r^2}{r^2 - 2GMr/c^2 + a^2}(dr)^2 - r^2\,(d\theta)^2 - \left[r^2 + a^2 + \frac{2GMa^2}{c^2 r}\right](d\phi)^2. \quad (3.41)$$

So the static region with timelike intervals (i.e., $g_{TT} > 0$) is here Schwarzschild-like:

$$r > r_0 = r_S \equiv \frac{2GM}{c^2}. \quad (3.42)$$

For any arbitrary angle θ, the static limit (from Equation 3.28) is evidently

$$r_0(\theta) = \frac{GM}{c^2} + \left(\left[\frac{GM}{c^2}\right]^2 - a^2\cos^2\theta\right)^{1/2}. \quad (3.43)$$

However, the second significant difference between the Kerr and Schwarzschild solutions—the mixing of ϕ with T—leads to the surprising conclusion that the static limit in the case that $a \neq 0$ is actually *not* the event horizon as previously defined. The interval ds can still be greater than zero (i.e., timelike) even when $r < r_0(\theta)$, due to the presence of the off-diagonal term $dT\,d\phi$ in Equation (3.28). We see that observers are permitted to exist at a fixed radius r inside $r_0(\theta)$ if the positive contribution to $(ds)^2$ from the term

$$2g_{T\phi}\,c\,dT\,d\phi = 2g_{T\phi}\,(c\,dT)^2\,\frac{d\phi}{c\,dT} \quad (3.44)$$

is large enough to offset the negative value of $g_{TT}\,(c\,dT)^2$. Thus, observers may exist at fixed r inside $r_0(\theta)$ only if they're rotating along with the black hole at a rate $\dot\phi > 0$.

This phenomenon is commonly interpreted to mean that the spacetime itself swirls around the spinning object with a speed decreasing with distance from the center. This 'frame dragging' forces everything within that spacetime into co-rotation with the source of gravity, even if within that frame these objects are completely stationary. If correct, this interpretation of Kerr's solution implies that even if we could somehow place a particle with zero angular momentum in the vicinity of a spinning black hole, that particle would still appear to be moving laterally from the perspective of a distant observer.

This frame dragging gives rise to another critical radius called the *stationary* limit. One can see exactly how this arises by rewriting Equation (3.28) in the following form:

$$(ds)^2 = (c\,dT)^2 \left[-\left\{ \left(\frac{r^3 + a^2 r + 2GMa^2/c^2}{r} \right)^{1/2} \frac{d\phi}{c\,dT} - \right. \right.$$

$$\left. \left. \frac{2aGM/c^2}{r^{1/2}(r^3 + a^2 r + 2GMa^2/c^2)^{1/2}} \right\}^2 + \frac{4a^2(GM/c^2)^2}{r(r^3 + a^2 r + 2GMa^2/c^2)} + \right.$$

$$\left. \left(1 - \frac{2GM}{c^2 r} \right) \right] - \frac{r^2}{r^2 - 2GMr/c^2 + a^2}(dr)^2 + r^2\,(d\theta)^2. \qquad (3.45)$$

Evidently, $(ds)^2 > 0$ for special values of $\dot{\phi}$, even when $dr = d\theta = 0$. The extremal condition on this effect occurs when the factor inside the curly brackets of this equation reaches a minimum:

$$\left(\frac{d\phi}{dT} \right)_0 = \frac{2aGM/c^2}{r^3 + a^2 r + 2GMa^2/c^2}, \qquad (3.46)$$

which now produces the *stationary* limit, corresponding to the radius at which the coefficient multiplying $(dT)^2$ goes to zero:

$$r = r_+ \equiv \frac{GM}{c^2} + \left(\left[\frac{GM}{c^2} \right]^2 - a^2 \right)^{1/2} \quad \text{(for all } \theta) . \qquad (3.47)$$

Since $(d\phi/dT)_0$ is the minimum rotation rate permitted in this region, we interpret it to be the angular speed forced on a zero angular momentum object by the spin of the black hole.

The radius r_+ is the true event horizon for a Kerr black hole because inside it neither static nor stationary observers are allowed, regardless of the value of $\dot{\phi}$. Stationary observers are permitted within the 'ergosphere,' the region bounded by r_+ from below and $r_0(\theta)$ from above, but only if they have the requisite $\dot{\phi}$. Static observers are permitted anywhere outside $r_0(\theta)$. So the static horizon is a flattened sphere, whose semi-minor axis lies parallel to $\mathbf{J}$, whereas the stationary—or the actual event—horizon is a sphere with radius r_+. This horizon matches that of the Schwarzschild metric when the spin parameter, a, is zero, but is otherwise always smaller than r_S. And as we noted before, these relations make sense only so along as $a \leq GM/c^2$.

CHAPTER 4

The Cosmic Spacetime

T HE cosmic spacetime has elements in common with Schwarzschild and Kerr, but differs from them in several distinct ways. For example, it may not be possible to find a 'vacuum' solution for this purpose, since we are now interested in a metric *within* the medium, as opposed to one outside the source of gravity. That is, we cannot proceed as before by setting $T^{\alpha\beta} = 0$ in Equation (2.167). In addition, it is not always obvious that the most convenient coordinates to use in the metric are necessarily those for which the coefficients $g_{\alpha\beta}$ are time independent.[1] Indeed, it often makes sense to use the evident degree of symmetry and homogeneity in the Universe (Section 1.3) to adopt a set of coordinates that assume a spatially-independent (though not time-independent) expansion factor, thereby simplifying the equations, but at the cost of rendering the $g_{\alpha\beta}$ functions of time. As we shall see, when the cosmic spacetime metric is non-static like this, additional complexities may arise in the usual interpretation of redshift and causality. Nevertheless, several long-standing questions have now been answered, notably whether in cosmology we see a third type of redshift, distinct from the well understood gravitational and kinematic forms appearing in other applications of general relativity (see Section 3.1). These features will be developed later in the book, especially Chapters 7 and 8.

4.1 THE METRIC IN COMOVING COORDINATES

As we discussed in the introduction (Section 1.3), our Universe appears to be homogeneous and isotropic on large scales, in the sense that observations made from our vantage point are evidently representative of the cosmos as viewed from anywhere else. Homogeneity means that the physical conditions are the same at every spatial point within the medium. Isotropy means that the physical conditions are identical in all directions when viewed from any given location. Isotropy at every point automatically guarantees homogeneity, but the reverse is not always true. The *Cosmological Principle*, which assumes both homogeneity and isotropy, is essential to any attempt at using what we see from Earth as a basis for testing cosmological models.

---

[1]As it turns out, there does exist a special category of cosmic spacetimes with time-independent metric coefficients. But as we shall show in Chapter 5, none of these appears to be directly relevant to the actual Universe.

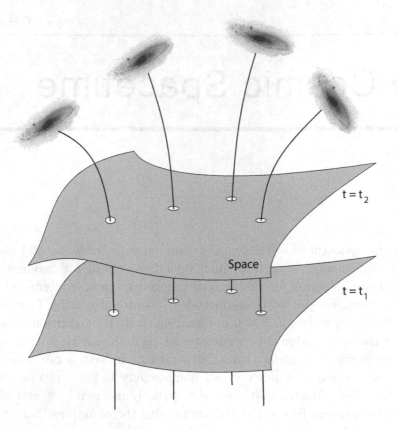

Figure 4.1 Non-intersecting particle worldlines consistent with Weyl's postulate. This regularity allows us to represent the evolution of the Universe as a time-ordered sequence of spacelike hypersurfaces (here shown at times t_1 and t_2) orthogonal to the worldlines.

The validity of the Cosmological Principle, however, does depend on the spatial scale we are considering. Light has been traveling towards us for up to 13.7 billion years, which means that the portion of the Universe observable to us today could be of order 4,200 Mpc (1 Mpc $\approx 3.26 \times 10^6$ light years).[2] But we also see structure in the form of galaxies, clusters, and superclusters on scales of 100 Mpc or less. Thus, although the Cosmological Principle is valid on a universal scale, it should not be taken too literally when considering the local matter distribution.

On large scales, at least, Hubble's observations indicate that the Universe is expanding in an orderly fashion, with galaxies moving apart from one another (except for the odd collision or two due to some peculiar motion on top of the 'Hubble flow'). Galactic trajectories on a spacetime diagram would therefore show worldlines forming a funnel-like structure in which the separation between any two paths is steadily increasing with time (see Figure 4.1).

[2]We shall see in Section 7.5, however, that the actual maximum *physical* distance traveled by any detectable light signal is only half of this value, since the most distant sources we see today have been receding from us at near lightspeed for half the age of the Universe.

Figure 4.2 Hermann Klaus Hugo Weyl was one of the most influential mathematicians of the twentieth century. In this regard, he very ably continued the tradition established by his doctoral adviser, David Hilbert (see Section 2.7 and Figure 2.3). His postulate concerning the regularity of cosmic expansion was a crucial ingredient in the development of the FLRW metric. (Image courtesy of ETH-Bibliothek Zurich, Bildarchiv; licensed under the Creative Commons to share and distribute the work freely)

Homogeneity and isotropy give rise to this type of regularity, and together suggest that the evolution of the Universe may be represented as a time-ordered sequence of 3-dimensional space-like hypersurfaces, each of which satisfies the Cosmological Principle.[3] This intuitive picture of regularity is often expressed formally as *Weyl's postulate*, after the mathematician Hermann Weyl (1885–1955) (see Figure 4.2), who did much of the early work on this subject in the 1920's [54].

The essential thinking here is that the worldlines of galaxies form a 3-bundle of non-intersecting geodesics *orthogonal* to a foliated sequence of spacelike hypersurfaces. We shall now advance the description of these worldlines by considering the full

---

[3]At this point, we should also mention that an even stronger version of this principle is sometimes used, in which the Universe is not only homogeneous and isotropic at any one time, but also has a structure invariant with respect to time. The observational evidence argues strongly against such a 'steady state' universe, so we shall not be using the strong version of the Cosmological Principle in this book.

impact of Weyl's postulate on the coordinates and metric of spacetime. Since it is not clear *a priori* whether our description will be based on observations from a Cartesian inertial frame, or from an accelerated frame—a distinction that bears on our notation in Chapter 2—we shall simply denote our coordinates as x^μ for now. Thus, on any spacelike hypersurface orthogonal to a given particle worldline in the Hubble expansion, Weyl's postulate and the Cosmological Principle require that $x^0 =$ constant.

Unlike the Kerr metric, this means that no mixing of the spatial and time coordinates is permitted in the cosmic spacetime, and therefore here we expect that

$$g_{0i} = 0 , \tag{4.1}$$

for $i = 1, 2, 3$. For reasons that will become clearer shortly, we shall choose our spatial coordinates such that they remain constant from one hypersurface to the next, allowing the dynamics associated with the universal expansion to emerge through the metric coefficients. Now, the worldlines are geodesics, so the coordinates must satisfy Equation (2.41), and since the spatial coordinates are constant, that means that

$$\Gamma^i{}_{00} = 0 \quad (i = 1, 2, 3) . \tag{4.2}$$

Thus, using the definition of the Christoffel symbols (Equation 2.45), with Equation (4.1), it is trivial to see that

$$\frac{\partial g_{00}}{\partial x^i} = 0 \quad (i = 1, 2, 3) . \tag{4.3}$$

The metric coefficient g_{00} must therefore depend on x^0 only, and without loss of generality, we can redefine x^0 in order that

$$g_{00} = 1 . \tag{4.4}$$

Let us pause for a moment and consider the impact of what we have just done. The symmetries present in the Cosmological Principle (Section 1.2) apparently require a constant *lapse* function, g_{00}. But we have arrived at this result without considering whether an ansatz for the metric that includes this coefficient is even consistent with the choice we shall make later for the stress-energy tensor $T^{\alpha\beta}$ in order to solve Einstein's Equations (2.167). Yet by forcing the condition $g_{00} = 1$, we are precluding any possibility of measuring a time dilation on our clock relative to the proper time in a local inertial frame (see Section 2.6). Does this mean we are excluding accelerated expansion of the Universe from our framework? This is a profound question that requires detailed analysis beyond what we are concerned with right now. Its answer constitutes one of the most important foundational elements of this book and will occupy our attention for much of Chapter 10. For now, we proceed in the traditional way by refining the metric in order to satisfy the Cosmological Principle without regard to the consequences on $T^{\alpha\beta}$. But we shall return to this topic and thoroughly complete the analysis later (Section 10.1).

The line element consistent with the Cosmological Principle and Weyl's postulate is therefore

$$ds^2 = (dx^0)^2 + g_{ij} \, dx^i \, dx^j . \tag{4.5}$$

We now define the *cosmic time t*, such that $x^0 = ct$, and therefore

$$ds^2 = c^2 dt^2 + g_{ij}\, dx^i\, dx^j \ . \tag{4.6}$$

The spacelike hypersurfaces in Weyl's postulate are membranes of simultaneity with respect to this time coordinate. However, we emphasize that although t is the proper time kept by each observer on his or her clock, this simultaneity is not adjudged by any one individual, because although each galaxy is in its own free-fall frame, the Universe as a whole is not a *global* inertial frame. This definition of cosmic time is unfortunately often a source of misunderstanding, and we shall therefore revisit the physical meaning of t on many occasions—later in this chapter, but also in subsequent applications throughout the remainder of this book.

In finding a mathematical description for the spacelike hypersurfaces constituting the $g_{ij}\, dx^i\, dx^j$ contribution to the line element ds, we realize that in three dimensions, there exist only three independent translations and three rotations. These permit only three types of homogeneous and isotropic spaces: (i) flat space, (ii) a 3-dimensional sphere of constant *positive* curvature, and (iii) a 3-dimensional hyperbolic surface of constant *negative* curvature.

It is not always easy to visualize these immediately; however it is quite helpful to first consider their analogues in two-dimensional space, as viewed from our 3-dimensional world. Let us therefore consider the plane and the two-sphere, both of which can be embedded in 3-dimensional Euclidean space with the usual Cartesian coordinates x, y, and z.

The surface of a two-dimensional sphere is given by the equation

$$x^2 + y^2 + z^2 = a^2 \ , \tag{4.7}$$

where a is its radius. In order to relate this to the metric, we must find a way of calculating the contribution to ds by an arc inscribed on the sphere's surface. In 3-space, the Euclidean metric is simply

$$dl^2 = dx^2 + dy^2 + dz^2 \ . \tag{4.8}$$

The additional condition imposed by the sphere is that not all of the coordinates x, y, and z are independent of each other if we constrain the motion to its surface. However, we do know from Equation (4.7) that

$$dz^2 = \frac{(x\, dx + y\, dy)^2}{a^2 - x^2 - y^2} \ . \tag{4.9}$$

Therefore, the distance between any two points located on the two-sphere may be expressed entirely in terms of just two independent coordinates, x and y, bounded by the condition $x^2 + y^2 \leq a^2$:

$$dl^2 = dx^2 + dy^2 + \frac{(x\, dx + y\, dy)^2}{a^2 - x^2 - y^2} \ . \tag{4.10}$$

Notice, in particular, that the constraint on the motion is now represented by the *radius of curvature a*.

Because of the spherical symmetry in this geometry, it is often more convenient to express the line element dl in terms of spherical coordinates (R, θ, ϕ), rather than the Cartesian (x, y, z), where

$$x = \eta \cos \phi$$
$$y = \eta \sin \phi ,\qquad (4.11)$$

in terms of the radius $\eta \equiv R \sin \theta$ projected onto the $x - y$ plane. Then, with

$$dx^2 + dy^2 = d\eta^2 + \eta^2 \, d\phi^2 , \qquad (4.12)$$

and

$$x \, dx + y \, dy = \eta \, d\eta , \qquad (4.13)$$

it is straightforward to see that

$$dl^2 = \frac{d\eta^2}{1 - (\eta^2/a^2)} + \eta^2 \, d\phi^2 . \qquad (4.14)$$

The dramatic increase in dl for a finite $d\eta$ as $\eta \to a$ is due to the fact that a plane tangential to the surface of the sphere becomes orthogonal to the $x - y$ plane at that radius.

In this elegant expression, we can see all of the homogeneous and isotropic surfaces represented by the three different kinds of curvature radius a. For a positive, finite value of a^2, we have the sphere with *positive* curvature. In the limit $a \to \infty$, dl is Euclidean, so this limit represents *flat* space with *zero* curvature. Finally, although it may not be immediately obvious, we may also formally take a^2 to be negative, in which case Equation (4.14) then describes a homogeneous, isotropic two-dimensional surface with constant *negative* curvature. However, unlike the previous two cases, this kind of surface cannot be embedded in Euclidean 3-dimensional space because a is imaginary, so in this instance a does not actually represent a radius of curvature in 3-dimensional space.

In order to standardize the usage of the curvature radius a, it is common to introduce the rescaled coordinate

$$r \equiv \frac{\eta}{\sqrt{|a^2|}} \qquad (4.15)$$

which, in the case of a sphere, is simply the fractional radius out to its surface. Notice, by the way, that r is dimensionless. Then, the two-metric (Equation 4.14) becomes

$$dl^2 = |a^2| \left(\frac{dr^2}{1 - kr^2} + r^2 \, d\phi^2 \right) , \qquad (4.16)$$

where the type of surface is now characterized by the value of k, which is $+1$ for the sphere $(a^2 > 0)$, 0 for the plane (two-dimensional flat space), and -1 for the hyperbolic space with negative curvature $(a^2 < 0)$.

The generalization to a 3-dimensional surface embedded in a four-dimensional Euclidean space is straightforward, and we may write the metric of a 3-dimensional space of constant curvature as

$$ds^2 = a^2 \left(\frac{dr^2}{1 - kr^2} + r^2 d\Omega^2 \right) \qquad (4.17)$$

where, as usual,

$$d\Omega^2 \equiv d\theta^2 + \sin^2 \theta \, d\phi^2 \, , \qquad (4.18)$$

and where it will be understood from now on that a^2 is a positive quantity. As before, k can take on the values 0, ± 1.

Putting this together with Equation (4.6), we therefore see that the most general line element satisfying Weyl's postulate and the Cosmological Principle is given by

$$ds^2 = c^2 \, dt^2 - a(t)^2 \left(\frac{dr^2}{1 - kr^2} + r^2 d\Omega^2 \right) , \qquad (4.19)$$

where the 3-spaces at constant t are Euclidean for $k = 0$, closed with positive curvature for $k + 1$, and open with negative curvature for $k = -1$. Notice that we have now explicitly written $a(t)$ as a function of time. In our discussion above, a was treated as a constant since we were considering hypersurfaces at a fixed cosmic time t. But though we have chosen to keep our spatial coordinates (r, θ, ϕ) constant from hypersurface to hypersurface in the expanding flow, there is no reason why a itself must remain constant. In fact, it will be the temporal behavior of the *expansion factor $a(t)$* (as this function is often called) that reflects the dynamics of the expanding universe. The line element in Equation (4.19) was first rigorously derived in the 1930's by Howard Percy Robertson (1903–1961) and (independently) Arthur Geoffrey Walker (1909–2001), and is commonly referred to as the *Robertson-Walker* metric [55, 56], though we shall shortly justify attaching both Friedmann's and Lemaître's names to it for cosmological applications.

4.2 FRIEDMANN, LEMAÎTRE, ROBERTSON & WALKER

With the line element written in Equation (4.19), we have taken our discussion of the cosmic spacetime as far as we can go without next introducing a dynamical theory to tell us how the scale factor $a(t)$ and the curvature constant k depend on the constituents of the Universe. For this, we need to calculate the Ricci tensor from derivatives of the metric coefficients drawn from Equation (4.19), and then solve the Einstein Equations (2.171) to determine how $a(t)$ varies with time. The identification of k is not so direct; its value must be determined from a comparison of the cosmological models with observations of the cosmic microwave background radiation and the redshift distribution of Type Ia Supernovae, among others. We shall revisit this important question in cosmology—whether the Universe is open or closed—in later chapters.[4]

To begin with, the metric tensor for the Robertson-Walker line element is

$$[g_{\alpha\beta}] = \begin{pmatrix} 1 & 0 & 0 & 0 \\ 0 & -a(t)^2/(1-kr^2) & 0 & 0 \\ 0 & 0 & -\{a(t)r\}^2 & 0 \\ 0 & 0 & 0 & -\{a(t)r\sin\theta\}^2 \end{pmatrix}, \qquad (4.20)$$

with a determinant

$$g = -r^4 \sin^2\theta \, a(t)^6/(1-kr^2). \qquad (4.21)$$

The corresponding inverse metric tensor (see Equation 2.50) is

$$[g^{\alpha\beta}] = \begin{pmatrix} 1 & 0 & 0 & 0 \\ 0 & -(1-kr^2)/a(t)^2 & 0 & 0 \\ 0 & 0 & -\{a(t)r\}^{-2} & 0 \\ 0 & 0 & 0 & -\{a(t)r\sin\theta\}^{-2} \end{pmatrix}. \qquad (4.22)$$

Given this simple diagonal form for $g_{\alpha\beta}$, only 13 Christoffel symbols are non-zero and, as before, may be evaluated directly from Equation (2.45):

$$\Gamma^r{}_{tr} = \Gamma^\theta{}_{t\theta} = \Gamma^\phi{}_{t\phi} = \frac{1}{c}\frac{\dot{a}}{a}$$

$$\Gamma^t{}_{rr} = \frac{a\dot{a}}{c(1-kr^2)}$$

$$\Gamma^t{}_{\theta\theta} = \frac{a\dot{a}r^2}{c}$$

$$\Gamma^t{}_{\phi\phi} = \frac{a\dot{a}r^2\sin^2\theta}{c}$$

$$\Gamma^r{}_{rr} = \frac{kr}{1-kr^2}$$

---

[4]We shall see in Section 4.3 that the value of k is directly related to the energy content of the Universe. One may therefore reformulate this question—not in terms of k—but also more physically in terms of whether the Universe has a net energy greater than, less than, or equal to, zero.

$$\Gamma^\theta{}_{r\theta} = \Gamma^\phi{}_{r\phi} = \frac{1}{r}$$
$$\Gamma^r{}_{\theta\theta} = -r(1 - kr^2)$$
$$\Gamma^r{}_{\phi\phi} = -r(1 - kr^2)\sin^2\theta$$
$$\Gamma^\theta{}_{\phi\phi} = -\sin\theta\cos\theta$$

and $$\Gamma^\phi{}_{\theta\phi} = \cot\theta \,, \tag{4.23}$$

where dot denotes a derivative with respect to t. From here it is straightforward to calculate the non-zero components of the Ricci tensor (Equation 2.152), which are

$$R_{tt} = \frac{3}{c^2}\frac{\ddot{a}}{a}$$

$$R_{rr} = -\frac{1}{c^2}\frac{a\ddot{a} + 2\dot{a}^2 + 2kc^2}{(1 - kr^2)}$$

$$R_{\theta\theta} = -\frac{r^2}{c^2}\left(a\ddot{a} + 2\dot{a}^2 + 2kc^2\right)$$

and $$R_{\phi\phi} = -\frac{r^2\sin^2\theta}{c^2}\left(a\ddot{a} + 2\dot{a}^2 + 2kc^2\right) . \tag{4.24}$$

Thus, the curvature scalar (Equation 2.154) is

$$R = \frac{6}{c^2}\left(\frac{\ddot{a}}{a} + \frac{\dot{a}^2 + kc^2}{a^2}\right) , \tag{4.25}$$

and therefore the essential components of the Einstein tensor (Equation 2.155) may be written

$$G_{tt} = -\frac{3}{c^2}\frac{\dot{a}^2 + kc^2}{a^2}$$

$$G_{rr} = \frac{1}{c^2}\left(\frac{2a\ddot{a}}{(1 - kr^2)} + \frac{\dot{a}^2}{(1 - kr^2)} + \frac{kc^2}{(1 - kr^2)}\right)$$

$$G_{\theta\theta} = \frac{1}{c^2}\left(a\ddot{a}r^2 + \dot{a}^2r^2 + kc^2r^2\right)$$

$$G_{\phi\phi} = \frac{1}{c^2}\left(2a\ddot{a}r^2\sin^2\theta + \dot{a}^2r^2\sin^2\theta + kc^2r^2\sin^2\theta\right) . \tag{4.26}$$

Finally, to complete the gravitational field equations (2.167), we need to identify the components of the energy-momentum tensor $T_{\alpha\beta}$. The Cosmological Principle per se does not require that the fluid be *perfect*, though it is certainly reasonable for us to assume that on large scales, not only are the rest mass density ρ_{m0} and pressure P_0 (as measured by an observer in a locally inertial frame comoving with the fluid at the instant of measurement) the same everywhere, but also no shear forces are present. Thus, we shall assume that the energy-momentum tensor (see Equation 2.126) may be written in the form

$$T_{\alpha\beta} = \left(\rho_{m0} + \frac{P_0}{c^2}\right)u_\alpha u_\beta - P_0\, g_{\alpha\beta} \tag{4.27}$$

where, as usual, $u_\alpha = dx_\alpha/d\tau$. As we learn more about the properties of the Robertson-Walker line element, it will become clear that the cosmic time t is also the local proper time τ for each observer at his or her location. Thus, the four-velocity may also be written $u_\alpha = dx_\alpha/dt$.

As we pursue our goal of finding the dynamical equations for $a(t)$, it is worth reminding ourselves of the important caveat preceding Equation (4.4). Our procedure here essentially replicates the historical approach of first simplifying the metric to comply with the Cosmological Principle—as we have done to arrive at Equation (4.19)—and then substituting it, along with the perfect-fluid stress-energy tensor, into Einstein's Equations (2.167). The danger here, however, is that we have preset $g_{tt} = 1$ without regard to the equation-of-state for ρ_{m0} and P_0. As we shall demonstrate more rigorously in Chapter 10, this approach is unique among the class of spherically-symmetric spacetimes, the majority of which instead retain g_{tt} as an unknown function of the coordinates that must be found—along with all the other metric coefficients—as part of the solution to the field equations. The Schwarzschild metric (Equation 3.19) is a prime example of a spherically-symmetric spacetime with $g_{tt} \neq 1$. The risk with this conventional approach is that a constant lapse function may not be consistent with the dynamics associated with every equation-of-state one may wish to consider. But mainly for pedagogical and historical reasons, we shall proceed through the rest of this chapter, and Chapters 5 and 6, to use the Robertson-Walker metric in Equation (4.19) to find $a(t)$ for various choices of ρ_{m0} and P_0. In Chapter 10 we shall repeat this procedure with a more rigorous approach and demonstrate that the applicability of the metric with $g_{tt} = 1$ in Equation (4.19) is actually very limited.

With this caveat in mind, we are now in a position to write down the individual components of the gravitational field equations (2.167) to arrive at a set of simultaneous expressions we must solve to find the expansion factor $a(t)$. We realize immediately that all three of the spatial components G_{rr}, $G_{\theta\theta}$, and $G_{\phi\phi}$ contain the same information, since we can easily show that $G^r{}_r = G^\theta{}_\theta = G^\phi{}_\phi$ using the metric coefficients in Equation (4.22) to raise the first index. From Equations (4.26) and (4.27), therefore, we get two independent equations, commonly written in the form

$$H^2 \equiv \left(\frac{\dot{a}}{a}\right)^2 = \frac{8\pi G}{3c^2}\rho - \frac{kc^2}{a^2} \, , \tag{4.28}$$

known as the *Friedmann* equation, and

$$\frac{\ddot{a}}{a} = -\frac{4\pi G}{3c^2}(\rho + 3p) \, , \tag{4.29}$$

called the *Raychaudhuri* (or *acceleration*) equation [57]. For convenience, we have here defined the Hubble parameter H in terms of a and $\dot{a}$, and the total energy density

$$\rho \equiv \rho_{m0}\, c^2 \, . \tag{4.30}$$

In addition, we have also relabeled the pressure as $p \equiv P_0$ to simplify the notation.

As we saw earlier, the gravitational field equations were derived in part to satisfy the zero covariant divergence of the energy-momentum tensor (see Equation 2.117), which expresses mathematically the conservation of energy and momentum. In principle, we can therefore add one more equation to this list, which is easily derived from the component $T^{00}{}_{;0}$:

$$\dot{\rho} = -3H \left(\rho + p \right) , \qquad (4.31)$$

the so-called *energy* equation. One often sees Equations (4.28), (4.29), and (4.31) written together as a complete set, though upon closer inspection we realize that only two of these are actually independent of each other. For example, it is straightforward to see that Equation (4.31) may be derived by differentiating Equation (4.28) with respect to t and then substituting for $\ddot{a}$ from Equation (4.29). In any case, the Friedmann, acceleration, and energy expressions form a complete set of dynamical equations that one can use, together with knowledge of ρ and p, to solve for the evolution of the expansion factor $a(t)$ as a function of cosmic time t.

Equation (4.28) is named in honor of Aleksandr Aleksandrovich Friedmann (1888–1925) (see Figure 4.3), a Russian mathematician who discovered the expanding-universe solution [14, 58] to Einstein's equations of general relativity in 1922, unmotivated by any experimental indication at the time, since the expanding universe idea would not be corroborated by Edwin Hubble (1889–1953) [59] for another seven years (Figure 8.2). In fact, his paper of 1924 (entitled 'On the Possibility of a World with Constant Negative Curvature of Space') introduced all three types of Friedmann models, describing a universe with positive, zero, and negative curvature, a decade before Robertson and Walker published their (independent) analysis of the metric that now bears their name. Today, the Friedmann equation, together with the acceleration and energy equations, has become the backbone for the standard model of cosmology, commonly referred to as the 'Big Bang' theory (Chapter 1).

Interestingly, however, his initial foray into general relativity was not met with immediate approval. The son of a ballet dancer (his father) and a pianist, Friedmann demonstrated a mathematical virtuosity expressed through many different areas of scientific inquiry, including meteorology and hydrodynamics. Einstein's general theory of relativity, though published in 1916, was not known right away in Russia due to World War I and the ensuing Civil War. As such, Friedmann did not learn of Einstein's field equations until the early 1920's. But within several years, he had written several books on the subject, one a textbook co-authored with Vsevelod Frederiks (1885–1944), with whom he would later lecture on the general theory of relativity at the Physical Institute of Leningrad University. Of this period, the Russian theoretician, Vladimir Fock (1898–1974), later reminisced [60] that 'They were in different styles: Frederiks had a deep understanding of the physical side of the theory, but did not like mathematical calculations, while Friedman put the emphasis not on the physics, but on the mathematics. He strove for mathematical rigor and gave great importance to the complete and precise formulation of the initial hypotheses.'

After mastering general relativity, Friedmann quickly proceeded to find his now famous solution [61] that he submitted to *Zeitschrift für Physik* on 29 June 1922. The expert referee for his paper was none other than Einstein himself, who quickly rejected

Figure 4.3 Aleksandr Aleksandrovich Friedmann circa 1922 or 1923 in Moscow, at the time he was developing his models for the expansion of the Universe. (Leningrad Physico-Technical Institute, courtesy AIP Emilio Segre Visual Archives)

the manuscript[5] on the basis that "... the results concerning the non-stationary world, contained in [Friedmann's] work, appear to me suspicious. In reality it turns out that the solution given in it does not satisfy the field equations." Undaunted, Friedmann wrote a lengthy rebuttal directly to Einstein, who eventually recognized his error and wrote to the journal expressing his support for the publication of Friedmann's work. In his retraction [62], Einstein wrote "I am convinced that Mr. Friedmann's results are both correct and clarifying. They show that in addition to the static solutions to the field equations there are time varying solutions with a spatially symmetric structure." But though Einstein accepted Friedmann's non-static solutions, he apparently did so only in a mathematical sense, suggesting that they were of no physical significance.

In retrospect, Friedmann's work had very little impact on the development of cosmology, probably because of his affinity to mathematical formalism and a reluctance to connect his results to observations. For example, he never referred to the measurements of redshift and appears not to have taken seriously the possibility that our Universe might have been created in a singularity, even though his calculations in some sense predicted it [63]. In his early papers on this subject, published in the world's leading journal of physics, Friedmann was very clear about the expansion of the Universe. Had this idea been accepted on the basis of Friedmann's prediction, rather than Hubble's observations years later, today we might be talking about the expanding universe in terms of Friedmann's law, rather than Hubble's law.

No doubt, Einstein's reluctance to accept such an awkward concept as the beginning of the Universe was largely responsible for Friedmann being ignored. His awareness of the young Russian's work certainly did not change his own views, partially revealed in a 1929 article appearing in the Encyclopedia Britannica,[6] for which he wrote "nothing certain is known of what the properties of the space-time continuum may be as a whole. Through the general theory of relativity, however, the view that the continuum is infinite in its time-like extent but finite in its space-like extent has gained in probability." Looking to Einstein for guidance, most astronomers probably considered Friedmann's work to be irrelevant. To be fair, Friedmann did not so much predict the actual expansion of the Universe, but rather that expanding solutions were possible. Without much interest in the observations, it was not feasible for him to even distinguish the relevant solutions from those unlikely to be consistent with nature. And it was his lack of interest in coupling his calculations to physics and astronomy that would frame the principal difference between his work and that of Georges Henri Joseph Lemaître (1894–1966) (see Figure 4.4), as we shall soon see.

Young physicists learning about the general theory of relativity sometimes wonder why someone as brilliant as Friedmann, who clearly made such an auspicious start with his mathematical contributions to cosmology, is not associated with ever greater achievements later in life. The sad truth is that, like Schwarzschild (and Boyer much later [50], his life was cut short in his prime. He apparently contracted typhoid fever during a vacation in Crimea and died at the young age of 37 on 16 September 1925.

---

[5]Einstein's report rejecting Friedmann's paper was received by *Zeitschrift für Physik* on 18 September 1922.
[6]Quoted from ref. [64].

Figure 4.4 Monsignor Georges Henri Joseph Édouard Lemaître in 1933, while he was visiting professor at the Catholic University of America. During this year, Lemaître achieved widespread recognition for his work in cosmology, particularly the model he proposed that eventually became known as the Big Bang theory of the origin of the Universe. Several media outlets around the world heralded him as the leader of the new cosmological physics. (Image courtesy of The American Catholic History Research Center and University Archives, The Catholic University of America, Washington, D. C.)

But Friedmann was not the only mathematically talented individual of that era interested in how the Universe behaves. The Belgian Catholic priest Lemaître was drawn to cosmology while a graduate student with Sir Arthur Eddington (1882–1944) at the University of Cambridge. He went on to complete his Ph.D. at MIT, and in 1927 published a paper entitled 'A Homogeneous Universe of Constant Mass and Growing Radius Accounting for the Radial Velocity of Extragalactic Nebulae' [65]. In this paper, preceding Edwin Hubble's landmark article by two years, Lemaître derived a relation linking the recession velocity of a source to its distance—an equation later known as Hubble's law.

Lemaître's interest in physics developed during his years (1914–1919) as a soldier in the Belgian army. Between heavy house-to-house fighting and surviving the first poison-gas attack in history, he made time to read texts on basic physics. After the war, he pursued his studies in the sciences, eventually enrolling for a PhD degree at the US's renowned Massachusetts Institute of Technology. One can easily understand the path he followed afterwards—combining an admiration for the purity of mathematics with the shear relevance of real observations—from what he wrote on his fellowship application to complete his doctorate. He proposed "to get practical knowledge of modern astrophysics work, especially in matters connected with the astronomical consequences of the Principle of Relativity."[7]

Not surprisingly, he was quick to notice the significance of Hubble's discovery of Cepheids in spiral galaxies and journeyed to both Lowell Observatory in Northern Arizona, where Slipher had begun the exploration of redshifts in faint nebulae (see Chapter 8), and Mount Wilson in Southern California, to learn about the latest redshift results from Hubble. It was very soon after this that his 1927 paper[17] appeared in the Annales Scientifique Bruxelles.

In a sense, Friedmann's and Lemaître's contributions were complementary, though made completely independently of each other. Whereas Friedmann developed the mathematical framework for describing the dynamics of an open or closed universe, Lemaître was the first to propose that the expansion explains the redshift of galaxies and, moreover, that an initial *creation-like* event must have occurred. He called this the 'hypothesis of the primeval atom' which, as we have seen, later became known as the Big Bang theory of the Universe.

Given Einstein's earlier skepticism of Friedmann's solution, however, it was not surprising to see him continue his refusal to accept the idea of an expanding universe. Lemaître later recalled [66] him commenting "Vos calculs sont corrects, mais votre physique est abominable" ("Your math is correct, but your physics is abominable"). But by the early 1930's, after Hubble's discovery was published, Einstein publically recanted his objection to Lemaître's theory during a visit to Pasadena and Mount Wilson Observatory. This period coincided with his growing dissatisfaction with the cosmological constant, which he had earlier introduced into his field equations of general relativity in order to maintain 'stability' in the cosmos (see Equation 2.171). But now that the expanding model of the Universe was gaining popular acceptance

---

[7]Lemaître to Shapley, 15 February, 1924. Harvard College Observatory: Records of Director Harlow Shapley, 1930–1940, box 11, Harvard University Archives.

among the growing body of relativists, he no longer found a need for it and removed it from his deliberations. In 1931, Einstein introduced a big-bang type of model of his own, though avoiding the question of what happened at the 'beginning.' He assumed that the distribution of matter for very small a(t) was inhomogeneous, and argued that the singularity could be explained away as a mathematical artifact [67].

The following year, Einstein and de Sitter produced one of the best studied cosmological models, based on the simplest assumptions (see Section 5.3): zero spatial curvature ($k = 0$), no cosmological constant ($\Lambda = 0$), and zero pressure ($p = 0$). As it turns out, the expansion factor in this universe has a sudden transition at $t = 0$, implying some type of explosive beginning. Perhaps because of this 'undesirable' feature, Einstein and de Sitter gave it only lukewarm support. As we shall see in Chapters 5 and 12, this cosmology—now known as the Einstein-de Sitter Universe—is used often in model comparisons due to its very basic foundation.

And so today we honor all four individuals who developed the framework for the standard model, based on the idea of an initial Big Bang, followed by an apparently perpetual expansion consistent with Weyl's postulate. Rather than simply Robertson-Walker, we should more appropriately refer to the solution of Einstein's equations that correctly describes the dynamics of this motion as the *Friedmann-Lemaître-Robertson-Walker (FLRW) metric*.

In Chapter 5 we shall exploit the formalism we have now developed, particularly the FLRW metric in Equation (4.19), and the dynamical Equations (4.28), (4.29), and (4.31), to explore three simple and well-known cosmologies that emerge from the adoption of certain specific assumptions concerning the universal constituents. These three cosmological models will form the basis for several illustrative applications we shall consider in subsequent chapters. We shall also revisit these in Chapter 10 when we probe more deeply into the dependence of the lapse function g_{tt} on the equation-of-state. Before we proceed, however, we shall take a moment to better understand the physical meaning of the Friedmann Equation (4.28), which will be quite useful in our interpretation of the various models.

4.3   PHYSICAL INTERPRETATION OF THE FRIEDMANN EQUATION

It is not difficult developing a simple heuristic argument for the physical meaning of the Friedmann equation, justified by the corollary to the Birkhoff-Jebsen theorem [68], a topic we shall revisit in Chapter 7. As we shall see there, the Birkhoff-Jebsen theorem allows us to understand why the influence of any *isotropic*, external source of gravity is zero within a spherical cavity. It has been noted on many occasions [69] that because of this symmetry, a limited use of Newtonian mechanics is permitted for some cosmological problems, such as the one that elicits our attention here.

Consider a sphere 'carved out' of a homogeneous, isotropic universal medium with radius $R_s(t) = a(t)r_s$. Adopting the Cosmological Principle, we assume that the density within this region is a function of time t only, and that every point within and without the sphere expands away from every other point in proportion to the time-dependent scale factor $a(t)$, which itself is the same everywhere. According to the Birkhoff-Jebsen theorem and its corollary, we only need to consider contributions to the energy from the contents enclosed within R_s to determine the local dynamics of this region extending out to R_s.

Relative to an observer at the center of this sphere, the kinetic energy of a shell with thickness dR at radius R is therefore

$$dK = 4\pi R^2 \, dR \, \frac{\rho(t) \dot{R}^2}{2c^2} \,, \tag{4.32}$$

and integrating this out from $r = 0$ to $r = r_s$, one easily gets the total kinetic energy of this sphere relative to the observer at the origin:

$$K = \frac{2\pi}{5} \frac{\rho(t)}{c^2} a^3 \dot{a}^2 \, r_s{}^5 \,. \tag{4.33}$$

Let us now calculate the corresponding gravitational potential energy of this spherical distribution (remember that this is a classical derivation). The potential energy of the shell at R is

$$dV = -4\pi R^2 \, dR \, \frac{GM(R)\rho(t)}{c^2 R} \,, \tag{4.34}$$

where

$$M(R) = \frac{4\pi}{3} R^3 \frac{\rho(t)}{c^2} \tag{4.35}$$

is the total mass enclosed within the radius R. And integrating this out from $r = 0$ to $r = r_s$, we see that the total potential energy of this sphere (as measured by the observer at the origin) is

$$V = -\frac{16\pi^2 G}{15} \frac{\rho(t)^2}{c^4} a^5 r_s{}^5 \,. \tag{4.36}$$

Thus, from a classical point of view, the observer measures a total energy of this expanding sphere given by

$$E = \frac{2\pi}{5} \frac{\rho(t)}{c^2} a^3 \dot{a}^2 \, r_s{}^5 - \frac{16\pi^2 G}{15} \frac{\rho(t)^2}{c^4} a^5 r_s{}^5 \,, \tag{4.37}$$

which may be re-arranged to cast it into a more recognizable form:

$$\left(\frac{\dot{a}}{a}\right)^2 = \frac{8\pi G}{3c^2} \rho(t) + \frac{5c^2 E}{2\pi \rho(t) \, a^5 \, r_s{}^5} \,. \tag{4.38}$$

Evidently, the local conservation of energy relative to the observer at the origin is actually the Friedmann Equation (4.28), when we interpret the spatial curvature constant as

$$k \equiv -\frac{10}{3 \, r_s{}^2} \left(\frac{\epsilon}{\rho}\right) \,, \tag{4.39}$$

in terms of the local *total* energy density

$$\epsilon \equiv \frac{3E}{4\pi R_s{}^3} \,. \tag{4.40}$$

Note that in Equation (4.39), which was derived classically, ϵ represents the *local* kinetic plus gravitational energy density, while ρ represents the energy density in the medium. Crucially, the ratio ϵ/ρ is therefore independent of volume, so the value of k is 'universal.' A universe with positive spatial curvature therefore corresponds to a net negative energy $(\propto -\epsilon)$, which means the system is bounded, whereas a negative curvature is associated with a positive total energy density $(\epsilon > 0)$, characterizing an unbounded universe. A universe with net zero energy[8] is spatially flat $(k = 0)$.

---

[8]As we shall see later in this book, particularly Chapter 6, the observations are apparently telling us that the Universe is flat, meaning that its total energy density ϵ is exactly zero.

II

The Metric

CHAPTER 5

Special FLRW Solutions

T HE identification of a spacetime metric relevant for cosmology does not complete the story. Thus far, our theoretical understanding is limited to identifying the essential symmetries encoded into the Cosmological Principle, that have helped us simplify the metric coefficients in FLRW (Equation 4.19). What is still missing is a physical basis for generating the expansion dynamics, characterized by the two principal unknowns in this metric: the spatial curvature constant k and the universal expansion factor $a(t)$. As we shall see in Chapter 6, this incompleteness is the reason why our current standard model is largely empirical, characterized by at least six unknown parameters that must be optimized using real data.

In this chapter, we shall use the dynamical equations (4.28, 4.29, and 4.31) derived within the FLRW framework to explore the range of expansion scenarios permitted by several historically important equations-of-state. In deriving Equation (4.19), we placed no constraint on the lapse function g_{tt} in connection with our choice of stress-energy tensor $T^{\alpha\beta}$. This has been the conventional approach followed since de Sitter's discovery (Figure 5.1) in 1917 of his famous cosmological solution to Einstein's equations, barely three years after general relativity was founded [70, 71]. At this stage, we are therefore free to choose any form of the energy density ρ and pressure p (see Equation 4.30) in the cosmic fluid, which directly impact the time dependence of $a(t)$.

In Chapter 10, we shall make significant progress beyond this point, concluding from a more in-depth analysis of the relationship between g_{tt}, ρ and p, that the applicability of the FLRW metric in fact does not extend to all possible equations-of-state. But for the moment, let us first re-visit some of the most influential historical solutions that have helped inform our understanding of how the Universe expands.

5.1 MILNE UNIVERSE

One of the simplest cosmologies we can think of is an empty universe with a corresponding equation-of-state $\rho = p = 0$. But though $\rho = 0$, empty does not necessarily mean trivial. This happens because even though $\rho = 0$, the Hubble constant in Equation (4.28) need not be zero, which means the 3-space curvature constant k must then be negative. This is the basis for the model introduced by Edward Arthur Milne

(1896–1950) in 1940. On first inspection, this situation seems somewhat paradoxical because an empty universe with a non-zero Hubble constant would then appear to have net positive energy (see Section 4.3), which would presumably be associated with the kinetics of expansion (since $\dot{a} \neq 0$). But we need to be careful here, because conclusions such as this depend strongly on the coordinates one chooses to describe the dynamics. We shall see shortly that a Milne type of cosmology is really just a re-parametrization of a static spacetime with zero (4-dimensional) curvature.[1]

In the heuristic argument we made in Section 4.3, the identification of k with energy ϵ was made entirely in terms of the coordinates associated with a *single* observer at the origin. However, the solutions to Equations (4.28), (4.29), and (4.31), written in terms of $R(t) = a(t)r$ and t, do not describe the behavior of the Universe solely on the basis of measurements of distance and time made by just one individual. This point deserves additional explanation because we shall encounter many instances throughout this book where the interpretation of the cosmological variables appears to be strongly dependent on the coordinate system one chooses.

Of course, none of the various interpretations is necessarily problematic because any coordinate system can be used with equal validity as long as it is associated with a legitimate solution of the dynamical equations. Having said this, there is no question that one must distinguish between coordinates measured on rulers and clocks at rest with respect to a single observer, and those that more appropriately ought to be called *community* coordinates associated with measurements made by two or more individuals. The Milne universe is a case in point, so this is a good place for us to think more deeply about this distinction before we continue with our development of the empty universe model.

Coordinates of the former type are quite easy to identify. They happen to be those with which we may write proper distances $(g_{ij} \, dX^i dX^j)^{1/2}$ extracted from Equation (2.31), and proper times $(g_{00} \, dX^0 dX^0)^{1/2}$, in terms of a single observer's measurements anywhere in the spacetime. An example of a community coordinate is the so-called 'proper' radius $R(t) \equiv a(t)r$ but, as we shall demonstrate shortly, whereas dR represents the proper differential distance for one observer, the integral $R = \int dR$ instead represents the total proper distance for a collaboration of observers (each contributing a segment dR to the total length).

Clocks and rulers at rest with respect to a single observer may be used, under the right circumstances, to represent the entire value of the line element ds. For example, the proper time in Equation (2.5) is the time measured on a clock at rest with respect to that observer, so that Δs (or ds in the infinitesimal limit) is entirely due to the passage of time. For a single observer, the proper distance is a length measured simultaneously from one spacetime point to another with rulers at rest with respect to her, such that $dX^0 = 0$.

---

[1]As we proceed through this chapter, it is worth reinforcing the distinction between 3-space curvature, characterized by the constant k, which depends on the total energy density in the Universe and ultimately says something about whether the Universe is bounded (closed) or unbounded (open), and 4-dimensional spacetime curvature, manifested through the Riemann-Christoffel tensor (Equation 2.147), a measure of the local acceleration due to gravity.

Insofar as the FLRW metric (Equation 4.19) is concerned, the differential element of proper radius (or, equivalently, distance when $d\Omega = 0$) is therefore

$$dR = a(t)\,\frac{dr}{\sqrt{1 - kr^2}}\,, \qquad\qquad (5.1)$$

or simply $dR = a(t)\,dr$ for a flat spacetime. The cosmic time t is the proper time for each observer measuring time on the clock locally at rest with respect to him. At all other radii, however, he sees the clocks moving and would therefore need to make a Lorentz transformation between those frames in which the other clocks are at rest and his own in order to get the correct time in his frame.

Yet each contribution dR to R is made at the same *local* time t, meaning that each segment dR is measured by a different observer—the person for whom t is in fact the proper time on the clock at rest with respect to him at his location. To find R, we integrate over a range of comoving radii r, evaluating $a(t)$ at the same local proper time t everywhere, not at the Lorentz-transformed time we would measure on our clocks. Therefore, the quantity $R(t)$ at best represents the sum of all the incremental segments measured by a collaboration of observers—each at time t in their own rest frame—stretched out from one spacetime point to another.[2] Thus, although $R(t)$ is usually referred to as the 'proper' distance, it is really only dR that represents a proper differential distance for one observer. The integrated value R is in fact not the physical distance measured between any pair of spacetime points by a single individual using rulers at rest with respect to her in her own frame. This is why R should more aptly be called a *community* coordinate.

Let us now see how the perceived behavior of the Milne universe changes depending on what type of coordinates we choose. A universe with $\rho = 0$ and $k = -1$ corresponds to a simple solution of the Friedmann Equation (4.28) with

$$a(t) = ct\,, \qquad\qquad (5.2)$$

i.e., a universe in which the scale factor grows linearly in time at a rate equal to the speed of light c. Since the 'acceleration' $\ddot{a}(t)$ is therefore zero in this cosmology (see Equation 4.29), first introduced by Milne [72] in the 1940's, one might expect such a universe to be flat and a mere re-parametrization of Minkowski spacetime.[3] Indeed, Milne intended this type of expansion to be informed by special relativity only, without any constraints imposed by the more general theory.

To demonstrate this formally, we shall now derive a straightforward transformation of the coordinates that allow the Milne metric to appear in its manifestly flat form. We first introduce the comoving distance variable χ, defined in terms of r according to

$$r = \sinh \chi\,, \qquad\qquad (5.3)$$

---

[2]This kind of description for the meaning of $R(t)$ initially appeared in ref. [69].

[3]In Minkowski spacetime, the three ordinary spatial dimensions are combined with a dimension of time to form a four-dimensional manifold, though strictly adhering to the metric of special relativity. In this spacetime, there is no acceleration due to gravity, so the components of the Riemann-Christoffel tensor (Equation 2.147) are all zero.

which allows us to write the FLRW metric for Milne in the form

$$ds^2 = c^2 dt^2 - c^2 t^2 [d\chi^2 + \sinh^2 \chi \, d\Omega^2] \, . \tag{5.4}$$

The transformation that brings Equation (5.4) into a stationary (and manifestly flat) form is

$$\begin{aligned} T &= t \cosh \chi \\ R &= ct \sinh \chi \, , \end{aligned} \tag{5.5}$$

for then

$$ds^2 = c^2 dT^2 - dR^2 - R^2 d\Omega^2 \, . \tag{5.6}$$

We now recognize that the coordinate $X^0 = cT$ represents the time measured by a single observer on clocks at rest in his frame, while $X^1 = R$ is the radial coordinate measured with rulers also at rest with respect to this observer. A comparison with the general form of the FLRW metric (Equation 4.19) shows that there is no evident expansion in this frame (since $a[t] = 1$) and, very importantly, $k = 0$ when the spatial curvature constant is associated with the coordinates (R, θ, ϕ). Writing the Milne solution this way, we recover our intuitive understanding of the physical meaning of k, in that an empty universe does in fact correspond to net zero energy after all. But we did not recognize this immediately when presented with the FLRW metric for the Milne universe written in terms of community coordinates. For this reason, one must be very careful when interpreting the value of k emerging from the general form of the FLRW metric, since our heuristic argument of Section 4.3 applies only when the coordinates are those of a single observer.

5.2   DE SITTER SPACE

Unlike the Milne universe, the de Sitter cosmology [70] has a nonzero spacetime curvature. In de Sitter space, $\rho \neq 0$ (when the cosmological constant is interpreted as an energy density), and objects not only recede from one another, but also accelerate under the influence of gravity. The de Sitter cosmology is a universe devoid of matter and radiation, but filled with a cosmological constant with an equation-of-state $p = -\rho$. According to the Friedmann Equations (4.28) and (4.29), the expansion factor in this model is therefore $a(t) = e^{Ht}$, and the FLRW metric in this case may be written

$$ds^2 = c^2 dt^2 - e^{2Ht}(dr^2 + r^2 d\Omega^2) \, , \tag{5.7}$$

in terms of the (constant) Hubble parameter H.

Though Willem de Sitter (1872–1934) (see Figure 5.1) is not credited with the discovery of an expanding Universe, de Sitter's 1917 papers were an important step in developing the theory behind it. His intention was to explore the possible astronomical implications of Einstein's static world model (see Section 2.1), but at the same time presented this alternative solution, which also has a time-independent spacetime curvature, as it turns out (see Section 5.4.3 below), though it may be written with a particular set of coordinates (ct, r, θ, ϕ) for which the medium expands.

Figure 5.1 The Dutch astronomer Willem de Sitter as he appeared around 1898. In 1917, he published a solution to Einstein's equations describing the spacetime within a uniform medium containing only a cosmological constant Λ (see Equation 2.171). (Image courtesy of Theodora Smit and the Center for History of Physics at the American Institute of Physics, from *God and the Astronomers* by Robert Jastrow)

Already by the time de Sitter's papers were published, it was known from the work of Vesto Slipher (see Chapter 8 and Figure 8.1) at the Lowell Observatory that the spectra of light from several spiral nebulae are redshifted, interpreted at that time as due to a Döppler effect. By the mid 1920's, a more enduring view took hold—that Slipher's catalog of redshifted galaxies was actually evidence in favor of an expanding Universe. Those developing this idea, which we shall explore in Chapter 8, remarked on the possible connection between these observations and de Sitter's solution. Of course, as more and more galaxies were discovered, it also became quite apparent that the Universe did not merely contain a cosmological constant, and could perhaps even be dominated by matter after all, though not in a static geometry, as Einstein had surmised in 1917. In the next section, we shall see how a subsequent collaboration between Einstein and de Sitter produced another influential cosmology, still used today for model comparison purposes.

When we examine the standard model in Chapter 6, we shall find that de Sitter's solution may represent the Universe's terminal state, and may also have corresponded to an early inflationary phase, during which it would have produced an accelerated expansion due to the strong dependence of the expansion factor on time.

5.3 EINSTEIN-DE SITTER UNIVERSE

By the early 1930's, Einstein had become convinced of the expansion of the Universe (see Section 4.2) and found reason to consider solutions to his field equations other than those producing steady-state conditions. In 1932, he collaborated with de Sitter, producing one of the most enduring cosmological models, which they based on the simplest possible assumptions consistent with the observations available then [73]. Called the Einstein-de Sitter universe, this world model used only what was known at that time—the mean density and expansion of the Universe. They simply took the curvature to be zero ($k = 0$), they assumed that there was no cosmological constant ($\Lambda = 0$), and that the pressure was also zero ($p = 0$). It is clear from Equations (4.28) and (4.29) that the density

$$\rho = \frac{3c^2}{8\pi G} H^2 \tag{5.8}$$

in this model is given directly in terms of the Hubble constant H and, moreover, that

$$\frac{\ddot{a}}{a} = -\frac{1}{2}\left(\frac{\dot{a}}{a}\right)^2 , \tag{5.9}$$

which has the simple polynomial solution

$$a(t) = a_0 + a_1 t^{2/3} . \tag{5.10}$$

Remarkably, Einstein and de Sitter avoided actually writing out this solution and discussing the implied Big Bang features of their model—i.e., that regardless of whether or not the coefficient a_0 is zero, this type of universe had an abrupt beginning at time $t = 0$.

But in spite of Einstein's and de Sitter's lukewarm support for this model, it has developed a historical significance due to its simplicity and ease of use. Hardly a candidate for the real structure of the Universe, it is nonetheless a good representation of the broader class of relativistic evolution cosmologies and is often employed as a basis for more sophisticated developments, some of which we shall consider in model comparisons described in Chapter 12. In a way, the Einstein-de Sitter cosmology paid homage over the years to the work of those, like Friedmann and Lemaître, whose inspiration eventually led to the idea of an expanding universe with an explosive beginning.

5.4 STATIC FLRW SOLUTIONS

Two of the solutions we have examined here—the Milne universe and de Sitter space—also happen to be members of a very special category of FLRW cosmologies. These are models with a constant spacetime curvature, sometimes also referred to as the 'static' solutions since their metric coefficients are independent of time. In a pivotal paper four decades ago, Florides presented a formal proof that there are exactly six, and only six, such solutions [74]. These are (1) the Minkowski spacetime, (2) the Milne universe, (3) de Sitter space, (4) the Lanczos universe, (5) a Lanczos universe with $k = -1$, and (6) anti-de Sitter space (a universe with negative mass density). We highlight these special solutions because, as we shall see in Chapter 8, they figure very prominently in our detailed study of cosmological redshift. It is worth mentioning that, although the spacetime curvature is constant in all these cases, it is generally *not zero*. Gravitational effects are present even when the FLRW metric is time-independent, as is well known from the analogous Schwarzschild and Kerr spacetimes (Equations 3.19 and 3.45).

5.4.1 Minkowski Spacetime

The Minkowski spacetime is spatially flat ($k = 0$) and is not expanding, $a(t) = 1$, so

$$ds^2 = c^2 dt^2 - dr^2 - r^2 \, d\Omega^2 \,. \tag{5.11}$$

The metric is already in static form, so there is no need to find an alternative set of coordinates for which the metric coefficients become time independent.

5.4.2 The Milne Universe (Revisited)

As introduced in Section 5.1 above, Milne's cosmology is based on the assumption of zero content ($\rho = 0$) and negative spatial curvature ($k = -1$), with a corresponding spacetime interval given in Equation (5.4). At least some of the metric coefficients written using comoving coordinates are time dependent, because the expansion factor $a(t)$ grows linearly with cosmic time t. With the transformation given in Equations

(5.5) and (5.6), however, the metric is brought into a manifestly flat form, with coefficients that are completely independent of the new time coordinate T. The form of the metric in Equation (5.6) is identical to the Minkowski spacetime (Equation 5.11), confirming the generally understood identity between these two models. Each of these is viewed as being a re-parametrization of the other. Neither Minkowski spacetime nor the Milne universe has any spacetime curvature, and may therefore be transformed into each other with an appropriate set of coordinates (as we have just seen).

5.4.3 de Sitter Space

Let us first present the transformation that casts the metric in Equation (5.7) into its static form, and then discuss the physical meaning of the new coordinate system. In terms of the new coordinates (cT, R, θ, ϕ), where R denotes the usual proper (or, community) distance, the required transformation may be written

$$
\begin{aligned}
R &= a(t)r \\
T &= t - \frac{1}{2H} \ln \Phi ,
\end{aligned}
\tag{5.12}
$$

where

$$
\Phi \equiv 1 - \left(\frac{R}{R_{\rm h}}\right)^2 ,
\tag{5.13}
$$

in terms of the Hubble radius

$$
R_{\rm h} \equiv \frac{c}{H} .
\tag{5.14}
$$

In this new coordinate system, the de Sitter metric becomes

$$
ds^2 = \Phi \, c^2 \, dT^2 - \Phi^{-1} dR^2 - R^2 d\Omega^2 .
\tag{5.15}
$$

Clearly, both the new metric coefficients $g_{TT} = \Phi$ and $g_{RR} = \Phi^{-1}$ are independent of the new time coordinate T.

As we shall study in some detail in Chapter 7, this form of the metric explicitly reveals the spacetime curvature most elegantly understood in terms of the Birkhoff-Jesben theorem and its corollary [68]. There we shall learn that, in a homogeneous and isotropic universe, an observer measures a spacetime curvature a proper distance R away from himself that depends solely on the mass-energy content of a sphere with radius R. Due to the spherical symmetry, the influence of the rest of the Universe is completely canceled locally. This may appear to contradict our assumption of isotropy, which could naively be interpreted to mean that the spacetime curvature should cancel because the observer sees mass-energy equally distributed in all directions. But the Birkhoff-Jesben theorem ensures that only the mass-energy between any pair of spacetime points affects the path linking them, so in fact the observer's worldlines are curved in *every* direction.

The metric in Equation (5.15) is how de Sitter himself first presented his now famous solution. One can understand the inspiration for it by considering the following heuristic argument, which will be justified more formally in Chapter 7. Consider the Schwarzschild solution for the spacetime surrounding an enclosed, spherically-symmetric object of mass M given in Equation (3.19). The de Sitter metric describes the spacetime surrounding a *radially-dependent* enclosed mass

$$M(R) = M(R_{\mathrm{h}})(R/R_{\mathrm{h}})^3 , \tag{5.16}$$

in a medium with uniform mass-energy density. Thus, the Schwarzschild factor $1 - (2GM/c^2R)$ transitions into $1 - (R/R_{\mathrm{h}})^2$, i.e., the function Φ (see Equation 5.13), which reproduces the static form of de Sitter's metric in Equation (5.15). In Chapter 7, we shall better understand why this equation implicitly contains the restriction that no mass-energy beyond R should contribute to the gravitational influence inside this radius, as required by the Birkhoff-Jebsen theorem.

When we study the nature of cosmological redshift in Chapter 8, it will be necessary for us to recognize that de Sitter's metric in Equation (5.15) is not only static in the new coordinates (cT, R, θ, ϕ), but that it also contains the effects of gravitational curvature through the factor Φ, analogously to the corresponding factor in the Schwarzschild metric (Equation 3.19).

5.4.4 The Lanczos Universe

The Lanczos universe [75] is described by the metric

$$ds^2 = c^2 dt^2 - (cb)^2 \cosh^2\left(\frac{t}{b}\right)\left[\frac{dr^2}{1 - r^2} + r^2 d\Omega^2\right] , \tag{5.17}$$

where b is a constant (though not the Hubble constant $H \equiv \dot{a}/a$) and $k = +1$. Evidently, the expansion factor is

$$a(t) = (cb)\cosh\left(\frac{t}{b}\right) , \tag{5.18}$$

so $H = (1/b)\tanh(t/b)$. The physical interpretation of this model is that it represents the gravitational field of a rigidly rotating dust cylinder coupled to a cosmological constant. The coordinate transformation that renders this metric in its static form is

$$R = cb\,r\cosh\left(\frac{t}{b}\right)$$

$$\tanh\left(\frac{T}{b}\right) = \left(1 - r^2\right)^{-1/2}\tanh\left(\frac{t}{b}\right) . \tag{5.19}$$

In terms of these new coordinates, the spacetime interval may be written in the form

$$ds^2 = \left[1 - \left(\frac{R}{cb}\right)^2\right]c^2 dT^2 - \left[1 - \left(\frac{R}{cb}\right)^2\right]^{-1} dR^2 - R^2 d\Omega^2 . \tag{5.20}$$

Clearly, the metric coefficients in Equation (5.20) are independent of the new time coordinate T, so this FLRW solution has constant spacetime curvature.

5.4.5 The Lanczos Universe with $k = -1$

The fifth static FLRW metric is simply the Lanczos universe with $k = -1$, for which the interval written in comoving coordinates is

$$ds^2 = c^2 dt^2 - (cb)^2 \sinh^2\left(\frac{t}{b}\right)\left[\frac{dr^2}{1+r^2} + r^2 d\Omega^2\right], \qquad (5.21)$$

where now

$$a(t) = (cb)\sinh\left(\frac{t}{b}\right). \qquad (5.22)$$

This metric may be written in static form with the coordinate transformation

$$R = cbr\sinh\left(\frac{t}{b}\right)$$

$$\tanh\left(\frac{T}{b}\right) = \left(1+r^2\right)^{1/2}\tanh\left(\frac{t}{b}\right). \qquad (5.23)$$

These new coordinates allow us to write the interval in the form

$$ds^2 = \left[1 - \left(\frac{R}{cb}\right)^2\right]c^2 dT^2 - \left[1 - \left(\frac{R}{cb}\right)^2\right]^{-1} dR^2 - R^2 d\Omega^2, \qquad (5.24)$$

identical (in terms of R and T) to the Lanczos metric in Equation (5.20). Note, however, that in spite of the identical form of these two metrics (Equations 5.20 and 5.24), they are not the same because the transformed coordinates T and R incorporate different curvature effects relative to the (same) comoving coordinates (ct, r, θ, ϕ).

5.4.6 Anti-de Sitter Space (Negative ρ)

The sixth, and final, FLRW metric with constant spacetime curvature is known as anti-de Sitter space—a universe with negative mass density and spatial curvature constant $k = -1$. This metric is given by

$$ds^2 = c^2 dt^2 - (cb)^2 \sin^2\left(\frac{t}{b}\right)\left[\frac{dr^2}{1+r^2} + r^2 d\Omega^2\right], \qquad (5.25)$$

where clearly the expansion factor is now

$$a(t) = cb\sin\left(\frac{t}{b}\right). \qquad (5.26)$$

The static form of the metric is produced with the coordinate transformation

$$R = cbr\sin\left(\frac{t}{b}\right)$$

$$\tan\left(\frac{T}{b}\right) = \left(1+r^2\right)^{1/2}\tan(t/b), \qquad (5.27)$$

which yields

$$ds^2 = \left[1 + \left(\frac{R}{cb}\right)^2\right] c^2 dT^2 - \left[1 + \left(\frac{R}{cb}\right)^2\right]^{-1} dR^2 - R^2 d\Omega^2 . \tag{5.28}$$

We shall return to these six static FLRW cosmologies in Chapter 8.

CHAPTER 6

The Standard Model

T HE significant progress we have made thus far describing the Universe and its
history within the framework of general relativity and a combination of philosophical, empirical and theoretical arguments, is not yet sufficient to identify a specific, well-motivated cosmological spacetime. Certainly the adoption of the Cosmological Principle (Section 1.3) eliminates numerous possible solutions to Einstein's
field equations. And an insistence on simplicity reduces the expansion dynamics to
very manageable analytic forms for the expansion factor $a(t)$ (as we saw in Chapter 5). But in the end, most of the essential physics resides in the stress-energy tensor,
$T^{\mu\nu}$, defined in Equation (2.126). As we shall show throughout the remainder of this
book, the choice of cosmology is ultimately dictated by our selection of this $T^{\mu\nu}$.
In the standard model, known as ΛCDM [76], this tensor's components are largely
empirical, without much input from theory, of which we are reminded by the rather
emphatic inscription of its contents across its name, 'Λ plus cold dark matter.'

Unlike many other subjects in science, the Universe is unique, limiting one's ability
to compare objects in the same category in order to find patterns and reproducibility.
Because of this, it is often difficult to discuss what is probable or necessary. And the
fact that we must incorporate its history into our current understanding compels us
to abduce the most likely evolutionary scenario, rather than to make an overarching
physical argument and prediction and then test them against the data. When one
attempts to reconstruct the conditions in the early Universe, the physical boundaries
one may use are imposed by what is believed to have occurred at later epochs, based
on the physical laws available today. Not to mention that this situation also has the
tendency of blurring the distinction between what is actually due to the physics versus
the initial conditions. Even now, after a century of scientific development, cosmology
cannot completely avoid some reliance on philosophical biases.

The difficulty with properly identifying the correct cosmological spacetime is com-
pounded by two other unavoidable factors. The first is that Einstein's field equations
cannot be solved under fully generalized conditions. These represent a system of 10
coupled, non-linear partial differential equations for the metric coefficients, so one
may find exact, analytic solutions for the simplest assumptions concerning the symmetries of the spacetime and the choice of $T^{\mu\nu}$, but not at all for even a basic blend
of contributions to the energy density ρ and pressure p. And in spite of the growing

refrain that we are now conducting 'precision' cosmology, the second factor is that the data are not good enough for us to fully identify all of the equations-of-state shaping the stress-energy tensor. Recall that all of the observations we make today are confined to our past lightcone, so some assumptions concerning the rest of the Universe are unavoidable. In addition, source catalogs are always limited in magnitude and redshift coverage, their possible evolution must be conjectured, and we do not actually observe every single component in the cosmic fluid. This is most famously the case for the hypothesized dark matter, which we believe couples only gravitationally to the rest of the Universe.

One also needs to be vigilant about the tendency of lending too much credence to a model when its parameters continue to be optimized with ever improving accuracy. After all, these variables are defined within a specific framework and refining their measurement is subject to the limitations and biases inherent in the model itself. If the assumed cosmology is incorrect or incomplete, a high precision determination of its parametrization will not mitigate its inadequacies. In such circumstances, high precision measurements often do not shed any additional light on the physical nature of the underlying model. Quite famously, even as we continue to refine the value of the cosmological constant, we still have no clue why it misses the predicted quantum-field energy of the vacuum by 120 orders of magnitude (see Sections 1.2 and 6.1). Nor should one forget the lessons learned from history, in which 'precision' measurements came and went. Bishop James Ussher (1581–1656), the Archbishop of Armagh and Primate of All Ireland, was well-known for his detailed reading of the Old Testament and other ancient documents, that produced a precise dating of historical events, including Earth's creation on October 23, 4,004 B.C.E.[1]

The standard model thus derives from an unavoidable compromise: rather than aiming to find the ultimate description of everything, ΛCDM seeks to identify a metric that approximates the observable Universe as well as possible. The downside of this approach, which is becoming a debilitating factor as the precision of our measurements continues to improve, is that—while this mostly empirical method accounts quite well for many large-scale properties of the cosmos—it can nonetheless still fail at small or large redshifts, as we shall chronicle in Chapters 9–12. There is sometimes a tendency to then disqualify the model, labeling it as incorrect, but that is not appropriate either, because it was never the intention of ΛCDM to claim that it provided the final verdict. By its very essence, it is a work in progress, an approximation to what we hope will eventually emerge as the correct cosmic spacetime.

Indeed, an important question one should keep asking is "Why, in the face of so many assumptions and poorly motivated physical inputs, does it work so well?" No doubt, the growing tension between ΛCDM and the data today are making us wonder whether we are relying too heavily on its overly-naive internal structure. But without preempting too many of the details from Chapters 9–12, we shall find later in this book

---

[1] Follow this delightful analysis in *Annales Veteris Testamenti, a Prima Mundi Origine Deducti, una cum Rerum Asiaticarum et Aegyptiacarum Chronico, a Temporis Historici Principio Usque ad Maccabaicorum Initia Producto* (Annals of the Old Testament, Deduced from the First Origins of the World, the Chronicle of Asiatic and Egyptian Matters Together Produced from the Beginning of Historical Time up to the Beginnings of Maccabees) [77].

that its parametrization turns out to be quite flexible, allowing an optimization that mimics what appears to be a very special equation-of-state in general relativity—the zero active mass condition $\rho + 3p = 0$, for the total energy density ρ and pressure p. Later, we shall explore the physics behind this emerging property of the cosmic spacetime, but shall make no further reference to this constraint for the time being.

6.1 STANDARD MODEL BASICS

It is helpful in discussing the structure and evolution of the Universe to have at hand several basic, measurable parameters and the functional form of the expansion factor under special circumstances. The Friedmann equation (4.28) may be used to define a critical energy density corresponding to a spatially-flat ($k = 0$) Universe,

$$\rho_c \equiv \frac{3c^2 H_0^2}{8\pi G} . \tag{6.1}$$

For the most recent *Planck* parameter optimization [29], with $H_0 = 67.4 \pm 0.5$ km s^{-1} Mpc^{-1}, the implied critical energy density is

$$\rho_c = 8.13 \times 10^{-9} \text{ erg cm}^{-3}$$
$$= 5.07 \times 10^{-6} \text{ GeV cm}^{-3} . \tag{6.2}$$

One typically defines the total energy density parameter Ω as the ratio of ρ to ρ_c,

$$\Omega \equiv \frac{\rho}{\rho_c} \tag{6.3}$$

so that, according to Equation (4.28), the Universe is closed ($k = +1$) when $\Omega > 1$, open ($k = -1$) when $\Omega < 1$, and spatially flat ($k = 0$) when $\Omega = 1$.

As best as one can tell from the observations (see Chapter 1), the cosmic fluid in the local Universe consists of pressureless baryonic matter, cold (and also pressureless) dark matter, radiation and dark energy, which may or may not be in the form of a cosmological constant Λ. The situation changes somewhat in the early Universe, where ρ may also have had a contribution from relativistic neutrinos (see Equation 6.39 below) and, if inflation actually took place, a very brief dominance by an inflaton field, about which we shall have much more to discuss in Sections 6.5–6.7.

The approximate, largely empirical nature of ΛCDM makes it necessary for us to simplify ρ and its equation-of-state as much as possible in order to have any chance of solving the Friedmann Equations (4.28) and (4.29) for the expansion factor $a(t)$. It therefore makes sense to subdivide the energy density into its constituent parts, which we shall call ρ_r, ρ_m and ρ_{de}, for the radiation, (baryonic plus dark) matter, and dark energy, respectively. We do the same for the individual pressures, p_r, p_m and p_{de}. With an intuitively obvious definition of the scaled energy densities, one therefore has

$$\Omega = \Omega_r + \Omega_m + \Omega_{de} . \tag{6.4}$$

According to the most recent *Planck* optimizations (see above), $\Omega_r = (5.14\pm0.001) \times 10^{-5}$, $\Omega_m = 0.315\pm0.007$ and $\Omega_\Lambda = 1.0 - \Omega_m - \Omega_r$. This last relation is an expression of the inferred spatial flatness of the Universe, $k = 0$.

But notice that, whereas writing the total energy density

$$\rho = \rho_r + \rho_m + \rho_{de} , \tag{6.5}$$

is trivial (and correct), the same cannot be said of the total pressure p. The reason is that when two or more species in the cosmic fluid are interacting, the pressure is not just $p = p_r + p_m + p_{de}$, in terms of the respective, individualized pressures p_i for $i = $ 'r', 'm', and 'de'. To overcome this hurdle, ΛCDM therefore usually assumes that either the Universe is dominated by just one component at a time, or that co-existing species are not interacting and p is simply a sum of their independent contributions. An important exception occurred prior to recombination, when radiation and baryonic matter are believed to have remained coupled, producing acoustic waves in the expanding fluid (see Section 6.4 and Chapter 13). In either case, it is sometimes helpful to also define a so-called equation-of-state parameter $w \equiv p/\rho$, in terms of the energy density and pressure of each component.

When a single species dominates the energy density (and pressure), the solution to the Friedmann and Raychaudhuri equations is very easy to find. The equation-of-state parameter is then constant, and Equation (4.31) may be written

$$\dot{\rho} = -3\frac{\dot{a}}{a}(1+w)\rho , \tag{6.6}$$

which is easily integrated to yield

$$\rho = \rho_0 \, a^{-3(1+w)} . \tag{6.7}$$

Thus, for $w \neq -1$ and $k = 0$, one may insert this density into Equation (4.28) and find that

$$a(t) = a_0 \, t^{2/[3(1+w)]} . \tag{6.8}$$

In these expressions, a quantity with subscript '0' denotes a normalization constant or amplitude, though not necessarily evaluated at the present time.

As we shall see shortly, the radiation energy density is a stronger function of $a(t)$ than that of either matter or dark energy, so in ΛCDM the dynamics of the early Universe (corresponding to $a \to 0$) was dominated by radiation—or relativistic particles, if we include neutrinos (see Section 6.2)—for which $w = +1/3$. In that case,

$$\rho_r = \rho_{r0} \, a^{-4} , \tag{6.9}$$

and

$$a(t) = a_0 \, t^{1/2} \quad \text{(Radiation)} . \tag{6.10}$$

The radiation energy density drops off by the additional factor of $a(t)$ in Equation (6.9) compared to matter (Equation 6.11) for the simple reason that cosmological redshift reduces the relativistic particle energy as the Universe expands (see Equation 8.8).

Non-relativistic matter eventually dominates the energy density (see Section 6.5) and one typically assumes that $p_m = 0$ during this phase, for which $w = 0$. The dilution in ρ then follows the inverse volume relation one would expect for a fixed particle number,

$$\rho_m = \rho_{m0}\, a^{-3}\,,\tag{6.11}$$

and

$$a(t) = a_0\, t^{2/3} \quad \text{(Matter)}\,.\tag{6.12}$$

None of the formalism in Chapter 4 requires (or even implies) the existence of dark energy, but we abduce from the apparent expansion history of the Universe that such an entity is necessary to prevent Ω falling below 1 (see, e.g., Sections 12.3.1–12.3.2). The aforementioned *Planck* optimizations suggest that $w_{de} = -1.03 \pm 0.03$, very close to -1, though the error still leaves some room for a different value. Of course, we cannot be entirely sure, but setting $w_{de} = -1$ has the advantage of allowing us to assume a cosmological constant Λ, for which $p_\Lambda = -\rho_\Lambda$. In that case, Equation (4.31) gives $\rho_{de} = \rho_{de0}$ (constant), which means

$$H_{de} \equiv \left(\frac{8\pi G}{3c^2}\,\rho_{de0}\right)^{1/2}\tag{6.13}$$

is constant as well (as long as $k = 0$ remains valid). Thus, according to the equation $\dot{a}/a = H_{de}$, we should expect the Universe to expand exponentially when dark energy (in the guise of a cosmological constant) dominates the energy density:

$$a(t) = a_0\, e^{H_{de}t} \quad \text{(Cosmological constant)}\,.\tag{6.14}$$

In the context of our discussion in Section 5.2, this result immediately suggests that a ΛCDM Universe with the density decomposition in Equation (6.5) enters a de Sitter phase at late times.

A schematic diagram showing these principal stages in the cosmic expansion is shown in figure 6.1, produced by the WMAP Science Team. The qualitative features of the history shown here have not changed with the improved *Planck* instrument capabilities, though some of the optimized parameters, e.g., H_0, Ω_m and several others, have been updated and their errors have been reduced. This diagram illustrates the key stages believed to have traced the ΛCDM evolution, starting with the inflationary expansion and the generation of quantum fluctuations very soon after the Big Bang (at $t \sim 10^{-35}$ sec), and the decoupling of baryonic matter and radiation at $t \sim 380,000$ yrs, that freed the CMB blackbody photons. In ΛCDM, the Universe then entered a period of 'darkness' lasting several hundred Myr, until the formation of Population III and Population II stars after $t \sim 400$ Myr, that initiated the Epoch of Re-ionization and the subsequent formation of structure. A more detailed account of the interesting and important first ~ 1 Gyr of this history will be provided in Sections 9.2–9.4, and the formation of structure will be discussed in Chapter 13. This diagram also shows graphically how the rate of expansion changed from exponential during inflation, to deceleration during the radiation and matter dominated eras, followed by late-time acceleration as the influence of dark energy became more

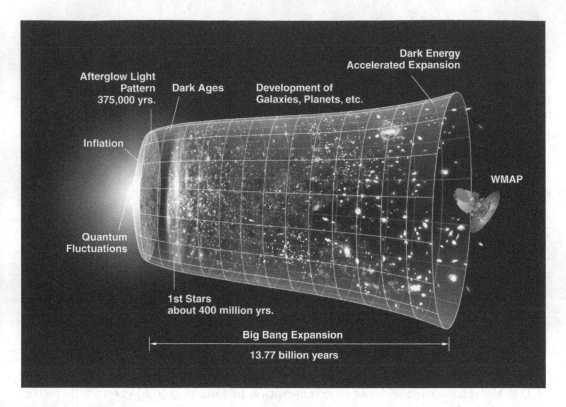

Figure 6.1 Schematic diagram showing the evolution of the Universe over its 13.77 Gyr history in ΛCDM, from the Big Bang on the far left, to recent times (with the WMAP satellite making its observations) on the right. The 'size' of the Universe is depicted by the vertical extent of the grid. At $t \sim 10^{-35}$ sec, a period of inflation produced an exponential growth. Decoupling occurred at $\sim 380,000$ yr, when the CMB was produced, after which the Universe entered its Dark Ages, lasting until ~ 400 Myr. The cosmic expansion then slowed down over the next several Gyr, but started accelerating again at mid age, due to the influence of dark energy. The assembly of structure started with the formation of Population III and then Population II stars near the end of the Dark Ages, and the assembly of galaxies and clusters subsequently persisted throughout the next ~ 13 Gyr. (Courtesy of NASA/WMAP Science Team)

pronounced. As noted in the previous paragraph, the simple (perhaps overly naive) formulation of the density and equation-of-state in ΛCDM predicts that this phase will grow in significance as the Universe ages, eventually turning the growth into purely de Sitter expansion.

6.2 THE EARLY UNIVERSE

Particle interactions are characterized by a reaction rate, Γ. As is well known from thermal physics, these particles can be maintained in thermodynamic equilibrium as long as Γ is much larger than any other rate of change in the system. In the context of an expanding Universe, one may safely assume thermodynamic equilibrium of a particular particle species 'i' at some temperature T_i as long as Γ_i greatly exceeds the Hubble expansion rate. In such a situation, these particles may be treated as either a perfect Fermi-Dirac ($+$) or Bose-Einstein gas ($-$), with distribution

$$F_i(E, T_i) = \frac{g_i}{(2\pi)^3} \frac{1}{e^{(E-\mu_i)/k_B T_i(t)} \pm 1} \equiv \frac{g_i}{(2\pi)^3} f_i(E, T_i) \,, \qquad (6.15)$$

where k_B is the Boltzmann constant, g_i is the degeneracy factor for this species and μ_i is its chemical potential. The energy E is given by the relativistic expression in Equation (2.29), and the normalization for f_i is chosen so that $f_i = 1$ for the maximum phase space density permitted by the Pauli exclusion principle for a fermion. By symmetry, T_i can be a function only of t, and interacting species are characterized by the same temperature. As long as the radiation is also in local thermodynamic equilibrium with these particles, T_i equals the photon temperature T, which may then be thought of as the temperature of the Universe at that epoch, since the photon energy density is dominant during this period.

For a particle species in thermodynamic equilibrium, its number density may be written

$$n_i = \int F_i(\mathbf{q}, T_i) \, d^3\mathbf{q} \,, \qquad (6.16)$$

and if we further assume isotropy in momentum space, then

$$n_i = \frac{g_i}{2\pi^2} \int f_i(q_i, T_i) \, q_i^2 \, dq_i \,, \qquad (6.17)$$

where we use the symbol q_i to designate the particle momentum to avoid confusion with the pressure p. Correspondingly, the energy density of species 'i' may be written

$$\rho_i = \int F_i(\mathbf{q}, T_i) \, E(\mathbf{q}) \, d^3\mathbf{q} \,, \qquad (6.18)$$

or

$$\rho_i = \frac{g_i}{2\pi^2} \int f_i(q_i, T_i) \, E_i(q_i) \, q_i^2 \, dq_i \,, \qquad (6.19)$$

and the pressure is given as

$$p_i = \int F_i(\mathbf{q}, T_i) \, \frac{q^2}{3E} \, d^3\mathbf{q} \,, \qquad (6.20)$$

or

$$p_i = \frac{g_i}{2\pi^2} \int f_i(q_i, T_i) \frac{q_i^2}{3E_i(q_i)} q_i^2 \, dq_i \, . \tag{6.21}$$

When the particles are ultra-relativistic, as one would expect for the extremely high temperatures in the early Universe, we also have $E_i \gg m_i c^2$ and $\mu_i \ll k_B T_i$. The integrals in Equations (6.17), (6.19) and (6.21) are then rather easy to evaluate, and we have

$$\rho_{\rm rel}(T) = g_*(T) \left(\frac{\pi^2 k_B^4}{30 c^3 \hbar^3} \right) T^4 \, , \tag{6.22}$$

where g_* is the effective number of relativistic degrees of freedom at this temperature:

$$g_*(T) = \sum_{i={\rm boson}} g_i \left(\frac{T_i}{T} \right)^4 + \sum_{i={\rm fermion}} g_i \frac{7}{8} \left(\frac{T_i}{T} \right)^4 . \tag{6.23}$$

The factor 7/8 arises from the different signs in the Fermi and Bose distribution functions (Equation 6.15).

For pure radiation, with $g_i = 2$ for the two photon polarization states, Equation (6.22) recovers the well-known result

$$\rho_{\rm r} = \left(\frac{\pi^2 k_B^4}{15 c^3 \hbar^3} \right) T^4 \, . \tag{6.24}$$

Similarly, from Equations (6.21) and (6.24), one gets

$$p_{\rm r} = \frac{1}{3} \rho_{\rm r} \, . \tag{6.25}$$

Very importantly, a comparison of Equations (6.9) and (6.24) shows that, as long as the radiation distribution is the Planck function (Equation 6.62, derived from Equation 6.15), its temperature traces the cosmic expansion according to

$$T = T_0 \, a^{-1} \, . \tag{6.26}$$

Perhaps more practically, we may also write

$$T = T_0(1 + z) \tag{6.27}$$

using Equation (8.9), if T_0 is chosen to be the CMB temperature measured today.

At very high temperatures in the early Universe, particles with mass m_i would have been produced and added to the thermal bath when $m_i c^2 \ll k_B T$. Thus, one may compute the energy density $\rho_{\rm rel}$ as a function of temperature by including only those species satisfying this inequality at each value of T. As an approximation, one may also assume that $T_i \approx T$ in this regime, allowing a straightforward determination of $g_*(T)$ in Equation (6.23) once the standard $SU(3) \times SU(2) \times U(1)$ model of particle physics is adopted. The complexity in this network, however, necessitates the use of lattice quantum chromodynamics (QCD) calculations [78]. The results are summarized in Table 6.1 and Figure 6.2. Two key points in this plot correspond to the

TABLE 6.1 Effective Number of Degrees of Freedom at High Temperature

Particle Mass	Temperature (GeV)	New Particles	$4g_*(T)$
$0 - m_e$	$0 - 0.0005$	γ, ν	29
$m_e - m_\mu$	$0.0005 - 0.11$	$e^\pm$	43
$m_\mu - m_\pi$	$0.11 - 0.14$	$\mu^\pm$	57
$m_\pi - E_c^\dagger$	$0.14 - 0.15$	π	69
$E_c - m_{\text{charm}}$	$0.15 - 1.32$	$\pi, u, \bar{u}, d, \bar{d}, s, \bar{s}$, gluons	247
$m_c - m_\tau$	$1.32 - 1.78$	$c, \bar{c}$	289
$m_\tau - m_{\text{bottom}}$	$1.78 - 4.24$	$\tau^\pm$	303
$m_b - m_{\text{W,Z}}$	$4.24 - 91.2$	$b, \bar{b}$	345
$m_{\text{W,Z}} - m_{\text{Higgs}}$	$91.2 - 159.5$	$W^\pm, Z$	381
$m_{\text{Higgs}} - m_{\text{top}}$	$159.5 - 173.1$	H^0	385
$m_t - 300$ GeV	$173.1 - 300$	$t, \bar{t}$	427

$^\dagger E_c$ corresponds to the hadron deconfinement transition energy.

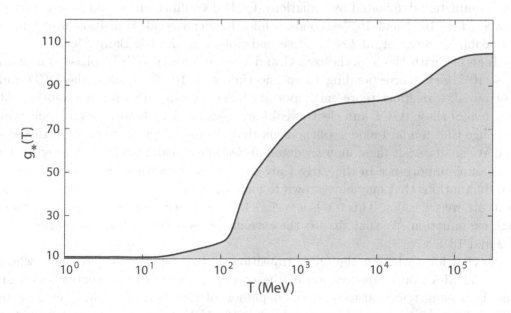

Figure 6.2 Effective number of relativistic degrees of freedom of the energy density ρ_{rel} at high T. The steep drop near the middle of the plot corresponds to the QCD transition at $T \sim 150$ MeV, which would have occurred at $t \sim 10^{-6}$ sec. The electroweak phase transition, associated with the turning on of the Higgs field, would have occurred at $T = 159.5$ GeV, corresponding to $t \sim 10^{-11}$ sec (see Section 11.4). (Adapted from ref. [78])

QCD transition at $T \sim 150$ MeV and the electroweak phase transition at $T = 159.5$ GeV, towards the right-hand side of this figure.

One may readily estimate the cosmic time at which these events took place using Equations (6.27), (9.4) and (9.5). An even simpler, somewhat cruder, approach in finding the temperature-age relation in the early Universe is merely to assume a radiation-dominated energy density, $\rho = a_{rad}T^4$, where a_{rad} is the radiation constant,

$$a_{rad} \equiv \frac{\pi^2 k_B^2}{15\, c^3 \hbar^3} \approx 7.5657 \times 10^{-15} \text{ erg cm}^{-3} \text{ K}^{-4} . \tag{6.28}$$

The Friedmann Equation (4.28) with $k = 0$, together with Equation (6.10), then gives

$$T(t) = \left(\frac{45\, c^5 \hbar^3}{32\, \pi^3 G k_B^4} \right)^{1/4} t^{-1/2} , \tag{6.29}$$

so that

$$T(t) \approx (1.52 \times 10^{10} \text{ K}) \, t_{sec}^{-1/2} . \tag{6.30}$$

Equivalently, we may estimate the temperature using energy units:

$$T(t) \approx (1.31 \text{ MeV}) \, t_{sec}^{-1/2} . \tag{6.31}$$

For the *Planck* optimized parameters (meaning that the early Universe would have been completely dominated by radiation), QCD deconfinement would have occurred between $t \sim 10^{-6}$ and 10^{-4} seconds, while the electroweak transition would have been completed earlier, at $t \sim 10^{-11}$ seconds following the Big Bang [79].

Together with the hypothesized Grand Unified Theory (GUT) phase transition at $\sim 10^{15}$ GeV, corresponding to cosmic time $t \sim 10^{-36}$ seconds, the QCD and electroweak transitions represent important milestones in early cosmic evolution. At least two of these (GUT and electroweak) are associated with horizon problems that challenge the standard model, phenomena that are too extensive to cover in this section. We shall revisit them in great detail in Sections 6.5 and 6.6 and Chapter 11. Of paramount importance in the early Universe is also the nature and origin of quantum fluctuations that may have grown to form the CMB anisotropies and large-scale structure we see today. This too is a very rich subject that warrants a separate, thorough examination. We shall discuss the current status of this concept in Sections 6.7, 11.2 and 13.1.

Moving forward from the QCD transition at $t \sim 10^{-6} - 10^{-4}$ seconds, where $T \sim 150$ MeV, our consideration of the cosmic expansion must contend with an ever decreasing temperature and the depletion of the thermal bath. According to Table 6.1, as the temperature dropped below 130 MeV, beyond $\sim 10^{-4}$ seconds after the Big Bang, the only particles that could have been present appreciably in thermal equilibrium were $e^{\pm}$, $\mu^{\pm}$, ν_e, $\bar{\nu}_e$, ν_μ, $\bar{\nu}_\mu$ and photons (γ). The photons are massless, so their behavior was different from that of the others. Of the rest, some of the neutrinos may also be massless, and could therefore have evolved more like radiation than matter as the Universe continued to expand.

The neutrino distribution would have been maintained via a network of weak interactions, including the following absorption, emission and scattering events:

$$e^- + \mu^+ \leftrightarrow \nu_e + \bar{\nu}_\mu$$
$$e^+ + \mu^- \leftrightarrow \bar{\nu}_e + \nu_\mu$$
$$\nu_e + \mu^- \leftrightarrow \nu_\mu + e^-$$
$$\bar{\nu}_e + \mu^+ \leftrightarrow \bar{\nu}_\mu + e^+$$
$$\nu_e + e^- \leftrightarrow \nu_e + e^-$$
$$\bar{\nu}_e + e^+ \leftrightarrow \bar{\nu}_e + e^+ . \qquad (6.32)$$

For $T < m_\mu c^2 / k_B \approx 110$ MeV, the typical reaction has a cross-section

$$\sigma_W = \alpha_W^2 \left(\frac{k_B T}{c^2 \hbar^2} \right)^2 , \qquad (6.33)$$

where $\alpha_W = 1.4 \times 10^{-49}$ erg cm^{-3} is the weak interaction coupling constant.

At least for the electrons and positrons, whose rest-mass energy is below 110 MeV, the particle number density is given by Equation (6.17) (with the Fermi-Dirac distribution) and is found to be

$$n_{e\pm} = 0.183 \left(\frac{k_B T}{ch} \right)^3 . \qquad (6.34)$$

On the other hand, the expansion rate H may be approximated in the radiation-dominated phase as

$$H = \left(\frac{8\pi G a_{\text{rad}}}{3c^2} \right)^{1/2} T^2 , \qquad (6.35)$$

or

$$H \approx (2.167 \times 10^{-21} \text{ s}^{-1}) T_K^2 , \qquad (6.36)$$

where T_K is the temperature in degrees Kelvin. The ratio of the neutrino-electron reaction rate to the Hubble expansion rate is therefore

$$\frac{c\sigma_W n_{e\pm}}{(2.167 \times 10^{-21} \text{ s}^{-1}) T_K^2} \approx \left(\frac{T}{2.3 \times 10^{10} \text{ K}} \right)^3 . \qquad (6.37)$$

The neutrino-muon reaction rate is smaller than $c\sigma_W n_{e\pm}$ because the muon concentration is depleted relative to that of the electrons as a result of $m_\mu c^2 > k_B T$ in this temperature range.

Thus, as the temperature dropped through the $\sim 1 - 2$ MeV threshold (Equation 6.31), reactions involving neutrinos started to run at a slower rate compared to the rate of expansion of the Universe. They ceased to interact with the rest of the matter and fell out of thermal equilibrium. Neutrinos are not charged so they do not interact directly with the radiation. The temperature $T \sim 1$ MeV, at roughly $t \sim 1$ second after the Big Bang, is therefore viewed as the neutrino decoupling (or freeze-out) temperature. After this point, the Universe would have become transparent to neutrinos and their energy and temperature would have continued to redshift with the cosmic expansion, so that $T_\nu \sim (1 + z)$.

Note, however, that although T_ν and T both would have evolved in the same way with redshift, their values were not quite the same. The reason is that prior to neutrino decoupling, there were 4 fermionic states (e^- and e^+, each having $g_e = 2$) and 2 bosonic states (the two photon projections, with $g_\gamma = 2$) all in equilibrium. From Equation (6.23), this gives $g_* = 11/2$ prior to ν-decoupling. But at even lower temperatures, the equilibrium would have been dominated by photons alone, so that $g_* = 2$. After $e^\pm$ annihilations were completed, transferring their energy to the radiation field, the photon temperature would therefore have been slightly higher than that of the already frozen out neutrinos, with a proportionality

$$T = \left(\frac{11}{4}\right)^{1/3} T_\nu . \tag{6.38}$$

According to Equations (6.22) and (6.23), we thus see that, as long as the neutrinos remained relativistic—which is certainly always true for the massless species—the total relativistic energy density would have been

$$\rho_{\rm rel}(T) = \left[2 + \frac{7N_\nu}{4}\left(\frac{T_\nu}{T}\right)^4\right]\left(\frac{\pi^2 k_{\rm B}^4}{30c^3\hbar^3}\right)T^4 , \tag{6.39}$$

where N_ν is the number of massless neutrino flavors. For example, if $N_\nu = 3$, then $\rho_{\rm rel} \approx 1.68\rho_{\rm r}$. When this calculation is done more carefully [80], the neutrinos do not all decouple at $e^\pm$ annihilation, so the effective value of N_ν is somewhat different from three. (It is actually closer to ~ 3.05.)

The caveat with this estimate, though, is that it ignores a possible non-zero neutrino mass. Neutrino oscillations have been observed [81], and the current best estimates require at least one neutrino eigenstate to have a mass exceeding 0.05 eV. Moreover, oscillation experiments only measure mass differences, so the overall mass content of neutrinos in the cosmic fluid may be somewhat larger than that inferred on the basis of 0.05 eV alone. But even with a neutrino number density estimated from Equation (6.39), such a tiny mass would result in a scaled neutrino mass density today $\Omega_\nu \ll \Omega_{\rm m}$ (see Section 6.1). In addition, these masses are too small to alter the validity of our relativistic approximation for the neutrino energy whenever the radiation content of the Universe was dynamically significant, thereby not impacting Equation (6.39) in any meaningful way.

During the epoch leading up to neutrino decoupling, another crucial development took place that is still largely wrapped in mystery. In Section 1.2 we pointed out that 95.1% of all the energy density in the Universe is dark and unknown, save for its gravitational influence on the expansion dynamics and condensations producing growth. But we cannot ignore the deep, unanswered question concerning the origin of the baryonic component as well, about which we supposedly know much more from experimental physics. We shall demonstrate shortly that, in standard cosmology, baryons and their antiparticles should have annihilated almost completely in the early Universe, leaving only a negligible abundance today, far from the level actually observed. So where did baryons come from? An understanding of this process, commonly referred to as *baryogenesis*, is still being sought in quantum field theories of elementary particles at early times.

When $T < T_i \equiv m_i c^2 / k_{\mathrm{B}}$, the distribution function for particles with mass m_i in Equation (6.15) simplifies to

$$F_i(E, T_i) \approx \frac{g_i}{(2\pi)^3} e^{-E/k_{\mathrm{B}} T_i(t)} . \tag{6.40}$$

The integral to Equation (6.17) is then

$$n_i = 0.0915 g_i \left(\frac{T_i}{T}\right)^{3/2} \left(\frac{k_{\mathrm{B}} T}{c\hbar}\right)^3 \exp\left(\frac{-T_i}{T}\right) , \tag{6.41}$$

showing a strong exponential suppression when the particle rest-mass energy becomes a significant fraction of the energy E. This limiting behavior is usually referred to as the 'non-relativistic' approximation. (Note that $T_i = 0$ for photons and massless neutrinos, so this approximation never applies to them.)

Thus, for baryons with mass $m_{\mathrm{b}} c^2 \sim 1$ GeV (and $g_i = 2$), the ratio of baryon to photon density in equilibrium at $T \lesssim 1$ GeV is

$$\frac{n_{\mathrm{b}}}{n_\gamma} \approx \frac{n_{\bar{\mathrm{b}}}}{n_\gamma} \approx \left(\frac{m_{\mathrm{b}} c^2}{k_{\mathrm{B}} T}\right)^{3/2} \exp\left(-\frac{m_{\mathrm{b}} c^2}{k_{\mathrm{B}} T}\right) . \tag{6.42}$$

The baryon density continues to drop relative to n_γ as along as their annihilation rate $\Gamma_{\mathrm{b}\bar{\mathrm{b}}}$ exceeds the Hubble expansion rate. Once H overtakes $\Gamma_{\mathrm{b}\bar{\mathrm{b}}}$, however, the baryons and antibaryons cease annihilating because their density is too low to sustain the required number of interactions.

The baryon-antibaryon annihilation rate at these energies may be written

$$\Gamma_{\mathrm{b}\bar{\mathrm{b}}} \approx n_{\mathrm{b}} \langle \sigma_{\mathrm{b}\bar{\mathrm{b}}} v \rangle , \tag{6.43}$$

where $\langle \sigma_{\mathrm{b}\bar{\mathrm{b}}} v \rangle$ is the thermally averaged cross section. Since $v \approx c$, its dependence on temperature is actually quite weak, and one may simply put

$$\langle \sigma_{\mathrm{b}\bar{\mathrm{b}}} v \rangle \sim \frac{\hbar^2}{m_\pi^2 c} , \tag{6.44}$$

in terms of the pion mass m_π. Thus, equating $\Gamma_{\mathrm{b}\bar{\mathrm{b}}}$ with H from Equation (6.36), one can easily show that the baryon-antibaryon annihilations freeze out at $T_{\mathrm{b}} \approx 20$ MeV. And using this value in Equation (6.42), one therefore finds that

$$\frac{n_{\mathrm{b}}}{n_\gamma} \approx \frac{n_{\bar{\mathrm{b}}}}{n_\gamma} \approx 10^{-18} . \tag{6.45}$$

As we shall see in Section 6.3 (notably Equation 6.48), however, this value is many orders of magnitude smaller than that required by the standard model to produce the light elements. It appears, therefore, that a new mechanism is needed to copiously produce baryons in the early Universe, while at the same time violating baryon-number conservation, since direct observational evidence shows that the Universe today contains no appreciable primordial antimatter.

Although several interesting mechanisms for generating the baryon asymmetry have been proposed over the past half century, we still do not know the answer to this mystery. What has been established quite firmly, starting with Sakharov's seminal paper in 1967 [82], is that at least three conditions must have been satisfied in order to comply with the required baryon creation rate. These are C and CP violation, and a departure from thermal equilibrium. C violation means that changing all particles in a given interaction to particles of opposite charge—including electric charge, but also other kinds of charge, and even neutral particles being replaced by their antiparticle partners, such as neutrinos switched to antineutrinos and neutrons to antineutrons—produces a different reaction rate. CP violation has an analogous result when, in addition to switching the particles for C, one also creates a mirror image of the Universe, effected by inverting all three spatial axes. It should be intuitively obvious why these two conditions must have existed during baryogenesis, for otherwise baryon-producing processes would have proceeded at the same rate as baryon-destroying processes, so the net baryon number would have always been zero. The third condition would have been necessary because the equilibrium state is symmetric under time reversal, but would have been less demanding than the other two in a rapidly evolving Universe, where the conditions pertaining to baryon creation could have changed more rapidly than the Hubble expansion rate (Equation 6.36).

All three of these conditions could be incorporated into a mechanism based on the out-of-equilibrium decay of a massive GUT particle [83]. Another possibility for the copious production of baryons is via scalar quarks and leptons in a supersymmetric version of GUT following inflation (Section 6.6) [84]. But the most commonly pursued mechanism appears to be electroweak baryogenesis, in which the asymmetry begins with the leptons and is then transferred to the baryons at the electroweak phase transition (Section 11.4). All three Sakharov conditions can be satisfied with the standard model of particle physics. Baryon number violation has never been observed, but the model predicts that baryons can turn into leptons via a quantum anomaly. We currently exist in a Universe with a high vacuum expectation value of the Higgs field, which has broken the $SU(2) \times U(1)$ symmetry (i.e., separated the weak and electromagnetic forces), making baryon number violating processes extremely rare. But this symmetry was unbroken prior to the electroweak phase transition at $T = 159.5$ GeV, approximately 10^{-11} seconds after the Big Bang (see Chapter 11), so baryon number violation would not have been suppressed earlier than that.

In the standard model of particle physics, C is maximally violated by weak interactions, while CP is also violated through quark and neutrino mixing. Together with the possibility that the rapid early expansion of the Universe maintained a state of disequilibrium in the baryon producing interaction network, all three Sakharov conditions could have been satisfied via the electroweak baryogenesis channel [85, 86]. Nevertheless, a major complication has emerged with the discovery of the Higgs boson in 2012 [87]. This mechanism is now known not to work with the minimal standard model because the measured mass of the Higgs boson implies that the electroweak phase transition was not of first order; it must have been a smooth crossover between the two phases. In addition, the measured CP violation would have been too weak anyway. Thus, to make electroweak baryogenesis feasible, some new physics beyond the standard model is required. One possibility is the existence of undiscovered TeV-scale particles that violate CP and make the electroweak phase transition stronger.

These possibilities notwithstanding, our lack of understanding concerning the origin of baryon asymmetry in the early Universe represents a major flaw in cosmological theory today. The problem may reside with the standard model of particle physics, rather than the standard model of cosmology. In either case, there is not much more we can do right now within the framework of ΛCDM before moving to the next step, which is to calculate the equilibrium density of protons and neutrons in order to set the stage for the nucleosynthesis of light elements, that we shall discuss in the next section.

6.3 NUCLEOSYNTHESIS

In the standard model of cosmology, the Universe was radiation-dominated throughout the epoch of baryogenesis and the subsequent creation of the light elements, a process that lasted several minutes beyond the neutrino freeze-out at $t \sim 1$ second. Thus, the total energy density during nucleosynthesis could be written in the form of Equation (6.39), though modified at the start by the inclusion of electrons and positrons prior to their annihilation. Taking these contributions into account, one would then have

$$\rho_{\text{rel}}(T) = \left[2 + \frac{7}{2} + \frac{7N_\nu}{4} \left(\frac{T_\nu}{T} \right)^4 \right] \left(\frac{\pi^2 k_{\text{B}}^4}{30 c^3 \hbar^3} \right) T^4 , \tag{6.46}$$

while the temperature exceeded ~ 1 MeV. Of course, this approximation is reasonable only as long as the concentration of baryons (and antibaryons) was dynamically insignificant, which we must assume in the absence of a working model for baryogenesis. Nevertheless, we must now consider the neutrons and protons because, no matter how dilute they were compared to the radiation, these were the particles directly involved in the formation of heavier elements.

Unfortunately, one must then rely on a poorly motivated assumption to ensure that the ratio $n_{\text{n}}/n_{\text{p}}$ met the requirements for producing the observed abundances. At $T \lesssim 1$ GeV, the neutron and proton distributions would have followed the non-relativistic approximation in Equation (6.41) as long as they remained in equilibrium with the rest of the cosmic fluid. One could then assume that

$$\frac{n_{\text{n}}}{n_{\text{p}}} \approx \exp \left(-\frac{[m_{\text{n}} - m_{\text{p}}] c^2}{k_{\text{B}} T} \right) . \tag{6.47}$$

But it was Equation (6.41) that led to the extremely low densities in Equation (6.45) in the first place, far smaller than what is required, as we shall confirm shortly. Yet one must fix the neutron to proton ratio using the Boltzmann factor in Equation (6.47) to make things work. The rather obvious weakness with this argument is that the (presumably non-thermal) mechanism for producing the baryons would not necessarily have been constrained by equilibrium concentrations prior to the baryon-antibaryon freeze-out at $T \sim 20$ MeV.

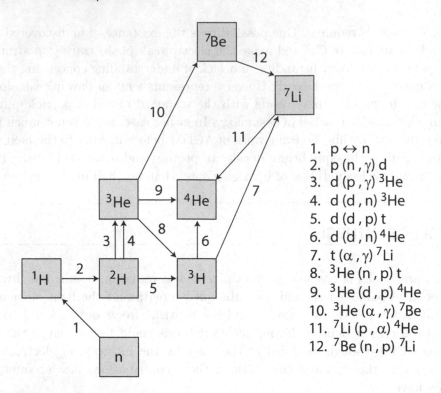

Figure 6.3 The minimal nuclear reaction network with 12 interactions needed to compute the primordial abundances up to lithium. (Adapted from ref. [88])

Let us ignore this difficulty for the moment, and simply claim that the neutron-to-proton ratio at the weak-interaction freeze-out ($T \sim 1$ MeV) was $n_{\rm n}/n_{\rm p} \sim 1/5$. For the sake of making order-of-magnitude estimates, it will be helpful for us to anticipate a key result we shall obtain shortly, viz. the inferred primordial baryon-to-photon ratio

$$\eta \equiv \frac{n_{\rm b}}{n_\gamma} = (6.10 \pm 0.04) \times 10^{-10} . \tag{6.48}$$

The first link in the nuclear chain is the formation of deuterium (d), via $p+n \to d+\gamma$. Deuterium's binding energy is quite small, $E_{\rm bind}^d = 2.2$ MeV, but the much larger concentration of photons relative to nucleons ($\eta^{-1} \sim 10^9$) would have destroyed it faster than it was produced, creating the so-called deuterium bottleneck. Big Bang nucleosynthesis (BBN) was thus delayed until $\eta^{-1} \exp(-E_{\rm bind}^d/k_{\rm B}T) \sim 1$, after which the deuterium destruction rate fell below its production rate, and the deuterium density could begin to rise. In the context of ΛCDM, subject to the above approximations, one therefore expects BBN to have started at $T_{BBN} \sim E_{\rm bind}^d/\ln(\eta^{-1}) \sim 0.1$ MeV which corresponds to a cosmic time $t \sim 2.9$ minutes, according to Equation (6.31).

The measured neutron lifetime is $\tau_{\rm n} = 880.3 \pm 1.1$ seconds [89]. Thus, some of the neutrons would have decayed by the time the deuterium bottleneck was broken, further reducing the $n_{\rm n}/n_{\rm p} \sim 1/5$ ratio at freeze-out by the additional factor $\sim e^{-2.9\,{\rm min}/\tau_{\rm n}} \approx 0.82$. A more detailed calculation shows that, after freeze-out, the

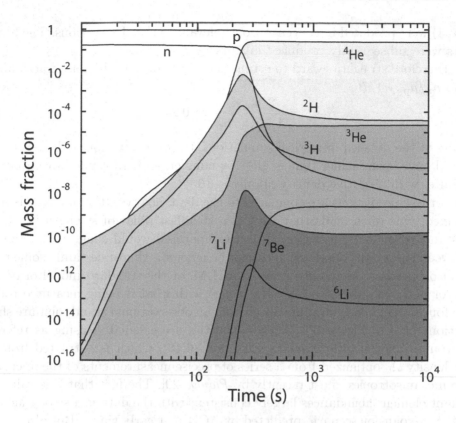

Figure 6.4 Big Bang nucleosynthesis in the standard model, showing the time evolution of neutrons, protons and the lightest elements, including deuterium, tritium, ^{3}He, ^{4}He, ^{6}Li, ^{7}Li, and ^{7}Be. The shaded areas highlight the four observationally most important elements and isotopes, i.e., deuterium, ^{4}He, ^{3}He and ^{7}Li (see also Figure 6.5). Below $T \sim 0.01$ MeV, i.e., after $t \sim 10^4$ seconds, the relative densities would have been frozen, becoming imprinted as the primordial abundances. (Adapted from ref. [91])

partial neutron decay would have resulted in a density ratio $n_n/n_p \sim 1/7$ at the time BBN actually began [90].

As deuterium started to form, essentially all of the neutrons would have ended up in ^{4}He, because several constraining factors would have placed strict limits on how far BBN could have proceeded in ΛCDM. The combination of a rapidly cooling Universe, insurmountable Coulomb barriers, and the mass gap at atomic number $A = 8$, would have strongly inhibited the production of elements beyond lithium. In the temperature range 0.05 MeV $\lesssim T \lesssim$ 0.6 MeV (i.e., cosmic time 5 sec $\lesssim t \lesssim$ 11 min), the synthesis of light elements would have proceeded solely via two-body reactions, accelerating as the deuterium concentration increased. The predicted relative abundances are therefore readily found solving the minimal series of nuclear reactions shown in Figure 6.3, with time-dependent source and sink terms. The state-of-the-art BBN codes today include up to 424 nuclear reactions in their network [91], but elements beyond lithium were produced only as trace amounts, far too small to be measurable. In standard BBN, only deuterium (or ^{2}H), ^{3}He, ^{4}He and ^{7}Li (see

Figure 6.4) were produced in concentrations relevant to the observations. The heavier elements were subsequently produced in stars.

It is therefore straightforward to estimate the primordial ^{4}He mass fraction from the ratio $n_n/n_p \sim 1/7$:

$$Y_p = \frac{2(n_n/n_p)}{1 + (n_n/n_p)} \approx 0.25 \,. \tag{6.49}$$

According to the detailed numerical simulations, the concentration of deuterium and ^{3}He would have been smaller than $\sim 10^{-5}$ by number, and ^{7}Li would have been even more dilute, with a relative density of only $\sim 10^{-10}$.

When these predictions are compared to the data, however, the outcome is somewhat mixed. One often reads that BBN is a steadfast pillar of strength in support of the standard model, but a more objective appraisal would suggest that the tension between theory and observation cannot be ignored. Almost certainly some major modification needs to be made, either to ΛCDM, or the standard model of particle physics, or both, in order to bring BBN in line with what is being measured today.

The four most important abundances from an observational standpoint are shown as functions of η in Figure 6.5. These quantities are calculated using as reference the baryon to photon ratio given in Equation (6.48), which is extracted from the baryon density Ω_b optimized from a series of precise measurements of the microwave background anisotropies, most recently by *Planck* [29]. The fact that this value of η yields light element abundances largely consistent with the data is a strong factor in favor of the expansion scenario predicted by ΛCDM at early times. But already this η creates significant tension with the standard model, either in cosmology, particle physics or both. As we saw earlier in Equation (6.45), the protons and neutrons produced in thermal equilibrium with the rest of the Universe would have been far too few to account for such a relatively large η. If one attempts to fix this problem by introducing a separate baryogenesis channel, it is then necessary to explain why the neutron to proton number ratio $n_n/n_p \approx 1/7$, required to yield the correct ^{4}He mass fraction, is based on the Boltzmann factor in Equation (6.47), which is strongly indicative of baryons in equilibrium with the ambient thermal pool.

These caveats aside, the predicted ^{4}He abundance is another strong factor in favor of the standard model. Observationally, the ^{4}He abundance determination is based primarily on a measurement of the emission lines in highly ionized gas in nearby extragalactic HII regions. Still, one must be wary of the fact that an ionization model must be used to extract this information, containing eight physical parameters that together yield a set of 10 H and He emission-line ratios [93]. Unfortunately, degeneracies among the parameters lead to relatively large uncertainties, but a recent measurement [94] has determined that

$$Y_p^{\text{obs}} = 0.2465 \pm 0.0097 \,, \tag{6.50}$$

in very good agreement with even just the simple estimation in Equation (6.49).

In contrast, ^{3}He is produced and destroyed in stars throughout galactic history, so the evolution of its abundance cannot be determined reliably as a function of time

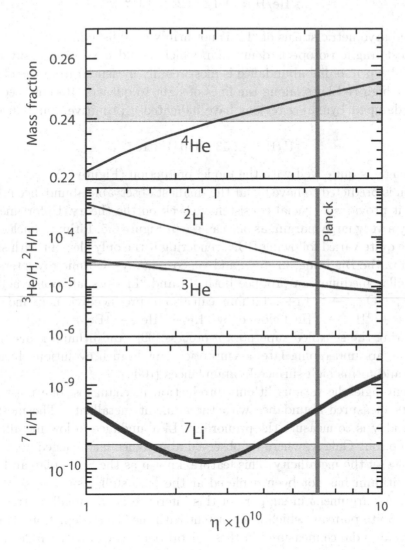

Figure 6.5 The four observationally most important abundances, deuterium (or ^{2}H), ^{4}He, ^{3}He, and ^{7}Li, as functions of η, calculated with a full network of over 400 nuclear reactions. The vertical bar indicates the *Planck* optimized value of η, while the horizontal shaded rectangles indicate the 1σ confidence regions associated with the measured abundances. The ^{7}Li abundance predicted by the standard model (Li/H $= 4.65 \times 10^{-10}$) is an outlier, missing the measured value (Li/H $= [1.6 \pm 0.3] \times 10^{-10}$) by $\sim 10\sigma$. (Adapted from ref. [92])

and is subject to large uncertainties. We do have a locally determined value from within our Galaxy [95],

$$^{3}\text{He/H} = (1.1 \pm 0.2) \times 10^{-5} \,, \tag{6.51}$$

though the baryometric status of ^{3}He is not firmly established.

The most fragile isotope is deuterium, which would have been easily destroyed after BBN. Its primordial abundance is most readily measured from the observation of clouds at high redshifts, along the lines-of-sight to quasars. Recent precise observations of damped Lyman-α regions have indicated a primitive abundance [96, 97]

$$^{2}\text{H/H} = (2.53 \pm 0.04) \times 10^{-5} \,, \tag{6.52}$$

which also agrees quite well with the model prediction (Figure 6.5).

Lithium is predicted to have by far the smallest observable abundance in standard BBN, but it provides a crucial consistency check on the theory. Unfortunately, this test fails by a very large margin, as one can see in Figure 6.5. Lithium nucleosynthesis sites can be quite varied following BBN, rendering it the only element with significant production in the Big Bang, in stars, and via cosmic rays. Cosmic-ray interactions in the interstellar medium can produce both ^{7}Li and ^{6}Li via spallation reactions [98], such as $p_{\text{cr}} + {}^{16}\text{O}_{\text{ism}} \rightarrow {}^{6,7}\text{Li} + \ldots$ Lithium can also be produced via neutrino spallation reactions, $\nu + {}^{4}\text{He} \rightarrow {}^{3}\text{He}$ followed by $^{3}\text{He} + {}^{4}\text{He} \rightarrow {}^{7}\text{Be} + \gamma$, and their mirror version, during the so-called supernova ν-process [99]. And lithium is also formed in lower mass stars undergoing late, asymptotic giant branch evolution, during which ^{3}He burns and forms high surface Li abundances [100].

To disentangle the various lithium production mechanisms, one must therefore compare its measured abundance with the ambient metallicity. The most reliable approach today is to measure the primordial ^{7}Li abundance in low metallicity stars in the halo of our Galaxy, where its observed abundance has reached an asymptotic independence of the metallicity. This feature, known as the Spite plateau [101], suggests that lithium has not been depleted at the halo stellar surface, so it must be primordial. An argument in support of this inference is the small scatter of values around the Spite plateau, which is consistent with inefficient depletion. The primordial lithium abundance measured in these metal poor halo stars is [102]

$$\text{Li/H} = (1.6 \pm 0.3) \times 10^{-10} \,. \tag{6.53}$$

But if we take the precision of this measurement at face value, we conclude from Figure 6.5 that the standard model prediction of the ^{7}Li abundance disagrees with the data by $\sim 10\sigma$. The frustrating aspect of this so-called *Lithium problem* is that the tension has actually gotten worse with time, as the measured value has tended to drift downwards with increased precision, rather than upwards towards the much higher theoretical prediction. Many solutions have been proposed to resolve this major inconsistency with the standard model, but none of them have been satisfactory thus far [103].

6.4 THE COSMIC MICROWAVE BACKGROUND

In the standard model, Big Bang nucleosynthesis ended by $t \sim 10^4$ seconds (Figure 6.4), and the relative nuclear densities would have become frozen, imprinted as primordial abundances. But the Universe continued to expand and cool, further suppressing any remaining interactions, especially the most important one relevant to the next major milestone—hydrogen photoionization:

$$p + e \leftrightarrow H + \gamma \,. \tag{6.54}$$

While $k_B T$ remained large compared to the hydrogen ionization energy,

$$E_{ion} = m_p + m_e - m_H = 13.6 \text{ eV} \,, \tag{6.55}$$

baryonic matter would continue to be ionized, sustaining a coupling between the ambient radiation field and electrons via Compton scattering. Given the overwhelming imbalance of the photon to lepton number ratio (Equation 6.48), the electron kinetic temperature would have remained close to the radiation temperature, and frequent inelastic Coulomb collisions between protons and electrons thermally equilibrated all of the particle species in Equation (6.54). At lower temperatures, however, the photoionization reaction would have been thermodynamically skewed to the right-hand side of this reaction equation. And as the concentration of neutral atoms grew, Compton scattering became less frequent, allowing the radiation to decouple from matter. It would eventually have started streaming freely across the Universe, creating the fossilized record we see today in the form of a cosmic microwave background.

While the photoionization reaction was able to maintain equilibrium, the relative abundances of the protons, electrons and hydrogen atoms would have been fixed by an expression governing chemical equilibrium, known as the Saha equation [104], which we may derive as follows. Analogously to Equation (6.47), the ratio of proton to hydrogen number densities may be written

$$\frac{n_p}{n_H} = \frac{g_p^*}{g_H} \exp\left(-\frac{E_{ion} + q_e^2/2m_e}{k_B T}\right) , \tag{6.56}$$

where g_H and g_p^* are the statistical weights of, respectively, hydrogen in its ground state and ions (i.e., protons) in their upper ionization state, a product of the ground state weight g_p and the number of possible states of the free electron within its available volume. The quantity $E_{ion} + q_e^2/2m_e$ is the latter's energy with respect to the ground state of the ion, written in terms of its momentum q_e.

The differential number of cells available to free electrons with momentum between q_e and $q_e + dq_e$ is $4\pi q_e^2 \, dq_e / n_e h^3$, where $1/n_e$ is their volume (per particle) in ordinary space and h is Planck's constant. Thus, taking all these factors into account, including the two-spin degeneracy of the electrons, and integrating over all the available cells, one finds that

$$\frac{n_p}{n_H} = \frac{g_p}{g_H} \frac{2}{n_e h^3} \int_0^\infty \exp\left(-\frac{E_{ion} + q_e^2/2m_e}{k_B T}\right) 4\pi q_e^2 \, dq_e \,. \tag{6.57}$$

With a straightforward evaluation of the integral, this expression yields

$$\frac{n_p n_e}{n_H} = \left(\frac{m_e k_B T}{2\pi\hbar^2}\right)^{3/2} \exp\left(-\frac{E_{\text{ion}}}{k_B T}\right) , \tag{6.58}$$

which includes the ratio $2g_p/g_H = 1$ for the ionization of hydrogen. Defining the ionization fraction

$$X_e \equiv \frac{n_e}{n_p + n_H} , \tag{6.59}$$

and putting $n_e = n_p$ for charge neutrality, and $n_b = n_p + n_H$, one arrives at the final result,

$$\frac{X_e^2}{1 - X_e} = \left(\frac{m_e k_B T}{2\pi\hbar^2}\right)^{3/2} \frac{1}{n_b} \exp\left(-\frac{E_{\text{ion}}}{k_B T}\right) . \tag{6.60}$$

If one continues to use the baryon density implied by Equation (6.48), it is clear from the Saha equation that recombination—defined, say, by the condition $X_e \sim 0.01$—would have occurred at a temperature $T \sim 4,000$ K. From Equation (6.27), with a CMB temperature today $T_0 \approx 2.73$ K (see Section 1.1), one therefore concludes that the microwave background started streaming freely by redshift $z_{\text{rec}} \sim 1,400$. Of course, this estimate ignores several ancillary features, including the recombination of helium, and a complete treatment based on the Boltzmann equation, which should also take into account electron-photon scattering events. Another important consideration affecting the neutral hydrogen formation rate is that the 13.6 eV photons couldn't really 'escape' from the plasma. The process that took photons out of the loop was actually the $2s \rightarrow 1s$ transition, which proceeds via two-photon emission to ensure angular momentum conservation. As such, neutral hydrogen could not form instantly. Also, the very tiny baryon to photon ratio means that ionizing photons did not have to come from the center of the Planck distribution. Even the Wien tail had enough high-energy photons to ionize all of the hydrogen atoms. The radiation temperature at decoupling was therefore not as tightly constrained as we have implied here. Finally, recombination on its own is not sufficient to determine precisely when the radiation completely decoupled from matter, since photons continued to collide with electrons until the scattering optical depth dropped to negligible values.

This optical depth τ may be estimated from the ionization fraction as a function of z around decoupling, and is given by the expression

$$\tau = \int n_e X_e \sigma_T \, dr , \tag{6.61}$$

where σ_T is the Thomson scattering cross-section and r is the comoving radius. As one may see from Equation (6.60), X_e (and therefore τ) varied rapidly around $z \sim 1,000$, so the aptly named 'visibility function' $g(z) = e^{-\tau} \, d\tau/dz$, related to the probability of a photon scattering between z and $z + dz$, is highly peaked. A practical definition of the 'last scattering surface' (LSS) is then the redshift at which this visibility function reached its maximum. A more detailed analysis [88] than the one carried out here thus refines the redshift at which the CMB was produced, from $z_{rec} \sim 1,400$ to $z_{cmb} \sim 1081$ which, according to Equation (9.4), corresponds to a cosmic time $t_{cmb} \sim 380,000$ yrs. In the context of ΛCDM, the Universe would have become transparent after this point, with CMB photons propagating freely in all directions. The LSS may thus be thought of as a two-dimensional sphere centered on us, with a comoving radius (Equation 8.5) evaluated at z_{cmb}.

The scenario we have just described makes several predictions, some of which have by now been confirmed by three satellite missions and ground-based followup observations. The first of these is that the CMB spectrum, formed in a hot cauldron of particles in approximate thermal equilibrium, should be well described by the Planck (blackbody) function,

$$B_\nu(T) = \frac{2h\nu^3/c^2}{\exp(h\nu/k_B T) - 1} , \qquad (6.62)$$

derived from the distribution in Equation (6.15) for a Bose-Einstein gas with zero chemical potential. And this is precisely what was measured by the Cosmic Background Explorer (COBE) satellite [105], later confirmed by NASA's Wilkinson Microwave Anisotropy Probe (WMAP) [106], and ESA's *Planck* mission [29]. In awarding two of COBE's principal investigators, George Smoot and John Mather, the 2006 Nobel Prize in Physics, the Nobel committee noted that "the COBE project may be regarded as the starting point for cosmology as a precision science."

Second, since $z_{cmb} \gg 1$, one would expect T to have closely followed the redshift dependence in Equation (6.27). At least in the nearby Universe, i.e., at $z \lesssim 3$, one should thus be able to test the CMB's cosmological origin by measuring the changing strengths of emission and absorption lines from radiating systems. For example, the CMB radiation excites the rotational levels of some interstellar molecules, including carbon monoxide (CO), which thereby function as cosmic 'thermometers.' Figure 6.6 shows several measurements of T, based on a variety of physical processes, including the CO rotational states [108] and, independently, the Sunyaev-Zeldovich (S-Z) effect [109], due to Compton scattering of the CMB photons by hot intracluster gas, producing a small change in the CMB spectral intensity. The steep frequency dependence of this effect allows T to be estimated at the redshift of the cluster, since the ratio of the intensity shift at two different frequencies is virtually independent of the cluster's properties, yielding a rather clean estimate of the temperature. Other techniques used to measure $T(z)$ include the analysis of the fine structure of atomic carbon [110], and a multi-transition excitation analysis of various molecular absorption lines towards radio-mm molecular absorbers, such as PKS 1830-211 [107]. The results shown in Figure 6.6 yield a CMB temperature evolution $T(z) = (2.72548 \pm 0.002) \times (1+z)^{1-\beta}$ K, with $\beta = -0.007 \pm 0.027$, in excellent agreement with the expected relation in Equation (6.27).

Perhaps more indicative of future research directions, the satellite missions designed to study the cosmic microwave background uncovered something even more profound than its pure blackbody spectrum, as illustrated quite dramatically in the recently published map of the CMB temperature distribution shown in Figure 1.3. The principal focus of CMB data analysis today is the angular variation

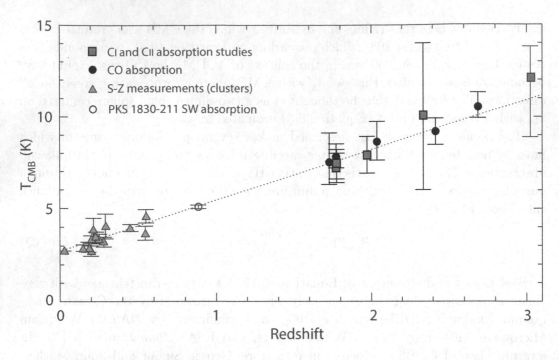

Figure 6.6 Measured blackbody temperature of the CMB as a function of redshift. Squares represent the analysis of the fine structure of atomic carbon; filled circles are based on the rotational excitation of CO molecules; triangles show the S-Z effect towards galaxy clusters, and the open circle is derived from the PKS 1830-211 SW absorption. The dotted line corresponds to the temperature dependence in Equation (6.27). (Adapted from ref. [107])

in temperature (or intensity) across the sky and, to a growing extent, fluctuations in polarization. These anisotropies were first detected by COBE, and have been mapped at increasing levels of sensitivity and angular resolution by WMAP, *Planck* and ground-based experiments, particularly the Atacama Cosmology Telescope (ACT) [111] and the South Pole Telescope (SPT) [31].

High-precision maps of the CMB reveal temperature anisotropies with an amplitude of 1 part in 100,000, and polarization anisotropies at the 10^{-6} level, spanning a broad range of angular scales (see Figure 1.3). By convention, the microwave temperature $T(\hat{\mathbf{n}})$ in every direction $\hat{\mathbf{n}}$ is assumed to be a Gaussian random field in the plane of the sky. It may then be written as an expansion in spherical harmonics $Y_{lm}(\hat{\mathbf{n}})$,

$$T(\hat{\mathbf{n}}) = \sum_{\ell m} a_{\ell m} Y_{lm}(\hat{\mathbf{n}}) \,, \qquad (6.63)$$

whose random coefficients a_{lm} have zero mean. Since the expansion coefficients are statistically independent, they must satisfy the condition

$$\langle a_{\ell m}^* a_{\ell' m'} \rangle \propto \delta_{\ell\ell'} \, \delta_{mm'} \,. \qquad (6.64)$$

Further, statistical isotropy ensures that the constant of proportionality depends only on ℓ:

$$\langle a_{\ell m}^* a_{\ell' m'} \rangle = C_\ell \, \delta_{\ell\ell'} \, \delta_{mm'} \, . \tag{6.65}$$

The quantity

$$C_\ell = \frac{1}{2\ell + 1} \sum_m |a_{\ell m}|^2 \tag{6.66}$$

in Equation (6.65) is called the *angular power* of multipole ℓ.

The largest measured anisotropy is in the $\ell = 1$ first spherical harmonic (dipole), due to the Doppler boosting of the average temperature (the 'monopole' component) arising from Earth's motion with respect to the isotropic blackbody field. It is not difficult to derive the angular dependence of this multipole component from the kinematic Doppler boost formula,

$$\epsilon'_{cmb} = \epsilon_{cmb} \gamma (1 - \vec{\beta} \cdot \hat{k}) \, , \tag{6.67}$$

where ϵ_{cmb} and ϵ'_{cmb} are the photon's energy in the observer's frame and Hubble frame, respectively, $\gamma \equiv (1 - \beta^2)^{-1/2}$ is the Lorentz factor, $\vec{\beta}$ is the relative velocity between frames in units of c, and $\hat{k}$ is the photon's direction of propagation. With $\epsilon'_{cmb} \propto T(\theta)$, one expects to see a temperature

$$T(\theta) = T_0 \frac{(1 - \beta^2)^{1/2}}{(1 - \beta \cos \theta)} \tag{6.68}$$

(save for the much smaller intrinsic anisotropic signal) of the blackbody spectrum in the direction θ relative to Earth's motion [112]. Its measured variation with angle implies a velocity 370.09 ± 0.22 km s^{-1} for the solar system barycenter, in the Galactic coordinate direction $(l, b) = (264.00° \pm 0.03°, 48.24° \pm 0.02°)$. This motion, and its induced dipole moment, are normally removed from the temperature map for the purpose of studying the *intrinsic* CMB anisotropies.

In ΛCDM, the higher multipole ($\ell > 1$) variations are largely due to the classical-ization and subsequent growth of quantum fluctuations seeded in the early Universe, manifested as density perturbations at the epoch of last scattering. We shall address this concept more fully in Section 6.7 below, and later consider updated developments on this thinking in Section 13.1. Figure 6.7 shows the measured CMB multipole power, represented by the quantity $D_\ell \equiv \ell(\ell+1)C_\ell/2\pi$, as a function of ℓ. The superscript TT is conventionally used to designate the temperature angular power spectrum, while TE usually denotes the temperature-polarization cross-power spectrum. The solid curve represents the optimized ΛCDM fit, which is calculated taking several physical influences into account. Some of these effects act preferentially at large angles (i.e., on fluctuations with size $\theta \gg 1 - 2°$), while others, e.g, baryon acoustic oscillations (BAO), contribute primarily on smaller scales [113, 114, 115, 116, 117].

At large angles, corresponding to $\ell \lesssim 30$, the anisotropies are produced primarily by the Sachs-Wolfe (SW) effect [118], constituting metric perturbations related to scalar fluctuations in the matter field. The large anisotropies observed in the CMB temperature today are mainly due to inhomogeneities of the metric fluctuation ampli-tude on the LSS.

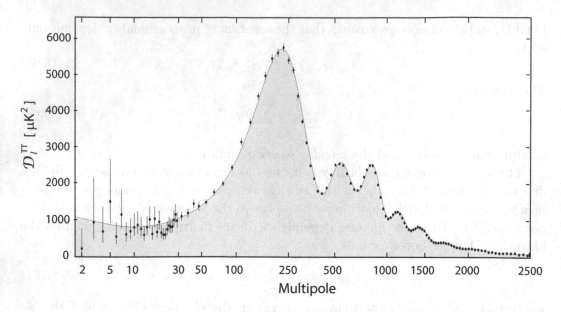

Figure 6.7 The foreground-subtracted, frequency-averaged, cross-half-mission angular power spectrum of the CMB temperature, based on the *Planck* 2018 data release. The grey curve shows the ΛCDM best-fitting model. The uncertainties in this spectrum are dominated by sampling variance, rather than by noise or foreground residuals. The superscript TT designates the temperature angular power spectrum, distinguishing it from the temperature-polarization cross-power spectrum labeled TE. (Adapted from ref. [29])

On sub-degree scales, however, corresponding to $\ell > 30$—including the very prominent (and now famous) peaks in the power spectrum—the anisotropies are due to the aforementioned gravity-driven acoustic oscillations propagating as sound waves before recombination. During this period, the proton-electron plasma remained tightly coupled to the photons via Compton scattering, as noted earlier in our discussion leading up to Equation (6.60), making these components behave as a single radiation-baryon fluid. Any dark matter present was uncharged and therefore not coupled electromagnetically to the other particle species.

The latter dominated perturbations in the gravitational potential, however, due to its inferred larger contribution to the total energy density ρ. Based on the latest *Planck* parameter optimization [29], in which $\Omega_m = 0.315 \pm 0.007$ and $\Omega_b = 0.0493 \pm 0.0002$ (67.4 km s^{-1} Mpc$^{-1}/H_0)^2$ (see Section 6.1), we estimate a factor ~ 5 greater contribution to ρ from dark matter than from baryons. But while gravity tended to compress the photon-baryon fluctuations, radiation pressure provided most of the restoring force, producing sound waves that grew the fluctuations until the Universe became neutral and the radiation was released.

The amplitude of these perturbations was quite small (of order $\sim 10^{-5}$), so each Fourier mode evolved independently of the others, with a frequency directly related to the sound speed in the fluid (via the usual dispersion relation). The acoustic oscillations constituted both density and velocity variations, with a $\pi/2$ phase difference, and both effects produced temperature perturbations. After recombination, these phases became 'frozen,' and we now see them projected on the sky as a harmonic series of peaks. The principal peak, at $\ell \sim 220$, is the mode that oscillated through $1/4$ of a period, attaining maximal compression, producing by far the strongest signal in the temperature fluctuations, while the troughs correspond to the velocity maxima. The overall pattern in the power spectrum results from this combination of gravitational and Doppler redshifts produced by the undulations.

The location of these peaks in terms of ℓ is decided by the physical scale created by the so-called 'sound horizon' at last scattering—basically the maximum comoving distance traveled by the sound waves prior to recombination, which depends critically on how the sound speed c_s evolved with redshift prior to that time. Using the comoving distance defined in Equation (6.73) below as basis, one simply defines the sound horizon by replacing the speed of light c in this expression with c_s:

$$r_s \equiv \int_0^{t_{\rm rec}} c_s(t')[1 + z(t')]\, dt' , \qquad (6.69)$$

where $t_{\rm rec}$ is the time corresponding to $z_{\rm rec}$, and we have made use of Equation (8.9) to replace the $a(t')$ in the denominator of the integrand with the redshift factor $[1+z(t')]^{-1}$. From this, one may then infer the proper distance $R_s(z_{\rm rec}) = r_s/(1+z_{\rm rec})$ (Equation 6.74), and finally the angular size θ_s corresponding to r_s at the LSS using the angular diameter distance (Section 12.4).

To determine θ_s carefully, one therefore needs to know how the sound speed c_s evolves with time. In a relativistic fluid, $c_s = c/\sqrt{3}$, but the early Universe contained matter as well as radiation. The strong coupling between photons, electrons and baryons allows one to consider the plasma a single fluid for dynamical purposes, though the contribution of baryons to the equation-of-state affects the dependence of c_s on z. A careful treatment of this quantity in the context of ΛCDM must take into account its evolution with time, showing that differences amounting to a factor ~ 1.3 could reduce the sound speed. These effects are typically rendered through the expression [119]

$$c_s = \frac{c}{\sqrt{3(1 + 3\rho_{\rm b}/4\rho_{\rm r})}} . \qquad (6.70)$$

Clearly, c_s reduces to $c/\sqrt{3}$ when $\rho_{\rm b}/\rho_{\rm r} \to 0$, as expected.

The three CMB satellite missions have identified the acoustic horizon and measured its angular size $\theta_s = (0.596724 \pm 0.00038)°$ on the LSS. The acoustic fluctuations are expected to have a characteristic size $\theta_{\rm fluc} \approx 2\theta_s$ because the sound waves expanded as spherical shells away from the dark-matter condensations. We see a cross section of this structure on the LSS, extending across twice the acoustic horizon. Thus, since the multipole number is defined as $\ell_s = 2\pi/\theta_{\rm fluc}$, one has $\ell_s = \pi/\theta_s$, which produces the location of the first peak at $\ell_1^{TT} \sim 300$, with an acoustic angular size $\theta_s \sim 0.6°$. But in fact several additional physical effects must be taken into account in order to produce an accurate measurement of ℓ_1^{TT}, including the decay of the gravitational potential and contributions from the Doppler shift of the oscillating fluid. These introduce a phase shift ϕ_1 into the spectrum [120, 121]. The general relation for all peaks and troughs is $\ell_m^{TT} = \ell_s(m - \phi_m)$, and since ϕ_m is typically $\sim 25\%$, the measured location of the first peak in the temperature power spectrum ends up at $l_1^{TT} \sim 220$, which we readily recognize in Figure 6.7.

The general agreement between the measured size of the acoustic horizon, which depends on $z_{\rm rec}$, and the value of the redshift at recombination implied by the astrophysical analysis leading up to Equation (6.60), is one of the most compelling successes of the expansion dynamics predicted by the standard model. Unfortunately, the story does not end there, however, so the conclusion is not that simple. This correspondence is based solely on the power spectrum at $\ell > 30$, associated exclusively with small-angle ($\theta \lesssim 1-2°$) acoustic fluctuations (see Figure 6.7). But we have known for several decades that—unlike the Sachs-Wolfe effect, which is quite sensitive to the expansion dynamics—the local physics where the CMB is produced appears to be generic to a broad range of evolutionary histories and/or model parameters [122]. So the fact that ΛCDM works beautifully to produce the exquisite multi-peaked power spectrum in Figure 6.7 does not ensure a correspondingly strong confirmation of its predictions at $\ell \lesssim 30$. On the contrary, the *Planck* data, in particular, now argue strongly that the expansion history in the standard model prior to recombination is inconsistent with the large-angle correlations measured in the CMB. We shall examine the framework and important consequences of this issue in Chapter 11.

6.5 THE TEMPERATURE HORIZON PROBLEM

The ΛCDM model described thus far is foundational and quite successful in accounting for many cosmological observations. Yet it hardly broaches the physical complexity inherent in the myriad high-precision measurements we have today. A pivotal turning point in the establishment of the 'standard model' was the serious attempt at addressing a persistent inconsistency—known as the 'horizon' problem—towards the end of the 1970's. Today we actually have two such problems: one arising from the apparent uniformity across the sky of the temperature in the cosmic microwave background; the second, analogously, arising from the observed uniformity of the Higgs vacuum expectation value throughout the cosmos. In this section, we shall describe the older, better known temperature horizon problem and explain how it arises in FLRW cosmology. We shall also introduce the inflationary paradigm invoked to address it. But we shall defer to Chapter 11 a more detailed discussion of this critical component of the theory, including a thorough explanation of the second (arguably more serious) issue associated with the Higgs field.

The older horizon problem arises from two features of the standard model, the first being that the Universe must have had a beginning (Chapter 1). As we have seen, the Big Bang initiated an expansion calculable from a set of rather simple differential equations (4.28, 4.29 and 4.31). Starting as a hot, dense plasma of thermalized matter and radiation, dynamically subjugated by its own self-gravity and the effects of an unknown dark energy, the Universe subsequently evolved via a gradual cooling and a drop in density and energy (Section 6.2).

The evidence that our Universe started at a finite time in the past is pervasive. In addition to the reasons introduced in Chapter 1, we must also acknowledge that distant galaxies recede from us at a speed proportional to their distance (see Equation 7.15 below); the relic microwave background radiation has a near-perfect Planck spectrum characterized by a highly-diminished temperature of 2.7° K (Section 6.4); and the relative cosmic abundance by mass of hydrogen ($\sim 75\%$) and helium ($\sim 25\%$) accords very well with one's expectation based on the physical conditions in the infant Universe (Sections 6.2 and 6.3).

The second factor is the observed uniformity of the cosmic microwave background (CMB), save for a spectrum of tiny fluctuations [28] at the level of one part per 100,000 (Section 6.4). This high degree of symmetry is certainly consistent with the Cosmological Principle, which posits that the Universe ought to be homogeneous and isotropic on scales[2] at least larger than 100 Mpc (Section 1.3). But these two aspects of the cosmic expansion are not fully consistent with each other.

To be more precise, the horizon problem is not necessarily viewed as a major shortcoming of the standard model per se but, rather, with the apparent requirement of highly improbable initial conditions. We measure a near identical CMB temperature in every direction yet, as we shall demonstrate shortly, regions on opposite sides of the sky lie beyond each other's causal horizon—here loosely defined as the distance light could have traveled prior to the time the CMB was produced. Opposite sides of the sky could not have come into thermal equilibrium after the Big Bang, because no physical process propagating at or below the speed of light could have causally connected them. It appears that some unknown mechanism had to produce a highly tuned spatial distribution of temperature.

The inflationary model of cosmology was invented in order to resolve this very serious problem [123, 124, 125, 126]. In this paradigm (see Chapter 11), the temperature horizon problem may be resolved if an inflationary spurt occurred from $\sim 10^{-35}$ seconds to 10^{-32} seconds after the Big Bang, during which the Universe expanded exponentially, stretching causally connected regions beyond the horizon.

To understand how this problem arises in the first place, let us begin by considering a much simpler analogy in flat spacetime (Figure 6.8). Suppose two frames of reference, A and C, are coincident with ours (frame B) at time $t = 0$, where t is measured in our frame, and then they recede at speed v for $t > 0$. If $v > c/2$, we infer that $2vt > ct$, and we would naively claim that A and C are beyond each other's 'horizon,' defined simply as ct. This statement is incorrect for several reasons, however.

At a very fundamental level in special relativity, we cannot use the coordinates in our frame B to determine who is causally connected in another frame, A or C, via the simple addition of velocities. For illustration, suppose that $v = 0.9c$. A's speed relative to C is then certainly not $2v = 1.8c$. The relativistic addition of velocities would tell us that A is instead moving at speed $0.994475c$ with respect to C, so it reaches a distance $0.994475ct'$ relative to C (which is less than ct') by time t' in C's frame. Hence, there is no violation of causality when A's distance from C is measured by an observer in C, rather than by the observer in frame B.

---

[2]On smaller scales, we obviously see evidence of mass and energy clumped in the form of galaxies and clusters of galaxies which, however, have negligible dynamical impact on the overall expansion of the Universe over much larger distances.

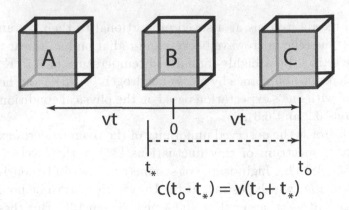

$$c(t_o - t_*) = v(t_o + t_*)$$

Figure 6.8 A simple analogy in flat spacetime, with two frames of reference (A and C) receding away from us (located at B) at speed v. If v is close to c, A and C are apparently beyond each other's horizon (here defined as ct) after a time t in our frame. A and C could still have communicated after $t = 0$ and before t_0, however, if light was emitted by A at time t_*, under the condition that $c(t_0 - t_*) = v(t_0 + t_*)$. (Adapted from ref. [127])

But even more importantly, no matter how close v is to c, we would still conclude that A and C had communicated with each other after separating at $t = 0$, as long as a light signal was transmitted (at time t_* in Figure 6.8) from one to the other shortly after their motion began. We would conclude that the light signal had reached C by the time t_0, if

$$c(t_0 - t_*) = v(t_0 + t_*) . \tag{6.71}$$

That is, we would see that A had communicated with C if the signal had been sent prior to the time

$$t_* \approx \frac{t_0}{2\gamma^2} , \tag{6.72}$$

where $\gamma \equiv (1 - [v/c])^{-1/2}$ is the Lorentz factor. It is true that the emission time t_* may be small when $v \to c$, but it is never zero. Thus, it makes no difference whether we view the interaction between A and C from our perspective in B, or from that of either A or C. Everyone agrees that A and C can be in causal contact after $t = t' = 0$, as long as A, B, and C were all coincident at the beginning. It does not matter if $v > c/2$ in our frame. This example is very illuminating because a similar effect emerges when we consider proper distances between us and patches of the CMB on opposite sides of the sky (Figure 6.9).

From the standard form of the FLRW metric in Equation (4.19), assuming spatial flatness with $k = 0$, we infer that the comoving distance traversed by a photon propagating radially from time 0 to t is

$$r = \int_0^t \frac{c\,dt'}{a(t')} . \tag{6.73}$$

Thus, the general definition of proper distance may be written

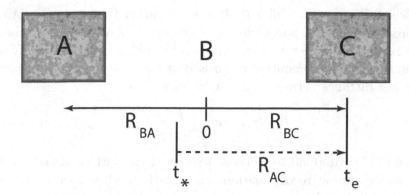

Figure 6.9 Light was emitted at cosmic time t_e by two patches (A and C) in the CMB, and is now reaching an observer at B. At the time of the emission, the patches were at proper distances $R_{BA}(t_e)$ and $R_{BC}(t_e)$, respectively. The two patches were in equilibrium when they emitted this light if A (say) emitted an earlier light signal at time t_* that could traverse the proper distance $R_{AC}(t_e)$ by time t_e. (Adapted from ref. [127])

$$R(t) \equiv a(t)r = a(t) \int_0^t \frac{c\,dt'}{a(t')} \,. \tag{6.74}$$

As we shall demonstrate in Section 7.4 below, it is straightforward to see from Equation (6.74) that the proper distance $R_\gamma(t)$ of a photon therefore satisfies the null geodesic Equation (7.27), written in terms of the gravitational (or Hubble) radius $R_h \equiv c/H(t)$ (see Section 7.3).

Both R_γ and R_h are functions of cosmic time t. When the equation-of-state is written as $p = w\rho$, with $w = $ constant, and p and ρ are, respectively, the total pressure and energy density in the cosmic fluid, the Friedmann Equations (4.28) and (4.29) give

$$\dot{R}_h = \frac{3}{2}(1+w)c \,. \tag{6.75}$$

The null geodesic linking a source of photons at proper distance $R_{\text{src}}(t_e) = R_\gamma(t_e)$ and cosmic time t_e, with an observer receiving them when $R_\gamma(t_0) = 0$, at time t_0, may then be found by simultaneously solving Equations (6.75) and (7.27).

This can sometimes be done analytically, e.g., when radiation was dominant in the early ΛCDM universe. During that epoch, we could simply put $w = +1/3$. Let us see how this works in practice by considering a specific scenario analogous to our earlier flat spacetime thought experiment. With reference to Figure 6.9, consider an observer B who is exchanging light signals with patches A and C in the CMB. Photons emitted by A and C at time t_e reach B at a later time t_0. By symmetry, photons emitted by B at t_e also reach A and C at t_0—remember that in the FLRW spacetime, t is the coordinate registering how much proper time has elapsed since the

Big Bang *everywhere* in the Universe. But at an earlier time t_* ($< t_e$), A emitted a signal in order to communicate with C at or before t_e. And just as we concluded in our earlier example (Figure 6.8), here too A and C will have been causally connected by the time their light was emitted towards B at t_e.

The proper distance between A and C at time t_* is

$$R_{AC}(t_*) = a(t_*) \int_{t_*}^{t_e} \frac{c\, dt'}{a(t')} . \qquad (6.76)$$

Patches A and C correspond to emission regions at the surface of last scattering, so think of t_e as the time at recombination, when matter and radiation started separating. Approximating the Universe as being radiation dominated up to t_e, we then find from Equations (4.28) and (4.29) that

$$a(t) = (2H_0 t)^{1/2} , \qquad (6.77)$$

in terms of the current value of the Hubble constant, H_0. During this time, Equation (6.76) is easy to evaluate, yielding the result

$$R_{AC}(t_*) = 2ct_* \left[\left(\frac{t_e}{t_*} \right)^{1/2} - 1 \right] . \qquad (6.78)$$

But we must also have $R_{\gamma e}(t_*) = R_{AC}(t_*)$ in order for the light signal emitted by A at t_* to reach C at t_e. Thus, for any arbitrary time $t < t_e$,

$$R_{\gamma e}(t) = 2ct \left[\left(\frac{t_e}{t} \right)^{1/2} - 1 \right] . \qquad (6.79)$$

The additional label 'e' in the subscript of $R_{\gamma e}(t)$ emphasizes the point that this is the particular null geodesic reaching C at time t_e. One can easily confirm that Equation (6.79) satisfies Equation (7.27), using the fact that $R_{\rm h}(t) = 2ct$ in this case. In probing the geometry of this null geodesic, we see that $R_{\gamma e}(t) \to 0$ at the beginning and end of the trajectory, i.e., at $t \to 0$ and $t \to t_e$. In addition, $R_{\gamma e}(t)$ has a maximum at $t = t_e/4$, where $R_{\gamma e}(t_e/4) = ct_e/2$. These are the features we should expect of a null geodesic with the constraints described above, all of which are quite evident in the photon trajectories plotted schematically in Figure 6.10.

From this figure, we can understand why the observed properties of the cosmic microwave background conflict with the predictions of basic ΛCDM without inflation. For convenience, we shall discuss the geodesics from the perspective of an observer in patch C. As noted earlier, proper distances are independent of this choice because the cosmic time t is the same everywhere. Thus, if C receives a light signal from B during a certain time interval, B will likewise receive a reciprocal signal from C during that same interval.

Figure 6.10 shows the null geodesics (solid curves) reaching observer C at two specific times: the present, t_0, and the time, t_e, when photons were emitted in the CMB she observes today. In her frame, some of the photons propagating along the path $R_{\gamma 0}(t)$ were emitted by B at time t_e, a proper distance $R_{BC}(t_e) = R_{\gamma 0}(t_e)$

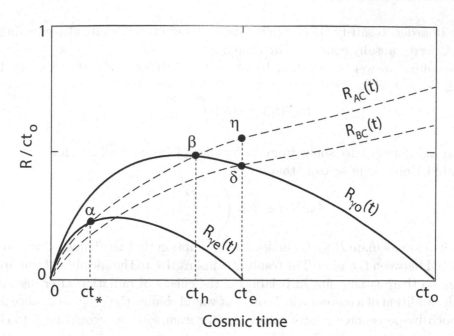

Figure 6.10 Null geodesics (solid curves) seen by an observer in C (see Figure 6.9). The light trajectory $R_{\gamma e}(t)$, with $t < t_e$, reaches C at time t_e from any source a proper distance $R_{\mathrm{src}}(t) = R_{\gamma e}(t)$ away. Correspondingly, the null geodesic labeled $R_{\gamma 0}(t)$ arrives at C at time t_0 ($> t_e$), though from sources at $R_{\mathrm{src}}(t) = R_{\gamma 0}(t)$, for any $t < t_0$. The other curves and symbols are discussed in the text. (Adapted from ref. [127])

away. The emission point is labeled δ in this figure. C is also receiving light emitted from patch A, radiated at time t_h, which is less than t_e because A was farther away, at a proper distance $R_{AC}(t_h) = R_{\gamma 0}(t_h)$. This point is labeled β in the figure. The subscript 'h' for the emission time at A denotes the fact that no light emitted there at $t > t_h$ could reach C by t_0. Clearly, patch A lies beyond C's current photon horizon for $t > t_h$. Of course, light emitted by A after t_h will become visible to C after t_0. Similarly, the observer at C will see in her future light emitted by B after t_e.

For a given observer, the null geodesic labeled $R_{\gamma 0}(t)$ is unique. If we knew the redshift at which the CMB was produced, our cosmological model would uniquely specify the time t_e at which the light was emitted. As such, there is only one point δ that satisfies the necessary conditions for C to receive the signal at t_0. According to our data, we see light emitted by patches A and C at $t_e \sim 380,000$ years, from equal proper distances on opposite sides of the sky. We thus know that $R_{AC}(t_e) = 2R_{BC}(t_e)$ in this diagram. Notice, however, that $R_{AC}(t_e)$ and $R_{BC}(t_e)$ are strongly dependent on the background cosmology. There is no guarantee that A and C were causally connected at arbitrary times. A necessary requirement would have been for A to emit a light signal at some $t_* < t_e$, and have it reach C at (or before) t_e. But could the ΛCDM universe have expanded at a rate that now permits us to identify such a time t_* in order to satisfy this proper-distance requirement, while also ensuring that A and C were causally connected at time t_e?

The simple answer is yes, though not for $t_e = 380{,}000$ years after the Big Bang. Writing

$$R_{AC}(t_e) = a(t_e) \int_{t_*}^{t_e} \frac{c\,dt'}{a(t')} \,, \qquad (6.80)$$

it is straightforward to show from Equations (4.28) and (4.29) for a radiation-dominated Universe prior to t_e that

$$R_{AC}(t_e) = 2ct_e \left(1 - \left[\frac{t_*}{t_e}\right]^{1/2}\right). \qquad (6.81)$$

We shall next calculate $R_{BC}(t_e)$ under the assumption that the Universe was matter-dominated between t_e and t_0. The results are insensitive to the specifics of this approximation, so there is no point in belaboring the effects of radiation after decoupling. Also, the addition of a cosmological constant would change the proper distance during this epoch by percentage points only, again far from what we would need to change the outcome of this comparison. Thus,

$$R_{BC}(t_e) = a(t_e) \int_{t_e}^{t_0} \frac{c\,dt'}{a(t')} \,, \qquad (6.82)$$

where $a(t) = (3H_0 t/2)^{2/3}$ in Einstein-de-Sitter space (Section 5.3), and so

$$R_{BC}(t_e) = 3ct_e \left(\left[\frac{t_0}{t_e}\right]^{1/3} - 1\right). \qquad (6.83)$$

Our constraint implies the equality $R_{AC}(t_e) = 2R_{BC}(t_e)$. Together with Equations (6.81) and (6.83), we therefore find that

$$\left(\frac{t_0}{t_e}\right)^{1/3} = \frac{5}{3} - \frac{2}{3}\left(\frac{t_*}{t_e}\right)^{1/2}. \qquad (6.84)$$

The earliest time at which A and C could have produced the CMB after having reached thermal equilibrium is therefore realized when $t_* \to 0$, for which $t_e \approx 27 t_0/125$, or roughly $1/5$ the current age of the Universe.

In basic ΛCDM without inflation, it would not have been possible for the temperature of the CMB to be equilibrated across the sky any earlier than approximately 2.7 Gyr after the Big Bang, much later than the redshift ($z \sim 1081$) we associate with the last scattering surface (Section 6.4), corresponding to a cosmic time $t_e \approx$ 380,000 years. This is the 'temperature' horizon problem which, however, would not exist if the CMB photons were produced within the past 10 billion years or so—even from opposite sides of the sky. It arises from the early deceleration of the Universe following the Big Bang, which is inevitable in a radiation-dominated environment, for which $a(t) \propto t^{1/2}$, as we have seen. In other words, this decreasing rate of expansion would not have permitted opposite sides of the sky to have reached as far as we see them today within only the first 380,000 years.

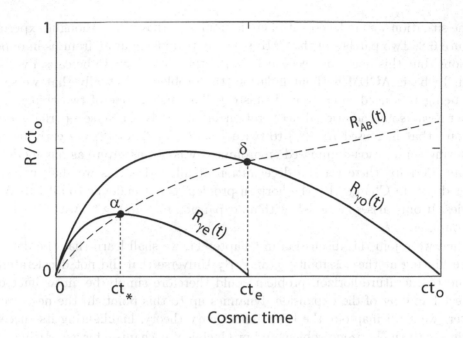

Figure 6.11 Null geodesics seen by an observer in patch B (see Figure 6.9). The null geodesic labeled $R_{\gamma e}(t)$ arrives at A or C at time t_e from B, which emitted a light signal at time $t < t_e$ from a proper distance $R_{AB}(t) = R_{\gamma e}(t)$ away. The corresponding null geodesic arriving at B at time t_0 ($> t_e$) is labeled $R_{\gamma 0}(t)$. The other curves and symbols have the same meaning as in Figure 6.10. (Adapted from ref. [127])

The only possible variation on the equilibration scenario we have just described is one in which the physical conditions in A and C are matched by a 'beacon' at B sending out signals that reach these patches before they emit at t_e. This geometry is shown schematically in Figure 6.11. The requirements are now

$$R_{AB}(t_*) = R_{BC}(t_*) = 2ct_* \left(\left[\frac{t_e}{t_*} \right]^{1/2} - 1 \right) , \qquad (6.85)$$

and

$$R_{AB}(t_e) = R_{BC}(t_e) = 3ct_e \left(\left[\frac{t_0}{t_e} \right]^{1/3} - 1 \right) \qquad (6.86)$$

(under the same conditions as before). Thus, putting

$$R_{AB}(t_e) = R_{BC}(t_e) = \frac{a(t_e)}{a(t_*)} R_{AB}(t_*) , \qquad (6.87)$$

we can easily show that the result for t_e is identical to that in Figure 6.10, i.e., the earliest t_e that would have permitted A and C to be causally connected—and yet have reached opposite sides of the sky—is about $1/5$ of t_0. We could have expected this outcome, of course, given the high degree of symmetry implicit in the FLRW metric.

The situation we are faced with here is similar to that of our thought experiment in Figure 6.8. Two patches of the CMB sky may be moving away from us in opposite directions, but this does not necessarily imply that they have to be causally disconnected. In basic ΛCDM without inflation, the problem is actually that we see the CMB being produced so early in its history. This means one of two things: either (1) there was some additional acceleration in the early Universe—perhaps due to inflation—that increased $R_{AC}(t_e)$ to twice $R_{BC}(t_e)$ by t_e; or (2) the early deceleration during a radiation-dominated environment was not as severe as our model now suggests. Perhaps there was no deceleration at all. Indeed, as we shall explore at greater depth in Chapter 11, the horizon problem is not endemic to all FLRW cosmologies. It only arises in models with an expansion rate $a(t) \sim t^\alpha$, with $\alpha < 1$, prior to decoupling at t_e.

When we rejoin this discussion in Chapter 11, we shall learn that the data may now be hinting at the possibility of an early Universe that did not decelerate after all. The temperature horizon problem could therefore simply be an artifact of an incorrect handling of the expansion dynamics up to this point. In the next section, however, we shall map out the basic inflationary theory, highlighting its success in resolving both the horizon problem and producing a mechanism for generating a near power-law fluctuation spectrum that may have seeded the anisotropies in the CMB and the subsequent formation of structure.

6.6 BASIC SLOW-ROLL INFLATION

To understand one of the principal motivations for the introduction of an inflaton field in the early 1980's, let us first consider a physical process we understand much better today—the Higgs mechanism for generating fermionic mass [128, 129]. As we shall see later in this book, the electroweak transition associated with this mechanism itself presents us with a serious horizon problem in cosmology, not unlike the one we are discussing here, but we shall defer the more detailed discussion of this topic to Section 11.3. There are actually two parts to this process: the first has to do with when (and if) the Higgs field acquired a non-zero 'vacuum expectation value,' commonly referred to as the 'turning on of the Higgs field;' the second has to do with the strength of the coupling between various elementary particles and the Higgs field. This interaction can spontaneously break the symmetry of the electroweak force, as we now describe. Prior to the Higgs turning on, all the 'messenger' particles, i.e., the gauge bosons carrying the electroweak force—in this case the photon and the $W^\pm$ and Z^0 particles—transferred the same amount of momentum per unit energy from one fermion to the next. In this regime, the relativistic energy, $E^2 = m^2c^4 + p^2c^2$, could not differentiate between them based on the value of p/E, since they all had $m = 0$, regardless of which particle was carrying the momentum. As such, there was only a single force.

The electric and weak forces separated, however, when the momentum per unit energy, p/E, changed due to the emergence of a non-zero mass. The symmetry was spontaneously broken because the Higgs acquired a non-zero vacuum expectation value and the particles had unequal coupling constants, and the consequent difference in mass m became more important relative to the momentum p as the temperature (and hence the energy E) dropped. Since in equipartition all the particles had the same energy distribution, the momentum distribution of different particle species had to change if their inertial masses were different. As we shall see in Chapter 11, we now know that the electroweak phase transition in ΛCDM must have occurred when the temperature dropped to the critical value $T = 159.5 \pm 1.5$ GeV, at $t_{\mathrm{ew}} \sim 10^{-11}$ seconds after the Big Bang.

The physics behind the Grand Unified Theory (GUT) concept—analogous to the electroweak unification—is that a similar spontaneous symmetry breaking could have occurred at much higher energies ($\sim 10^{15}$ GeV, as it turns out) to separate the electroweak force from the strong force. In fact, the thinking back then was that the same Higgs field might have been responsible for both transitions. But the discovery [87] of the Higgs particle at the CERN Large Hadron Collider (LHC) in 2012, revealed that its mass (~ 125 GeV/c^2) is much too large to permit the cosmic microwave radiation to appear as it does today. Since then, several proposals have been made to somehow include the Higgs field and its properties in a GUT-like transition, but with very little success thus far.[3]

Regardless of how/if a GUT transition was involved, however, the notion took hold that a scalar field may have emerged shortly after the Planck time[4] and briefly dominated the energy density in the cosmic fluid. With suitable characteristics, that we shall describe below, this field could have expanded the Universe at an exponential rate to overcome the temperature horizon problem. Perhaps even more importantly, the anisotropies in the CMB bear testimony to the creation of tiny fluctuations at early times, and suggest that these might have grown gravitationally to form the large-scale structure we see today. These perturbations could have started as quantum fluctuations in the inflaton field. Cosmologists now view this mechanism as an even more important consequence of inflation than the resolution of the horizon problem itself.

As we shall see in Section 6.7, the inflationary paradigm today posits that microscopic quantum fluctuations in the inflaton field got stretched during inflation to macroscopic scales, beyond the gravitational horizon, $R_{\mathrm{h}} = c/H$, where H is the Hubble parameter, at that time (see Chapter 7). These fluctuations oscillated while their amplitude grew until their wavelength λ exceeded $2\pi R_{\mathrm{h}}$, at which point they 'froze' with a constant amplitude until they re-entered the horizon much later, when the Universe was again dominated by radiation and, to a lesser extent, matter and dark energy. In the context of ΛCDM, this re-entry would have occurred at roughly 100,000 years, not too long before they imprinted themselves as acoustically induced anisotropies on the CMB at $\sim 380,000$ years.

---

[3]Some workers are considering the possibility that a 'little brother' to the Higgs may exist, with similar properties, though with a much smaller mass [130].

[4]See Chapter 13 for a thorough discussion of physics at the Planck scale.

To get started with a more formal description of the inflationary process, let us define the first of the so-called small parameters of inflation,

$$\epsilon \equiv -\frac{\dot{H}}{H^2} . \tag{6.88}$$

With it (and the definition $H \equiv \dot{a}/a$), one can trivially show that

$$\frac{\ddot{a}}{a} = H^2 \left(1 - \epsilon\right) . \tag{6.89}$$

Since we need the Universe to have accelerated during inflation, a minimal requirement is that $\epsilon < 1$. It will also be useful for us to introduce the number N of e-folds during the inflationary expansion,

$$dN \equiv H \, dt = d \ln a . \tag{6.90}$$

Its meaning stems from the fact that $a(t)$ is an exponential function when $H = $ constant. Understanding the constraints on the quantities ϵ and N will be critical to establishing whether or not a given model can produce the required expansion.

The simplest inflationary models assume a single scalar field, ϕ, which we shall always refer to as the *inflaton* field. Since the foundational nature of this field is still unknown, not much can be said about it except that its equation-of-state is constrained by inflationary requirements, and its evolution in time characterizes how the inflationary energy density changes. The inflaton field may or may not have been minimally coupled to gravity. If it was minimally coupled, its dynamics would have been governed by the action

$$S = \int d^4x \sqrt{-g} \; \mathcal{L}(\phi, \partial_\mu \phi) , \tag{6.91}$$

where $\sqrt{-g} = a^3(t)$ for the FLRW metric in Equation (4.19) with $k = 0$, and

$$\mathcal{L} = \frac{m_{\rm P}^2}{16\pi} \mathcal{R} + \frac{1}{2} g^{\mu\nu} \partial_\mu \phi \, \partial_\mu \phi - V(\phi) \tag{6.92}$$

is the Lagrangian density. In this expression, $\mathcal{R}$ is the Ricci scalar, which we have slightly relabeled in comparison to Equation (2.154) in order to avoid confusion with the proper radius R; $V(\phi)$ is the inflaton potential; and $m_{\rm P} \equiv G^{-1/2}$ is the Planck mass. Throughout this section, we use 'natural' units, in which $c = \hbar = 1$.

In Equation (6.92), the first term on the right-hand side is the Einstein-Hilbert Lagrangian density, while the other two terms constitute the Lagrangian density,

$$\mathcal{L}_\phi \equiv \frac{1}{2} g^{\mu\nu} \partial_\mu \phi \, \partial_\mu \phi - V(\phi) , \tag{6.93}$$

of a scalar field with kinetic and potential contributions. The function $V(\phi)$ describes the self-interactions of ϕ, which are necessary when we want the pressure it exerts to be negative (more on this below). If the inflaton field were not minimally coupled, $\mathcal{L}$ would contain an additional interaction term coupling $\mathcal{R}$ to ϕ.

The energy-momentum tensor $T^{\mu\nu}$ follows from an application of Noether's theorem (Figure 2.4) to S in Equation (6.91), which relates the invariance of the action to a constant translation in spacetime.[5] It may be found here by varying S with respect to the metric $g_{\mu\nu}$, which appears both in the kinetic term for ϕ and in the determinant g (Equation 2.66). The energy-momentum tensor for ϕ is

$$T^{\mu\nu} = \frac{2}{\sqrt{-g}} \frac{\delta(\sqrt{-g}\mathcal{L}_\phi)}{\delta g_{\mu\nu}} , \qquad (6.94)$$

whose evaluation will require the use of

$$\delta\sqrt{-g} = -\frac{\delta g}{2\sqrt{-g}} = \frac{1}{2}\sqrt{-g}\, g^{\mu\nu}\, \delta g_{\mu\nu} , \qquad (6.95)$$

and

$$\delta g^{\mu\nu} = -g^{\mu\alpha} g^{\nu\beta} \delta g_{\alpha\beta} . \qquad (6.96)$$

The first of these expressions uses Jacobi's formula $\delta g = g\, g^{\mu\nu}\, \delta g_{\mu\nu}$.

With Equations (6.95) and (6.96), the energy-momentum tensor then becomes

$$\begin{aligned} T^{\mu\nu} &= \frac{2}{\sqrt{-g}}\left(\sqrt{-g}\frac{\delta\mathcal{L}_\phi}{\delta g_{\mu\nu}} + \frac{1}{2}\sqrt{-g}\, g^{\mu\nu}\mathcal{L}_\phi\right) \\ &= \partial^\mu\phi\, \partial^\nu\phi - g^{\mu\nu}\left(\frac{1}{2}\partial_\alpha\phi\, \partial^\alpha\phi + V(\phi)\right) . \end{aligned} \qquad (6.97)$$

And since the background field in FLRW spacetime is homogeneous, we may ignore the spatial gradients, so the corresponding energy density ρ_ϕ $(= T_{00})$ and pressure p_ϕ $(= T_{ii})$ of the inflaton field are given simply as

$$\rho_\phi = \frac{1}{2}\dot{\phi}^2 + V(\phi) , \qquad (6.98)$$

and

$$p_\phi = \frac{1}{2}\dot{\phi}^2 - V(\phi) , \qquad (6.99)$$

in the form of a perfect fluid (Equation 4.27) [131].

The dynamics equation for the inflaton field may be derived in several ways, perhaps the simplest of which is to just substitute ρ_ϕ and p_ϕ for ρ and p into the energy Equation (4.31). This immediately yields the well-known Klein-Gordon equation,

$$\ddot{\phi} + 3H\dot{\phi} + \frac{\partial V}{\partial\phi} = 0 . \qquad (6.100)$$

Correspondingly, the Friedmann Equation (4.28) for a spatially-flat spacetime reduces to

$$H^2 = \frac{8\pi}{3m_{\mathrm{P}}^2}\left(\frac{1}{2}\dot{\phi}^2 + V[\phi]\right) . \qquad (6.101)$$

---

[5] For a concrete example of how this works in practice, see Section 6.4 in *Electrodynamics (Chicago Lectures in Physics)* by Melia (2001).

Once $V(\phi)$ is known, Equations (6.100) and (6.101) completely determine the evolution of the Universe while the inflaton field dominates the energy density and pressure in the cosmic fluid. The challenge, of course, is that we know very little about its underlying physical properties, so we must rely quite heavily on the observational constraints to guide us in shaping this potential. Unfortunately, no extension to the standard model of particle physics has yet been found to accommodate the scenario we are describing here.

In this spirit, then, we note that the equation-of-state of the inflaton field ϕ may be written

$$w_\phi \equiv \frac{p_\phi}{\rho_\phi} = \frac{\dot{\phi}^2/2 - V(\phi)}{\dot{\phi}^2/2 + V(\phi)} . \tag{6.102}$$

Thus, in order for ϕ to exert negative pressure ($w_\phi < 0$) and produce an accelerated expansion, its potential $V(\phi)$ must dominate over the kinetic energy $\dot{\phi}^2/2$. According to Equation (4.31), the acceleration may be written

$$\frac{\ddot{a}}{a} = -\frac{8\pi}{3m_P^2} \left(\dot{\phi}^2 - V[\phi] \right) . \tag{6.103}$$

Therefore, from Equations (6.89), (6.101) and (6.103), we find that

$$\epsilon = \frac{3}{2} (w_\phi + 1) = \frac{4\pi}{m_P^2} \frac{\dot{\phi}^2}{H^2} . \tag{6.104}$$

As we shall see in the following section, the near-perfect power-law fluctuation spectrum observed in the CMB requires so-called *slow-roll* inflation, in which H changes very slowly with respect to ϕ. According to Equation (6.88), this means that the *slow-roll parameter* ϵ must be much smaller than one. Thus, not only must $\epsilon < 1$ for accelerated expansion, we must also have $\epsilon \ll 1$ to produce the correct distribution of anisotropies in the CMB. This means, therefore, that $w_\phi \to -1$ in Equation 6.104, i.e., $p_\phi \to -\rho_\phi$, which takes us into the de Sitter regime (Section 5.2). And from Equations (6.98) and (6.99), we conclude that inflation must have happened while

$$\frac{1}{2}\dot{\phi}^2 \ll V(\phi) , \tag{6.105}$$

which produces an exponentiated expansion as long as this condition is maintained. Some of these features are shown schematically in Figure 6.12. This image also shows several other properties of the inflaton potential that are required by the observed fluctuation spectrum, which will be discussed in Section 6.7.

A quick inspection of Figure 6.12 suggests that a single constraint on $\dot{\phi}$ may not be sufficient to ensure that the temperature horizon problem is overcome. In order to achieve this goal, the inflated expansion must have lasted long enough to prevent the limitation on t_e appearing in Equation (6.84). This requirement translates into a minimum value of $\Delta\phi$ in Figure 6.12.

This requirement motivates the introduction of a second small parameter of inflation,

$$\eta \equiv -\frac{\ddot{\phi}}{H\dot{\phi}} . \tag{6.106}$$

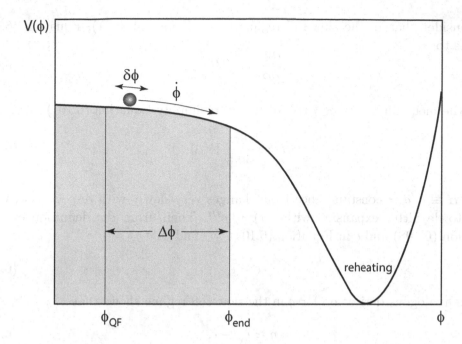

Figure 6.12 Schematic diagram showing a scalar field ϕ rolling down the shallow portion of a potential with 'speed' $\dot{\phi}$. During slow-roll inflation, the potential energy $V(\phi)$ dominates over the kinetic energy $\dot{\phi}^2/2$, and therefore changes very little while ϕ grows . The acceleration phase ends at ϕ_{end} when $\dot{\phi}^2/2 \sim V(\phi)$. The anisotropies in the CMB were created as quantum fluctuations (QF) at ϕ_{QF}, about 60 e-folds before inflation ended. At reheating, the inflaton energy density is dissipated into radiation, and the standard model takes over.

It is not difficult to show from Equations (6.88), (6.90) and (6.104) that

$$\eta = \epsilon - \frac{1}{2\epsilon} \frac{d\epsilon}{dN} , \tag{6.107}$$

so that the constraint $|\eta| < 1$ ensures a small fractional change in ϵ per e-fold. We may also see this directly in the Klein-Gordon Equation (6.100) and the Friedmann Equation (6.101), where the condition $|\eta| \ll 1$ implies $|\ddot{\phi}|^2 \ll |3H\dot{\phi}|$ and $|\ddot{\phi}|^2 \ll |\partial V/\partial\phi|$, so that $V(\phi)$ changes very slowly with respect to ϕ.

Sometimes it is more convenient to consider a second set of small parameters, equivalent to ϵ and η, though written in terms of the potential $V(\phi)$:

$$\epsilon_V(\phi) \equiv \frac{m_{\text{P}}^2}{16\pi} \left(\frac{1}{V} \frac{\partial V}{\partial \phi} \right)^2$$

$$\eta_V(\phi) \equiv \frac{m_{\text{P}}^2}{8\pi} \frac{1}{V} \frac{\partial^2 V}{\partial \phi^2} . \tag{6.108}$$

Clearly, the slow-roll regime as we have defined it also requires the shape of $V(\phi)$ to comply with the restrictions $\epsilon_V \ll 1$ and $|\eta_V| \ll 1$. Not surprisingly, the two sets of parameters are directly related to each other.

Consider that in the slow-roll regime ($\epsilon \ll 1$ and $|\eta| \ll 1$), Equation (6.100) reduces to

$$\frac{\partial V}{\partial \phi} \approx -3H\dot{\phi}\,, \tag{6.109}$$

while the inequality $\dot{\phi}^2/2 \ll V(\phi)$ allows us to simplify Equation (6.101) to

$$H^2 \approx \frac{8\pi}{3m_{\rm P}^2} V(\phi)\,. \tag{6.110}$$

Thus, $H \equiv \dot{a}/a \approx$ constant since $V(\phi)$ changes very slowly with respect to ϕ, which leads to de Sitter expansion with $a(t) \sim e^{Ht}$. Then, from the definition of ϵ_V in Equation (6.108) and ϵ in Equation (6.104), we find that

$$\epsilon \approx \epsilon_V\,, \tag{6.111}$$

and an analogous derivation for η in the slow-roll regime shows that

$$\eta \approx \eta_V - \epsilon_V\,. \tag{6.112}$$

For any given inflaton potential $V(\phi)$, we can tell when inflation would have ended based on the threshold condition

$$\epsilon(\phi_{\rm end}) = 1 \quad \text{or} \quad \epsilon_V(\phi_{\rm end}) = 1\,, \tag{6.113}$$

by which time the Universe would have expanded by the number of e-folds

$$N(\phi) = \ln\left(\frac{a_{\rm end}}{a}\right)\,, \tag{6.114}$$

starting with an inflationary phase at a. Thus,

$$N(\phi) = \int_{\phi}^{\phi_{\rm end}} \frac{H}{\dot{\phi}}\, d\phi \tag{6.115}$$

and, with the use of Equation (6.104), may be written as

$$N(\phi) \approx \int_{\phi_{\rm end}}^{\phi} \left(\frac{4\pi}{m_{\rm P}^2\epsilon}\right)^{1/2} d\phi\,. \tag{6.116}$$

Insofar as the temperature horizon problem is concerned, these equations are sufficient—at least in the slow-roll approximation—to test whether a given inflaton potential can overcome the proper distance deficit leading to Equation (6.84). In Chapter 11, we shall consider some of the latest observational data and examine how well slow-roll inflation is likely to fare going forward. We shall postpone this extended analysis until then in part because it will be beneficial for us to couple this probe with the equally problematic electroweak horizon problem which still has no workable solution in the context of ΛCDM.

6.7 QUANTUM FLUCTUATIONS IN THE INFLATON FIELD

The mechanism with which quantum fluctuations may have been seeded in the inflaton field and may have grown to form the primordial perturbation spectrum responsible for the CMB anisotropies and large-scale structure is at least as important as resolving the temperature horizon problem. This topic has received extensive coverage in the literature and several excellent accounts already exist. Our goal here is not to replicate or supplant these resources, but to provide a framework for understanding the physics and assumptions at the core of this work, with references to the more detailed treatments when needed.

At decoupling, the observed inhomogeneities in the CMB had an amplitude of order $\sim 10^{-5}$ relative to the otherwise uniform background (Figure 1.3). Perturbations arising in such a medium were therefore unquestionably very small, suggesting that linear perturbation theory should be adequate to handle the growth of these fluctuations. The common approach is therefore to write all relevant physical quantities $U(t, \mathbf{x})$ as the sum of a smooth time-dependent background component and a perturbation that locally also depends on space:

$$U(t, \mathbf{x}) = U_0(t) + \delta U(t, \mathbf{x}) . \tag{6.117}$$

It is understood that this approximation is to be made for all components of the energy-momentum tensor $T^{\mu\nu}$ and the metric coefficients $g_{\mu\nu}$. For example, we write for the inflaton field, which appears in the density ρ_ϕ and pressure p_ϕ in $T^{\mu\nu}$,

$$\phi(t, \mathbf{x}) = \phi_0(t) + \delta\phi(t, \mathbf{x}) , \tag{6.118}$$

with $|\delta\phi|/\phi_0 \ll 1$.

Already at the outset there is a complication with this approach from the fact that the decomposition in Equation (6.117) is not unique—it depends on the choice of coordinates, more commonly referred to as a *gauge choice*. For the smooth background, the 'natural' system to use is based on the comoving coordinates (ct, r, θ, ϕ) leading to the form of the FLRW metric in Equation (4.19). This makes sense when we consider how the spatial and time coordinates relate to each other in Figure 4.1. The cosmic time is the local proper time for every observer, no matter where they happen to be in the spacetime fabric. This allows for a straightforward *slicing* of the spacelike hypersurfaces of constant t on each timelike worldline defining the *threading*, corresponding to the trajectory of comoving observers with zero momentum density at their location (i.e., with zero 'peculiar' velocity). This is, of course, a geometric description of Weyl's postulate. The slicing is everywhere orthogonal to the threading, and each spacelike slice represents a homogeneous Universe at that time.

This situation changes when we introduce perturbations, however, because now the local proper time almost never corresponds to the cosmic time measured along the threads in Figure 4.1. The local, spatially-dependent fluctuations introduce curvature-related time dilations, albeit small, though not homogeneously distributed across each slice. One must therefore decide which coordinates to use for the description of the new spacetime when inhomogeneous perturbations are present, which in turn affects how the fluctuations $\delta U(t, \mathbf{x})$ are addressed.

This should make sense intuitively because the perturbation $\delta U(t, \mathbf{x})$ is the difference between $U(t, \mathbf{x})$ in the physical (i.e., perturbed) spacetime, and $U_0(t)$ in the smooth, homogeneous background. But we know from differential geometry that in order to compare tensors on two different membranes, one must first decide how to map the spacetime points on one hypersurface into the other. Choosing a gauge amounts to fixing this map. Unfortunately, there is no unique way of doing this. As we shall see below, a significant advance with cosmological perturbation theory was made when gauge-invariant combinations of various perturbed quantities were identified, allowing one to solve the dynamical equations independently of the gauge choice.

The partial differential equations we must solve for the various perturbed quantities $\delta U(t, \mathbf{x})$ depend on both $\mathbf{x}$ and t. It therefore makes sense to expand them using a Fourier decomposition, which at least allows us to evaluate the spatial derivatives immediately, thereby reducing these expressions to ordinary differential equations involving solely the time derivatives. Moreover, this approach will allow us to study each Fourier mode $\mathbf{k}$ *separately*, greatly reducing the complexity in the system.

But this decomposition also produces another important byproduct, resulting from the high degree of symmetry in the FLRW metric. The linearized equations allow us to identify independent scalar, vector and tensor components. Formally, these various designations have to do with how a given perturbation changes under rotation (by an angle ψ) of the coordinate system about the Fourier wavevector $\mathbf{k}$. The perturbation has a helicity m if its amplitude transforms according to

$$\delta U_k(t) \rightarrow e^{im\psi} \, \delta U_k(t) \, . \tag{6.119}$$

It is called a scalar, vector or tensor perturbation, depending on whether its helicity is, respectively, 0, ± 1, and ± 2. The benefit of this separation is that perturbations of each type evolve independently and differently of each other in the linear approximation.

Again, this can be understood quite easily at the intuitive level, even if the formal description may seem daunting. Scalar perturbations describe local, non-traveling inhomogeneities that grow isotropically due to their own self-gravity. Tensor modes, on the other hand, are simply traveling gravity waves that rely on spacetime oscillations within the cosmic fluid. But vector perturbations represent local peculiar motions of this fluid that must compete with the global Hubble expansion. Not surprisingly, they die off rather quickly and have no lasting impact on the evolving spacetime. The latter are therefore always ignored in cosmological applications of linearized perturbation theory.

The perturbed FLRW metric for the linearized scalar and tensor fluctuations is described by the line element [131, 132, 133, 134]

$$ds^2 = (1 + 2\Phi) \, dt^2 - 2a(t)(\partial_i B) \, dt \, dx^i - $$
$$a^2(t)[(1 - 2\Psi)\delta_{ij} + 2(\partial_i \partial_j E) + h_{ij}] \, dx^i \, dx^j \, , \tag{6.120}$$

where the indices i and j denote spatial coordinates, and Φ, Ψ, B and E describe the scalar metric perturbations, while h_{ij} represent the tensor perturbations. This form of the metric follows the notation of Mukhanov et al. (1992) and, as a reminder, we are using natural units with $c = \hbar = 1$ throughout this section.

When this metric is folded through Einstein's equations, the scalar and tensor components decouple to first order. The tensor fluctuations h_{ij} are always gauge-invariant, so any choice of coordinates is adequate for calculating them. The scalar equation, however, is gauge dependent. Fortunately, several gauge-independent combinations of the scalar perturbations have been identified. For example, in the comoving frame, the metric perturbation Ψ and the scalar field perturbation $\delta\phi$ may be combined to form the gauge-invariant *curvature* perturbation

$$\Theta \equiv \Psi + \left(\frac{H}{\dot{\phi}}\right)\delta\phi \qquad (6.121)$$

on hypersurfaces orthogonal to comoving worldlines.

Let us now expand Θ in Fourier modes,

$$\Theta(t, \mathbf{x}) = \int \frac{d^3\mathbf{k}}{(2\pi)^{3/2}}\, \Theta_k(t)\, e^{i\mathbf{k}\cdot\mathbf{x}}, \qquad (6.122)$$

where $\mathbf{k}$ is the *comoving* wavenumber. Using the linearized FLRW metric in Equation (6.120) with the linearized version (see also Section 13.2) of Einstein's Equations (2.167), one arrives at the perturbed equation of motion

$$\Theta_k'' + 2\left(\frac{z'}{z}\right)\Theta_k' + k^2\Theta_k = 0, \qquad (6.123)$$

where overprime now denotes a derivative with respect to conformal time $d\tau \equiv dt/a(t)$. This equation is satisfied individually by each mode $\mathbf{k}$.

The quantity z in Equation (6.123) is defined by the expression

$$z \equiv \frac{a(t)(\rho_\phi + p_\phi)^{1/2}}{H}. \qquad (6.124)$$

For an inflaton field satisfying Equations (6.98) and (6.99), this reduces to the simpler form

$$z = a(t)\frac{\dot{\phi}}{H}. \qquad (6.125)$$

The factor z'/z thus clearly depends on the background dynamics. For convenience, one typically rewrites Equation (6.123) in terms of the so-called Mukhanov-Sasaki variable [135, 136],

$$u_k \equiv z\,\Theta_k, \qquad (6.126)$$

in order to reduce the differential equation to a more manageable form with only two terms. Now represented by u_k, the scalar fluctuations evolve according to the expression

$$u_k'' + \left(k^2 - \frac{z''}{z}\right)u_k = 0, \qquad (6.127)$$

which is known as the Mukhanov equation. Note that it depends only on the magnitude of $\mathbf{k}$, so there is no need to keep the vector notation beyond this point.

To solve Equation (6.127) exactly, one needs to employ numerical techniques. Nevertheless, there are reasonable approximations one can make to simplify this expression and find simple analytic solutions, e.g., in the remote conformal past ($\tau \to -\infty$), or during de Sitter inflation in the slow-roll regime. The bigger issue, however, is *normalization*. The Mukhanov equation describes the time-dependent evolution of each mode u_k, but does not specify its initial amplitude. The inflaton field, however, functions in the quantum domain, where fluctuations such as this are necessarily constrained by *canonical quantization*.

To understand what this means, one typically introduces the quantum mechanical oscillator into the discussion. The essence of this constraint, however, actually exists at a more fundamental level—it is due to the Heisenberg Uncertainty Principle. We do not know what sets the value of the (reduced) Planck constant $\hbar$, though we know that it is not zero. And it apparently has not changed throughout the Universe's expansion history. We also know that in the laboratory, canonical variables, such as the position $\mathbf{x}$ and momentum $\mathbf{p}$, cannot be measured with arbitrary precision within a confined volume. The product of their uncertainties has a minimum value set by the Planck constant. If we assume that quantum mechanics was just as valid in the early Universe as it is now, we cannot avoid the implication that the fluctuation u_k and its canonical partner u'_k must also have been constrained by standard commutation relations

$$[u_k(\tau), u'_k(\tau)] = i \tag{6.128}$$

(where we have set $\hbar = 1$, as we do throughout this section). These clearly embody the spirit of the Heisenberg Uncertainty Principle in the sense that each canonical variable scales as the inverse of its partner.

Now consider the Hamiltonian $\mathcal{H}$ written in terms of u_k and u'_k. As long as the oscillation frequency one extracts from Equation (6.127) is constant (more on this below), $\mathcal{H}$ is just the sum of terms proportional to $|u_k|^2$ and $|u'_k|^2$, and since these scale inversely with each other, $\mathcal{H}$ has a minimum whose value depends on $\hbar$ and k or, equivalently, the mode's frequency ω. Thus, if a quantum fluctuation emerges in flat spacetime, where its frequency is independent of time, its amplitude must correspond to a minimization of the Hamiltonian, which simply means that it must have a minimum energy consistent with $\hbar$ expressed via the Heisenberg Uncertainty Principle. We therefore assume that the initial amplitude of each mode u_k was set by this constraint at the time it was created within the inflaton field.

But how does one actually do this in the early Universe, where the spacetime curvature was extreme? Notice that the factor multiplying u_k in Equation (6.127) is definitely not constant, negating any possibility of the frequency being time-independent. The step taken here is somewhat controversial because it apparently violates the Compton wavelength limit. A detailed discussion of this issue will be postponed to Section 13.1, where we shall consider the physics of quantum fluctuations at the Planck scale in concert with the latest constraints from the *Planck* observations of the CMB. For now, we shall merely introduce the approach taken to minimize the

Hamiltonian for the scalar modes u_k, so that later we may better appreciate some of the inconsistencies still plaguing the inflationary paradigm.

The argument is made that we can find the initial state of the quantum fluctuation u_k in the Minkowski vacuum of a comoving observer in the distant, conformal past, $\tau \to -\infty$, where comoving scales were all well inside the gravitational (or Hubble) horizon $R_{\rm h}$ (see Chapter 7). In this limit, known as the Bunch-Davies vacuum [137], the mode Equation (6.127) becomes

$$u_k'' + k^2 u_k = 0 \,, \tag{6.129}$$

which is merely the equation of a simple harmonic oscillator with a time-independent frequency. Then, using the minimum energy constraint, we can impose the initial condition

$$\lim_{\tau \to -\infty} u_k = \frac{e^{-ik\tau}}{\sqrt{2k}} \,, \tag{6.130}$$

which completely fixes the mode functions on all scales. As we shall learn in Chapter 13, however, this procedure assumes that we can track the wavelength of such a fluctuation well below its Compton wavelength, which is inconsistent with basic principles in quantum mechanics. This question has given rise to what is sometimes called the 'trans-Planckian' problem [138] associated with inflationary dynamics, and we shall address it more thoroughly in Section 13.1.

As inflation proceeds through its de Sitter phase, subject to the slow-roll conditions $\epsilon \ll 1$ and $H \approx$ constant (Equation 6.88), we then have

$$\frac{z''}{z} \to \frac{a''}{a} = \frac{2}{\tau^2} \tag{6.131}$$

(recalling that $a \propto e^{Ht}$), and the Mukhanov equation reduces to the simpler form

$$u_k'' + \left(k^2 - \frac{2}{\tau^2} \right) u_k = 0 \,. \tag{6.132}$$

The general solution to Equation (6.132), which may be confirmed by direct substitution, is

$$u_k = A \frac{e^{-ik\tau}}{\sqrt{2k}} \left(1 - \frac{i}{k\tau} \right) + B \frac{e^{ik\tau}}{\sqrt{2k}} \left(1 + \frac{i}{k\tau} \right) \,, \tag{6.133}$$

where A and B are unspecified constants. If we next impose the canonical quantization condition (Equation 6.128) and the Bunch-Davies limit (Equation 6.130) on this general form, the constants of integration take the unique values $A = 1$ and $B = 0$. The final mode functions subject to the assumptions and constraints we have discussed in this section are therefore

$$u_k = \frac{e^{-ik\tau}}{\sqrt{2k}} \left(1 - \frac{i}{k\tau} \right) \,. \tag{6.134}$$

One of the most important observational signatures associated with these modes is the predicted power spectrum of $\Theta(t, \mathbf{x})$ (Equation 6.121), derivable from u_k. The power spectrum for this quantity, expanded in Fourier space according to Equation (6.122), is derived from

$$\langle \Theta_k(t)\Theta_{k'}(t) \rangle \equiv (2\pi)^3 \delta^{(3)}(\mathbf{k} + \mathbf{k'})|\Theta_k(t)|^2 \,, \tag{6.135}$$

where $\langle ... \rangle$ is the ensemble average of the fluctuations. Thus,

$$\langle \Theta(t, \mathbf{x})^2 \rangle = \int \frac{d^3\mathbf{k}}{(2\pi)^3}\, |\Theta_k(t)|^2 \,. \tag{6.136}$$

The power spectrum of the fluctuations $\Theta(t, \mathbf{x})$ is then defined by the equation

$$\int \frac{dk}{k} \mathcal{P}_\Theta(k) \equiv \int \frac{d^3\mathbf{k}}{(2\pi)^3}\, |\Theta_k(t)|^2 \,, \tag{6.137}$$

which gives

$$\mathcal{P}_\Theta(k) = \frac{k^3}{2\pi^2}|\Theta_k(t)|^2 \,. \tag{6.138}$$

The scale-dependence of the power spectrum is defined by the scalar spectral index n_s, or *tilt*, where

$$n_s - 1 \equiv \frac{d \ln \mathcal{P}_\Theta(k)}{d \ln k} \,. \tag{6.139}$$

Therefore, *scale invariance* corresponds to the index $n_s = 1$. It is also useful to define a quantity—the so-called *running of the spectral index* n_s—that characterizes the dependence of n_s on k, which is given by

$$\alpha_s \equiv \frac{d\, n_s}{d \ln k} \,. \tag{6.140}$$

These theoretical predictions are compared to the measured spectrum, which is typically reported as an approximate power law:

$$\mathcal{P}_\Theta^{\text{obs}}(k) = A_s(k_*) \left(\frac{k}{k_*}\right)^{n_s(k_*)-1+\frac{1}{2}\alpha_s(k_*)\ln(k/k_*)} \,, \tag{6.141}$$

where the fixed wavenumber k_* is known as the *pivot* scale.

The astute reader will have noticed by now that, throughout this discussion, the quantities $\Theta(t, \mathbf{x})$, u_k, and $\mathcal{P}_\Theta(k)$ are all functions of cosmic time t. So where or when do we actually calculate $\mathcal{P}_\Theta(k)$ for comparison with the data (Equation 6.141)? The answer emerges from a closer inspection of the solution u_k in Equation (6.134), which reveals that the modes behave very differently on sub-Hubble horizon and super-Hubble horizon scales.

In de Sitter space,

$$\tau = -\frac{1}{Ha} \,. \tag{6.142}$$

Therefore we can rewrite Equation (6.132) in the form

$$u_k'' + k^2 \left(1 - 2 \left[\frac{Ha}{k} \right]^2 \right) u_k = 0 \, , \qquad (6.143)$$

showing its dependence on the ratio $\lambda_k / R_{\mathrm{h}}$, where

$$\lambda_k \equiv \frac{2\pi a}{k} \qquad (6.144)$$

is the wavelength of mode k. On sub-Hubble horizon scales, with $Ha/k \ll 1$, the Mukhanov equation reduces to

$$u_k'' + k^2 u_k = 0 \, , \qquad (6.145)$$

whose solution is the oscillating function

$$u_k = \frac{e^{-ik\tau}}{\sqrt{2k}} \quad (k \gg aH) \, . \qquad (6.146)$$

On super-Hubble horizon scales, we have instead

$$u_k'' - \frac{a''}{a} u_k = 0 \, , \qquad (6.147)$$

which is solved by the simple expression

$$u_k = D(k)a \quad (k \ll aH) \, . \qquad (6.148)$$

The constant amplitude $D(k)$ is fixed by equating the solutions in Equations (6.146) and (6.148) at $k = aH$, i.e., at $\lambda_k = 2\pi R_{\mathrm{h}}$, which gives

$$u_k = \frac{H}{\sqrt{2k^3}} \quad (k \ll aH) \, . \qquad (6.149)$$

The distinction between the mode solutions at sub-horizon and super-horizon scales is critical to understanding how quantum fluctuations in the inflaton field grew and formed the primordial spectrum. During slow-roll inflation, H would have changed very slowly, so the gravitational (i.e., Hubble) radius $R_{\mathrm{h}} = c/H$ would have remained almost constant during the corresponding de Sitter expansion. The mode wavelength λ_k (Equation 6.144), however, would have continued to grow with a. Thus, all sub-Hubble horizon modes would have oscillated and grown until $k \ll aH$, after which they *froze* with an amplitude given by Equation (6.149). The fluctuation spectrum would therefore have been an accumulation of all the modes that had crossed the Hubble horizon by the time they became relevant to the observations and froze until the Universe returned to its non-inflationary expansion.

From Equations (6.126), (6.134) and (6.138), we thus find that

$$\mathcal{P}_\Theta(k) = \frac{k^3}{2\pi^2} \left(\frac{H}{a\dot\phi} \right)^2 \frac{1}{2k} \left(1 + \frac{1}{(k\tau)^2} \right) \Bigg|_{k=a_c H_c} , \qquad (6.150)$$

where all quantities on the right-hand side are to be evaluated at the crossing of the Hubble horizon, when $k = a_c H_c$. But on super-horizon scales we also have $k\tau \ll 1$. In addition, τ may be replaced using Equation (6.142) during the de Sitter phase. Thus, all told,

$$\mathcal{P}_\Theta(k) = \left(\frac{H_c}{2\pi}\right)^2 \left(\frac{H_c}{\dot{\phi}_c}\right)^2 . \qquad (6.151)$$

Introducing Equation (6.104) into this expression gives us the final result,

$$\mathcal{P}_\Theta(k) = \left(\frac{1}{\pi m_{\mathrm{P}}^2}\right)\left(\frac{H_c^2}{\epsilon_c}\right) . \qquad (6.152)$$

This equation suggests that quantum fluctuations generated by a scalar field during slow-roll inflation are scale invariant with $n_s = 1$ (Equation 6.139). In reality, however, H is not completely constant, since $\dot{\phi}$ is not exactly zero, so different modes crossing the Hubble horizon at different times will be subject to slightly different values of a_c and H_c. With H_c decreasing monotonically with time, this results in a scalar spectral index slightly smaller than 1, its exact value depending on ϵ. And if ϵ itself is changing with time, then the running of the spectral index, α_s, is also not zero, meaning that n_s is somewhat dependent on the wavenumber k.

The most precise measurements of the scalar fluctuation spectrum we have to date were made by the *Planck* satellite [29]. They reveal a scalar spectral index, $n_s = 0.9649 \pm 0.0042$, in the power spectrum

$$\mathcal{P}_\Theta^{\mathrm{obs}}(k) = A_s(k/k_*)^{n_s-1} , \qquad (6.153)$$

with a corresponding amplitude

$$A_s(k_*) = (2.1 \pm 0.04) \times 10^{-9} . \qquad (6.154)$$

The pivot scale for these measurements is $k_* = 0.05 \ \mathrm{Mpc}^{-1}$. The agreement between this observed fluctuation spectrum and that expected from slow-roll inflation is one of the most significant successes of the inflationary paradigm. And yet, several problems persist. For example, very little is known about the inflaton field itself and its potential. There is no indication of how the standard model of particle physics should be extended in order to accommodate it. Worse, a third feature of $\mathcal{P}_\Theta(k)$ has now been measured in the *Planck* CMB data, and it does not fit well with the model we have described in these two sections. The data are telling us that $\mathcal{P}_\Theta(k)$ has a non-zero cutoff. How and why this presents a new challenge to inflation will be discussed at length in Sections 11.1 and 11.2. In the meantime, we have one more piece of the puzzle to consider: the metric fluctuations h_{ij} produced in parallel with Θ_k.

Besides the perturbed equation of motion for $\Theta(t,\mathbf{x})$, the use of the linearized FLRW metric in Equation (6.120), with the linearized form of Einstein's Equations (2.167), also produces the dynamics equation for the metric fluctuations h_{ij}:

$$\vec{\nabla}^2 h_{ij} - a^2 \ddot{h}_{ij} - 3a^2 H \dot{h}_{ij} = 0 , \qquad (6.155)$$

i.e., the wave equation, justifying the conventional designation of these tensor modes as *gravity waves*. The coefficients h_{ij} must therefore satisfy several constraints. The tensor they form must be symmetric, transverse and traceless [69]. As these waves propagate, the h_{ij} components fluctuate cyclically, so they may be represented by the decomposition

$$h_{ij}(t, \mathbf{x}) = h(t)e_{ij}(\mathbf{x}) , \qquad (6.156)$$

where $e_{ij}(\mathbf{x})$ is a polarization tensor satisfying the following conditions: $e_{ij} = e_{ji}$ (symmetric); $k^i e_{ij} = 0$ (transverse); and $e_{ii} = 0$ (traceless). Thus, starting with the six independent components of the symmetric tensor h_{ij}, the four traceless and transverse constraints reduce these to only two physical degrees of freedom. As the gravitational wave propagates, it therefore maintains precisely two independent polarizations, usually labeled $\Pi = +$ and $\Pi = \times$. The most general solution to Equation (6.155) may thus be written

$$h_{ij}(t, \mathbf{x}) = \sum_{\Pi=(+,\times)} h^\Pi(t)\, e_{ij}^\Pi(\mathbf{x}) . \qquad (6.157)$$

At this point, we are now faced with a similar hurdle we were confronted with in Equation (6.123) for the scalar fluctuations. As it turns out, Equations (6.123) and (6.155) are actually quite similar, especially if were to write the latter in terms of the conformal time $d\tau = dt/a$ rather than dt. The principal difference between them is the coefficient z'/z, that appears in front of the so-called Hubble frictional term for the scalar fluctuations, but not the tensor modes, which instead contain the factor $-3a^2 H$.

This similarity suggests that we should introduce the Mukhanov-Sasaki variable here as well in order to transform the differential Equation (6.155) into a more manageable form. In this case, however, the definition cannot be exactly the same due to this difference in the Hubble frictional term. With the definition

$$v_{ij} \equiv \frac{a\, m_P}{4\sqrt{2\pi}} h_{ij} , \qquad (6.158)$$

and an analogous expansion in Fourier modes,

$$v_{ij}(t, \mathbf{x}) = \int \frac{d^3\mathbf{k}}{(2\pi)^3} \sum_{\Pi=(+,\times)} v_k^\Pi(t)\, e_{ij}^\Pi(k)\, e^{i\mathbf{k}\cdot\mathbf{x}} , \qquad (6.159)$$

the dynamics equation for h_{ij} reduces to a form identical to Equation (6.132):

$$v_k^{\Pi\,\prime\prime} + \left(k^2 - \frac{2}{\tau^2}\right) v_k^\Pi = 0 , \qquad (6.160)$$

where we have again used the reduction $a''/a = 2/\tau$, valid during the slow-roll approximation (see Equation 6.131).

From here, our derivation of the power spectrum $\mathcal{P}_\Pi(k)$ for each single polarization of the tensor perturbations follows exactly the same reasoning that took us from

Equation (6.142) to (6.151) except, of course, that the coefficient multiplying the Hubble frictional term is different. Thus, we find that

$$\mathcal{P}_\Pi(k) = \left(\frac{H_c}{2\pi}\right)^2 \frac{32\pi}{m_P^2} , \qquad (6.161)$$

which should be compared directly with Equation (6.151). But since there are two independent polarization modes, the total tensor power is twice this value:

$$\mathcal{P}_T(k) = \frac{H_c^2}{m_P^2} \frac{16}{\pi} . \qquad (6.162)$$

The tensor fluctuation spectrum is conventionally normalized relative to the amplitude (Equation 6.154) of the scalar fluctuations, via the *tensor-to-scalar* ratio

$$r \equiv \frac{\mathcal{P}_T(k)}{\mathcal{P}_\Theta(k)} . \qquad (6.163)$$

Very importantly, notice from Equations (6.152) and (6.162) that, for slow-roll inflation,

$$r = 16\epsilon_c , \qquad (6.164)$$

the slow-roll parameter ϵ_c being evaluated when mode $\mathbf{k}$ crossed the Hubble horizon. This ratio therefore provides a powerful observational probe of inflation at, say, the pivot scale k_*, since $\mathcal{P}_T(k_*) \propto H_*^2$. An accurate measurement of $A_s(k_*)$ and $r(k_*)$ can therefore provide a direct indication of the energy scale of inflation, since $H^2 \propto V(\phi)$ (Equation 6.110). These primordial gravitational waves (tensor modes) reveal themselves via a curl-like ('B-mode') pattern in the CMB polarization. Unfortunately, even the latest satellite's (*Planck*) B-mode measurements have been noise and systematics limited, providing only a weak constraint on r. At the 95% confidence level, *Planck's* results show that $r_{0.002} < 0.41$. (The subscript '0.002' here refers to the wavenumber, $k = 0.002$ Mpc^{-1}, at which this observation was made.) As of today, the tightest constraints come from the Keck Array and BICEP2 Collaborations [139], when combined with the *Planck* high-frequency maps to remove the Galactic dust emission. The combined data sets yield the currently best available limit on r, specifically $r_{0.002} < 0.07$ at the 95% confidence level. These results tend to favor so-called *concave* inflaton potentials (with $\partial^2 V(\phi)/\partial\phi^2 < 0$) over *convex* potentials (with $\partial^2 V(\phi)/\partial\phi^2 > 0$), because these generally have smaller values of the inflation parameter ϵ.

It is safe to say, however, that we still have a long way to go before cosmological observations can clarify the issue concerning inflation-induced gravitational waves. At this stage, they may or may not exist, and unfortunately this lack of resolution is coupled to our inability to directly probe the inflaton energy. When we resume this discussion in Chapter 11, we shall see that other high-precision measurements of the CMB have introduced a new challenge to slow-roll inflation, making it even more difficult to put together the pieces we have assembled in this chapter.

For now, this brings us to the end of our brief survey of the standard (ΛCDM) model. There is much, much more to cover on this subject, of course, but our primary goal here has been to highlight the essential elements relevant to the cosmic spacetime—the principal focus of this book. The temperature horizon problem and the generation of a primordial fluctuation spectrum are at the front and center of any concerted effort at finding the metric that correctly yields the cosmic expansion history.

CHAPTER 7

The Gravitational Horizon

I N Chapter 5, we began to explore the properties of the FLRW spacetime viewed from the vantage point of different observers. To be sure, the form of the metric most commonly used in cosmology is rendered in terms of comoving coordinates introduced (separately) by Friedmann, Robertson and Walker. As we have seen, however, otherwise hidden properties of the spacetime emerge when we compare the metric coefficients written in various coordinate systems. This should not be too surprising given that an analogous situation exists in the case of the Schwarzschild metric (Equation 3.19). Consider, for example, that a free-falling observer in that system measures incremental proper distances as dr, compared to the modified form $(1 - r_S/r)^{-1/2}\, dr$ seen by her static (i.e., accelerated) counterpart. The difference depends on the proximity of their location to the Schwarzschild radius r_S (Equation 3.27), asymptotically trending to infinity as $r \to r_S$.

Like the Schwarzschild solution, the FLRW metric is spherically symmetric and it too possesses a horizon with a radius analogous to r_S, though with a somewhat different role. In general relativity, a gravitational horizon—more commonly known as the 'apparent horizon'—is a surface beyond which all null geodesics recede from the observer.[1] As is well known, the gravitational horizon in a Schwarzschild (or Kerr) metric is an event horizon because the spacetime is static (stationary) and r_S is fixed for a given mass. Here, we shall learn that the gravitational (apparent) horizon in the cosmos grows with time. Therefore, it may eventually turn into an event horizon—separating causally connected events from those that are not—depending on our choice of ρ and p in the cosmic fluid, which together determine the expansion factor $a(t)$.

In this chapter, we shall study how and why an apparent (gravitational) horizon manifests itself in the FLRW spacetime and, in one of the most important consequences of the role it plays in cosmology, we shall see in Chapter 8 how it helps us resolve the long-standing debate concerning the origin of cosmological redshift. Specifically, we shall investigate whether this redshift is due to a third form of time dilation, distinct from the better known gravitational and kinematic versions in other applications of general relativity, or whether it is merely a combination of the two,

---

[1] A 'null' geodesic is the trajectory of a massless particle through spacetime (curved or not), analogous to that calculated from the set of differential Equations (2.41) for an object with mass.

often called a 'lapse' function. Later, especially in Chapter 9, we shall learn why it is becoming so pivotal to our correct interpretation of many cosmological observations. In so doing, we shall conclude that it must now be an indispensable ingredient of any fundamental cosmological theory.

7.1 BRIEF HISTORICAL BACKGROUND

The term 'horizon' in cosmology has variously been used to denote (1) the maximum distance particles could have traveled relative to an observer since the Big Bang (the 'particle horizon'); or (2) a two-surface separating causally connected spacetime events from those that are not (the 'event horizon'), or (3) several other definitions with their own customized applications.[2] There is good reason to use them all, though the truth is that some are often applied incorrectly in the context of conformal diagrams, which are more difficult to interpret than concepts described in terms of proper distances and times. Some confusion is also due to the fact that several definitions employ comoving lengths, while others are based on proper distances. Of course, the problem is not the definitions themselves but, rather, possible errors in interpretation. As we shall see in Chapter 9, there is now compelling observational evidence that one definition (the gravitational, or apparent, horizon) stands out above all the others as the most relevant to our correct interpretation of the data. But to better understand where it fits into the pantheon of such definitions, we shall also compare it to some of the other horizons later in this chapter.

Early attempts at describing the nature of cosmological expansion and its consequences (e.g., its impact on the redshift of light reaching us from distant sources) have employed simplified approaches with various levels of success. Today, we have an extensive compilation of observational data that can be used to test our basic theoretical ideas. We introduced several of these data sets in Chapter 6, where we discussed how the current standard model is constructed more or less empirically from them. It is safe to say, however, that regardless of how a cosmological theory is assembled, a 'correct' understanding of the cosmos is best developed through the application of general relativity. This means that coordinate transformations, and their relevance to the spacetime metric, proper distances and times, and the manifestation of the aforementioned horizons due to the finite speed of light, cannot be ignored in the development of a comprehensive cosmological framework.

In pursuit of this goal, Nemiroff and Patla [142] discussed the possible existence of a gravitational horizon in the cosmos based on a 'toy' model of a universe dominated by a single, isotropic, stable, static, perfect-fluid energy (see Equation 2.126) and its associated pressure. They found that the Friedmann and Raychaudhuri equations (4.28, 4.29 and 4.31) allow a maximum scale length over which this energy can have

---

[2]Several horizons were first introduced by Rindler in ref. [140]. In this context, however, ref. [141] was especially helpful because, in avoiding unnecessarily complicated presentations, Ellis and Rothman made it easy to understand how misconceptions often arise from the misinterpretation of coordinate-dependent effects. They carefully differentiated stationary horizons from apparent horizons, extending in a clear, pedagogical manner Rindler's earlier definitions.

a gravitational effect on an object. However, they also concluded (incorrectly, as we shall see) that the observations do not limit this 'gravitational horizon' in our local Universe.

Our handling of the gravitational horizon in this chapter goes well beyond such a simple foray into what is—in reality—a much more elaborate physical process supported by an enormous body of empirical evidence. Our exposition includes the fact that the gravitational horizon in cosmology coincides with what is commonly called an 'apparent' horizon in general relativity.[3] As noted earlier, an apparent horizon separates regions in which null geodesics approach us, from those in which they recede, based on measurements of proper distance. The gravitational origin of an apparent horizon was the subject of Gautreau's study in ref. [144]. Other sources discussing the role of apparent horizons in cosmology include refs. [145, 146, 147]. Gautreau's approach was based on a pseudo-Newtonian description of gravity with a Schwarzschild-like curvature spatial coordinate. He showed that light travels a maximum distance along its geodesic through the Universe, and then turns around and reaches some remote origin. Our analysis below will show that this turning point is closely related to an apparent horizon, but to do so, we shall also demonstrate pedagogically that the best way to understand such features is to invoke and utilize the theorem of George David Birkhoff (1884–1944) (see Figure 7.1) [68]. As demonstrated in Birkhoff's remarkable book, this theorem is an important generalization of Newton's theory couched in the language of general relativity.

We shall see that the gravitational (or apparent) horizon has a radius (denoted $R_{\rm h}$) coincident with that of the better known Hubble sphere. Hubble's law, $\dot{R} = HR$ (see Chapter 6), predicts a radius c/H at which $\dot{R} \to c$. This radius is in fact $R_{\rm h}$, whose very existence is due to the presence of the gravitational horizon. The Hubble radius is simply a manifestation of the latter when the FLRW metric is written using comoving coordinates (Equation 4.19). Ironically, though the notion of a gravitational horizon in cosmology [148] was formally introduced in 2007, it also appeared as an unrecognized distance scale over a century ago in de Sitter's own papers [70, 71] announcing the discovery of his now famous metric (see also Section 5.2). Over the years, the coordinates for which $R_{\rm h}$ appears in de Sitter's metric were gradually abandoned as the use of Friedmann's comoving coordinates became more conventional.

In those rare instances when $R_{\rm h}$ has been discussed in the primary literature, the properties assigned to it were not always accurate, or even correct. One of our goals is to establish as clearly as possible the physical basis underlying its origin. The time-dependent gravitational horizon need *not* be a null surface, though it has often been confused with one. Some have even claimed that sources beyond the gravitational radius measured today are already observable to us now [149, 150, 151, 152], but this is absolutely not true.[4] The confusion stems from the misuse of proper and coordinate speeds in general relativity. Although coordinate speed has no limit, and may exceed c, the proper (or physical) speed is absolutely limited, and its

---

[3]An excellent discussion on the origin and meaning of apparent horizons in general relativity may be found in ref. [143].

[4]Some discussion concerning this topic has appeared in refs. [153, 154, 155].

Figure 7.1 George David Birkhoff (1884–1944), photographed in his Harvard office. Birkhoff was a very influential American mathematician who, in 1923, proved that the spacetime exterior to a spherically-symmetric concentration of mass is correctly given by the Schwarzschild metric, even if this object is expanding or contracting. (Courtesy of Harvard University Press: Birkhoff, George 6, Harvard University Archives)

value must be found using the curvature-dependent metric coefficients. If one ignores this distinction, the outcome can lead to the alarming conclusion that recessional speeds in the cosmos can exceed c, even within an observer's particle horizon [156]. On occasion, this 'superluminal' recession is claimed to contradict special relativity, rationalized with the argument that the limiting speed c is valid only within a non-expanding space, and that superluminal motion is due to the expansion [157, 158]. Such ambiguous, and often incorrect, statements can be easily discounted with the aforementioned distinction between coordinate and proper speeds. As we develop a deeper understanding of the role played by R_h, it will become quite clear that its properties are always fully consistent with such fundamental aspects of the metric in general relativity.

Once our theoretical understanding of R_h is complete, we shall discuss in Chapter 9 a truly amazing coincidence in nature—that the gravitational radius today satisfies the empirical relation $R_h(t) = ct_0$, where t_0 is the age of the Universe [159, 148, 160, 161, 162, 163]. But how can this be true, when the rate at which R_h grows is strongly dependent on the equation-of-state in the cosmic fluid? In the context of black-hole horizons, this result may seem familiar at first, given that an observer falling freely towards such an object sees the event horizon approaching him at speed c. However, R_h in cosmology is an apparent horizon, not an event horizon (at least not yet), so there is no *a priori* reason based on what we have assumed and learned thus far for it to be a null surface. Our principal goal in this chapter is therefore to prepare ourselves for the surprising assessment of the observational data in Chapter 9 by reiterating what the *physical meaning* of R_h is, to understand its properties at a deeper level than ever before, and to ascertain its role in delimiting the visible Universe via the subdivision of null geodesics that can, and cannot, physically reach us today at time t_0.

A very important by-product of this effort, when applied to the debate concerning the nature of cosmological redshift (Chapter 8), will be a resolution of another long-standing debate in cosmology, concerning whether or not space itself is expanding. Those who believe that cosmological redshift is distinct from its gravitational and kinematic counterparts seek to justify this position by conjuring up a mechanism whereby the stretching of wavelengths is due to an actual expansion of space. The other side of the debate has been arguing that such a poorly understood mechanism is probably unphysical. At the very least, they suggest, it would seem to violate the conservation of energy.[5]

7.2 THE BIRKHOFF-JEBSEN THEOREM

A gravitational radius is rather easy to comprehend in the context of a compact object, say a star or black hole, but not so much in cosmology. Since the mass in the former is confined to a finite volume, typically a sphere, and surrounded by vacuum, one can intuitively appreciate how its gravitational influence curves the surrounding

---

[5]A non-exhaustive set of informative papers on this very interesting debate includes refs. [164, 165, 166, 167, 168, 169, 170].

spacetime. A gravitational horizon appears for the observer outside the star when its radius shrinks sufficiently for the escape speed at the surface to equal or exceed the speed of light, c. The Schwarzschild and Kerr metrics describe the spacetime in the vacuum region surrounding such objects. Since they are time independent, their gravitational radii are static and actually define an event horizon. For these environments, the terms gravitational horizon and event horizon are often used interchangeably.

When the standard model is used to interpret the cosmological observations (see Chapter 6), we infer that the Universe is spatially flat (i.e., $k = 0$), or at least very close to it [171]. In this context, the Universe is thus open and will expand forever (because its energy density ϵ is zero; see Section 4.3). It is therefore also infinite, and has been infinite since the very beginning. As observers, we are embedded within it, and do not immediately recognize how to interpret the gravitational influence of the cosmic fluid as a function of distance. Nevertheless, the radius R_h is just as important in the cosmological metric as it is for Schwarzschild or Kerr, but to fully understand this, we need to use the Birkhoff-Jebsen theorem [68] and its corollary. An additional helpful discussion may also be found in refs. [69, 148]. Though this theorem has traditionally been attributed solely to Birkhoff, it was actually 'pre-discovered' by the Norwegian physicist Jørg Tofte Jebsen (1888–1922) (Figure 7.2) several years earlier [172], but his work was recognized only recently.

As a relativistic generalization of Newton's law of universal gravitation applied to spherical systems, the Birkhoff-Jebsen theorem states that the spacetime surrounding such an isotropic distribution of mass-energy—either static or time-dependent—is simply given by the Schwarzschild metric. In cosmology, the corollary provides an additional indispensable feature showing that, in an isotropic Universe, the spacetime curvature a proper distance R away from an observer depends solely on the mass-energy contained within a sphere of radius R centered at her location. The rest of the Universe outside has no influence on the metric within that radius due to complete cancellations as a result of the spherical symmetry.

This concept may be difficult to understand at first, since the Cosmological Principle ensures that the origin of the coordinates may be placed anywhere in the cosmos. Therefore it would seem that a radius R for one observer is different than the radius R' for another. But this argument does not rely on the *absolute* positioning of the origin. What matters is the *relative* spacetime curvature between two points. Each observer infers an identical gravitational influence felt by a test particle the same distance away. This effect may also be characterized as follows: any two points within a medium with non-zero energy density ρ (and/or pressure p) experience a net acceleration (or deceleration) towards (or away) from each other, depending only on how much mass-energy is present between them. This is why the Universe cannot be static, because local motions depend solely on local densities, even in an infinite cosmos. Einstein himself missed this point—and advanced the notion of a cosmological constant in his field Equations (2.171) in order to cancel the attraction due to $T^{\alpha\beta}$ and create a steady-state universe—because his thinking on this subject preceded the work of Birkhoff and Jebsen in the 1920's.

In a homogeneous universe, the observer measures a spacetime curvature at R that increases as this radius grows. Eventually, the spherical volume with radius R

Figure 7.2 An undated family portrait of the Norwegian physicist Jørg Tofte Jebsen (1888–1922) prior to his departure for Italy in 1920. Jebsen was one of the first to seriously work on Einstein's general theory of relativity. His life was cut short due to illness at the age of 33, but he managed to publish several works, including the first derivation (in 1921) of what used to be commonly referred to as the 'Birkhoff theorem' or, should more appropriately be called, the 'Birkhoff-Jebsen theorem.' (Image courtesy of N. Voje Johansen og F. Ravndal, Fra Fysikkens Verden 4, 96-103 (2004); licensed under the Creative Commons to share and distribute the work freely)

will contain sufficient mass-energy to create a gravitational horizon. We shall prove this below using the FLRW metric and demonstrate that the critical radius at which this happens is simply

$$R_{\mathrm{h}} = \frac{2GM}{c^2} , \tag{7.1}$$

where M is the *proper* mass contained within a sphere of proper radius R_{h}, i.e.,

$$M \equiv \frac{4\pi}{3} R_{\mathrm{h}}^3 \frac{\rho}{c^2} . \tag{7.2}$$

In some quarters, this quantity is also known as the Misner-Sharp mass [173]. This quantity is sometimes also referred to as the Misner-Sharp-Hernandez mass, to include the later contribution by Hernandez and Misner [174], which emerged from the pioneering work of Misner and Sharp [173] on spherical collapse in general relativity. Although the cosmological R_{h} may depend on time, it nonetheless defines a gravitational horizon that, at any time t, separates null geodesics approaching us from those receding, as we shall demonstrate more rigorously below.

The simplicity of Equations (7.1) and (7.2)—and the fact that the expression for $R_{\rm h}$ looks identical to $r_{\rm S}$—belies the complexity of identifying the physical mass-energy in a non-asymptotically flat geometry in general relativity [143]. This definition of $R_{\rm h}$, however, is actually required by the form of the metric coefficient g_{rr} in problems of spherically-symmetric expansion or collapse. In such cases, other possible definitions, such as the Hawking-Hayward quasilocal mass [175], also reduce completely to the Misner-Sharp-Hernandez definition. A similar simplification occurs for the Brown-York energy [176], defined as a two dimensional surface integral of the extrinsic curvature on the two-boundary of a spacelike hypersurface referenced to flat spacetime.

In spite of the difficulty in deriving $R_{\rm h}$ when a simplifying symmetry is not available, apparent horizons in general relativity are now very well understood. An essential condition for such a horizon is the aforementioned subdivision of the congruences of outgoing and ingoing null geodesics from a compact, orientable surface. Of course, when the spacetime is spherically symmetric, these are simply the outgoing and ingoing radial null geodesics from a two-sphere of symmetry [145, 146, 147, 143]. For compact objects, such a horizon is often more practical than stationary event horizons because the latter require knowledge of the entire future history of the spacetime in order to be located. These two horizons happen to coincide in the Schwarzschild and Kerr metrics precisely because the former are time-independent in these systems. Apparent horizons are also often preferred in other dynamical situations, such as the generation of gravitational waves during black-hole merging events, where one would need to know how the event unfolds into the distant future in order to find the event horizon in that scenario.

In the cosmological context, the situation is much simpler because the FLRW metric used in all theoretical considerations is always spherically symmetric. Thus, although our derivation of $R_{\rm h}$ below (leading to Equation 7.1) is designed for pedagogy, it turns out to be entirely consistent with all of the formal properties of an apparent horizon. This is because the Misner-Sharp-Hernandez mass is directly related to the apparent horizon in cosmology, thanks to the Birkhoff-Jebsen theorem and its corollary.

To put this another way, we are using the Birkhoff-Jebsen theorem and its corollary to define a 'gravitational horizon' in the FLRW spacetime which, however, turns out to be simply the more broadly defined 'apparent horizon' in general relativity when the metric is spherically symmetric.

At this point, the reader may be wondering why we are investing so much time and effort understanding the nature of $R_{\rm h}$, when so much development has already occurred in cosmology without even mentioning it. (Actually, this is not quite true either since the Hubble radius is often invoked as some generic measure of a 'horizon'—e.g., in inflationary physics—without knowing that it actually represents the gravitational, or apparent, horizon.) With the advent of so-called 'precision cosmology,' we shall see in Chapter 9 that the motivation for better understanding the role played by $R_{\rm h}$ in our interpretation of the data is now greater than ever. Its impact emerged after the optimization of model parameters in the standard model ΛCDM [106, 28, 177] revealed that $R_{\rm h}(t_0)$ equals ct_0 within the measurement error

[159, 148]. As we shall see, this observed equality cannot be a mere 'coincidence,' since it can be achieved only once in the entire (presumably infinite) history of the Universe, making it an astonishingly unlikely event.

Thus, in spite of the fact that the apparent horizon need not be a null surface, it does in fact turn out to be one in the real Universe. There may be several possible explanations for this unexpected constraint but, later in this book, we shall conclude that the simplest (and perhaps most likely) reason is that the Universe's gravitational horizon R_h is *always* equal to ct, and discuss the physics behind this totally unexpected outcome.

7.3   THE (APPARENT) GRAVITATIONAL HORIZON

For the Schwarzschild solution, we first become aware of the gravitational (or event) horizon r_S when we write the metric in terms of the spacetime curvature seen by a static, accelerated observer. This critical radius does not appear in the metric coefficients seen by the free-falling observer. An analogous correspondence exists for the FLRW metric, in which the coordinates (ct, r, θ, ϕ) represent the perspective of a *free-falling* observer (see Equation 4.19), so to see the gravitational horizon in the cosmological spacetime, we similarly must transform the metric into a form relevant to the *accelerated* observer—i.e., one at rest with respect to the source of gravity.

In Chapter 4, we introduced the proper radius, $R(t) \equiv a(t)r$, which is often used to express—not the comoving distance r between two points but, rather—the physical distance that increases as the Universe expands. This definition is actually a consequence of Weyl's postulate [54], which argues that no two worldlines can ever cross following the Big Bang (aside from local peculiar motion that may exist on top of the averaged Hubble flow) in order for the Cosmological Principle (Section 1.2) to be satisfied from one time-slice to the next. This condition requires every distance in an FLRW cosmology to be the product of a constant comoving length r, and a universal function of time, $a(t)$. R is sometimes referred to as the *areal* radius—the radius of a two-sphere of symmetry—defined by the relation $R \equiv \sqrt{A/4\pi}$, in terms of the area A of the two-sphere [178, 179]. When the FLRW metric is written in terms of R, the apparent horizon is located by the constraint $g^{RR} = 0$, which is equivalent to the condition $\Phi = 0$ in Equations (7.7) and (7.13) below.

Starting with the FLRW metric written in comoving coordinates (Equation 4.19), we set $k = 0$ to reflect the fact that the observations indicate the Universe is spatially flat (Chapter 6), and rewrite the expansion factor using the expression

$$a(t) = e^{f(t)} , \tag{7.3}$$

where $f(t)$ is itself a function only of cosmic time t [162]. Thus, putting

$$r = Re^{-f} , \tag{7.4}$$

it is straightforward to see that

$$ds^2 = \left[1 - \left(\frac{R\dot{f}}{c} \right)^2 \right] c^2 \, dt^2 + 2 \left(\frac{R\dot{f}}{c} \right) c \, dt \, dR - dR^2 - R^2 \, d\Omega^2 . \tag{7.5}$$

Here and elsewhere in this book, an overdot signifies a derivative with respect to cosmic time t. If we now complete the square, Equation (7.5) becomes

$$ds^2 = \Phi \left[c\,dt + \left(\frac{R\dot{f}}{c} \right) \Phi^{-1} dR \right]^2 - \Phi^{-1} dR^2 - R^2\, d\Omega^2 \qquad (7.6)$$

where $d\Omega^2$ is given in Equation (4.18). For convenience we have also defined a quantity that appears often in the FLRW metric coefficients:

$$\Phi \equiv 1 - \left(\frac{R\dot{f}}{c} \right)^2 , \qquad (7.7)$$

which happens to be the same function that appeared in Equation (5.10) for the static version of the de Sitter spacetime element. The interval ds may be cast into a more standard form by re-arranging its terms as follows:

$$ds^2 = \Phi \left[1 + \left(\frac{R\dot{f}}{c} \right) \Phi^{-1} \frac{\dot{R}}{c} \right]^2 c^2\, dt^2 - \Phi^{-1} dR^2 - R^2\, d\Omega^2 . \qquad (7.8)$$

The derivative $\dot{R}$ is to be understood as representing the proper velocity along the worldlines of observers having t for their proper time, i.e., the comoving observers.

Those familiar with the Oppenheimer-Volkoff equations for the interior structure of a star [180, 173] will see a great deal of similarity between their metric and Equation (7.8). The principal difference is that Oppenheimer and Volkoff assumed ab initio that the spacetime of a star in equilibrium is static, whereas $\dot{R}$ and $\dot{f}$ in Equation (7.8) are functions of time. Of course, the Universe is expanding so neither R nor a can be static. But this is not to say that the spacetime curvature must also vary with time in an FLRW universe. The six special cases we considered in Section 5.4 clearly demonstrate that, in spite of the Universe's expansion, some equations-of-state do permit a choice of coordinate systems for which the metric coefficients remain constant. For reference, we reiterate that the standard model (Chapter 6) is not a member of this set.

The physical meaning of the factor $\dot{f}/c$ and the function Φ may be understood through the use of the Birkhoff-Jebsen theorem and its corollary. From Equations (7.7) and (7.8), we infer that a threshold distance is reached when $R \to c/\dot{f}$. To see this, recall the dynamical Equations (4.28), (4.29) and (4.31) obtained with the FLRW metric. From Equations (7.1) and (7.2), we see that

$$R_{\rm h}^2 = \frac{3c^4}{8\pi G\rho} , \qquad (7.9)$$

which (with the flat condition $k = 0$ and the Friedmann equations) gives simply

$$R_{\rm h} = \frac{c}{H} = \frac{ca}{\dot{a}} . \qquad (7.10)$$

In other words,

$$R_{\rm h} = c/\dot{f} , \qquad (7.11)$$

consistent with the appearance of $c/\dot{f}$ in Equations (7.7) and (7.8). This expression for R_h is well known in the study of apparent horizons in cosmology [146, 143]. Thus, our transformed FLRW metric in Equation (7.8) may also be written

$$ds^2 = \Phi \left[1 + \left(\frac{R}{R_h}\right)\Phi^{-1}\frac{1}{c}\dot{R}\right]^2 c^2\,dt^2 - \Phi^{-1}dR^2 - R^2\,d\Omega^2\,, \qquad (7.12)$$

in which we now understand that the function

$$\Phi \equiv 1 - \left(\frac{R}{R_h}\right)^2 \qquad (7.13)$$

signals the dependence of the coefficients g_{tt} and g_{RR} on the proximity of the proper distance R to the gravitational radius R_h.

7.4 GEODESICS IN THE FLRW SPACETIME

To understand why the gravitational radius R_h is in fact the apparent horizon, we now consider the worldlines of comoving observers in the FLRW spacetime. From Weyl's postulate, we know that

$$\dot{R} = \dot{a}r\,, \qquad (7.14)$$

which quickly and elegantly reproduces Hubble's law:

$$\dot{R} = \frac{\dot{a}}{a}R \equiv HR\,, \qquad (7.15)$$

in terms of the previously defined Hubble constant H (Equation 4.28).

A particle moving along its geodesic follows the Hubble flow, with proper velocity $\dot{R}$. Therefore, the coefficient g_{tt} in Equation (7.12) simplifies to

$$g_{tt} = \Phi \left[1 + \left(\frac{R}{R_h}\right)^2\Phi^{-1}\right]^2\,, \qquad (7.16)$$

which reduces yet again to an even simpler form

$$g_{tt} = \Phi^{-1}\,. \qquad (7.17)$$

Along a particle worldline, the FLRW metric written in terms of the cosmic time t and the proper radius R is evidently

$$ds^2 = \Phi^{-1}c^2dt^2 - \Phi^{-1}dR^2 - R^2d\Omega^2\,. \qquad (7.18)$$

Further, since

$$dR = c\left(\frac{R}{R_h}\right)dt \qquad (7.19)$$

in the Hubble flow, Equation (7.18) reduces to its final form,

$$ds^2 = c^2 dt^2 - R^2 d\Omega^2 \ . \tag{7.20}$$

As a first application of Equation (7.20), consider an observer watching a particle moving radially with the Hubble flow (i.e., with $d\Omega = 0$). He measures a speed $\dot{R} = HR$ and a spacetime interval

$$ds = c\,dt \ . \tag{7.21}$$

This equation displays the behavior we would expect from our original definition of the coordinates. Notice that, according to Equation (7.21), the cosmic time t is the proper time in the local comoving frame seen by an observer anywhere in the Universe, regardless of the distance R from the origin. The clock with the free-falling observer reveals the local passage of time unhindered by any external gravitational influence and the time dilation it would induce.

Now consider what happens when we instead view the metric applied to a fixed radius, say R_0, with respect to the observer (at the origin). This situation is analogous to an (accelerated) observer measuring distances and times at a fixed radius within the Schwarzschild spacetime. Instead of R being associated with particles (e.g., galaxies) expanding with the Hubble flow, the radius at which the observations are made is now held fixed (i.e., $R \to R_0$) while the particles move through it. By analogy with the Schwarzschild case, we expect that gravitational effects must emerge in the metric since our measurements of distance and time are no longer made within the free-falling frame.

According to Equation (7.12), we have in this case $dR = 0$, so that

$$ds^2 = \Phi_0\, c^2 dt^2 - R_0^2 d\Omega^2 \ , \tag{7.22}$$

and if we again consider only radial motion (with $d\Omega^2 = 0$), then

$$ds^2 = \Phi_0\, c^2 dt^2 \ , \tag{7.23}$$

where $\Phi_0 \equiv 1 - (R_0/R_{\rm h})^2$. This elegant expression reproduces the time dilation we should have expected by analogy with the Schwarzschild metric, in which the flow of time at a fixed proper radius R_0 is no longer the proper time in a local free-falling frame. For any finite spacetime interval ds, we now have $dt \to \infty$ as $R_0 \to R_{\rm h}$, assigning characteristics to $R_{\rm h}$ quite similar to those of $r_{\rm S}$ in the Schwarzschild metric.

An apparent horizon emerges in the FLRW metric when we next consider the null geodesics in this spacetime. The derivative $\dot{r}$ cannot be zero for light, as we can readily verify from the FLRW metric in Equation (4.19). The interval for light satisfies the null condition, $ds = 0$, and therefore a ray of light moving radially in a flat universe obeys the equation

$$c\,dt = \pm a\,dr \ . \tag{7.24}$$

Thus, along an inwardly propagating radial null geodesic, we must have

$$\dot{r} = -\frac{c}{a} \ . \tag{7.25}$$

Letting R_γ designate the proper radius of our light signal, we may then describe its motion relative to an observer at the origin of the coordinates as

$$\frac{dR_\gamma}{dt} = \dot{a}r_\gamma + a\dot{r}_\gamma \,, \tag{7.26}$$

or

$$\frac{dR_\gamma}{dt} = c\left(\frac{R_\gamma}{R_{\rm h}} - 1\right) \,. \tag{7.27}$$

This is the null geodesic equation in an FLRW universe. The minus sign signifies that the ray of light is propagating towards the origin. A plus sign would be relevant for an outwardly directed light signal which, as we shall see later in this chapter, is central to the definition of a 'particle' horizon, distinct from the gravitational and apparent horizons we are concerned with now. It is trivial to confirm that replacing $\dot{R}$ with dR_γ/dt in Equation (7.27), and putting $dR = (dR_\gamma/dt)\,dt$, gives exactly $ds = 0$, as required for a null geodesic.

In our treatment, the behavior of light is always consistent with the properties expected of a null geodesic, regardless of which coordinates we use to write the metric. Moving beyond this straightforward confirmation of anticipated results, we also learn from Equation (7.27) that $dR_\gamma/dt = 0$ when $R_\gamma = R_{\rm h}$. That is, the spatial velocity of light measured in terms of the proper distance per unit cosmic time has two contributions that exactly cancel at the gravitational radius. The first is the propagation speed of light measured in the comoving frame:

$$\frac{dR_{\gamma\,\rm com}}{dt} = -c \,. \tag{7.28}$$

The second is due to the Hubble expansion:

$$\frac{dR_{\gamma\,\rm Hub}}{dt} = c\left(\frac{R_\gamma}{R_{\rm h}}\right) \,. \tag{7.29}$$

Given the definition of $R_{\rm h}$, we could also have simply written $dR_{\gamma\,\rm Hub}/dt = HR_\gamma$.

Further, Equation (7.27) tells us that when $R_\gamma > R_{\rm h}$, the photon's proper distance actually increases away from us, even though the ray of light is pointed in our direction as seen in the comoving frame (indicated by the negative sign in Equation 7.28). When we use its proper distance to locate it within the FLRW spacetime, the photon approaches us only if located within our gravitational horizon at $R_{\rm h}$. But note that since $R_{\rm h}$ changes with time, $\dot{R}_\gamma$ can flip sign as $R_{\rm h}$ overtakes R_γ, or vice versa. One can get a clear visual impression of this feature from Figure 7.3, in which the photon geodesic is shown as a solid, black curve (labeled $R_{\gamma 0}[t]$) and the gravitational radius $R_{\rm h}(t)$ is represented by the dashed curve.

The distinction between light rays approaching us when $R_\gamma < R_{\rm h}$ and receding when $R_\gamma > R_{\rm h}$ is the reason why the gravitational radius $R_{\rm h}$ is in fact an apparent horizon, as conventionally defined in general relativity [145, 146, 147]. From Figure 7.3, we see that a photon emitted at time $t_e < t_0$ beyond our apparent horizon

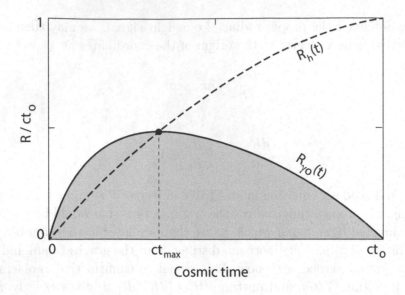

Figure 7.3 Schematic diagram showing the null geodesic (solid black curve) in an FLRW universe reaching us today (at cosmic time t_0), and the gravitational (apparent) horizon R_h. Only photons emitted along this curve can be seen by us at the present time, though there is no restriction on when they were created. This particular illustration is based on the optimized parameters in ΛCDM (see Chapter 6), though the qualitative features are generic to all cosmologies adopting the FLRW metric. Prior to t_{max}, at which the photon reaches its maximum proper distance relative to us, $R_{\gamma 0}$ is greater than R_h, so according to Equation (7.27), it must increase with time—even though the photon's velocity is pointed towards the origin. (Subscript "0" means this is the R_γ reaching us at t_0.) This motion reverses at t_{max}, after which $R_{\gamma 0} < R_h$, so $\dot{R}_{\gamma 0} < 0$. Of course, $R_{\gamma 0} \to 0$ by the time the photon reaches us. (Adapted from ref. [181])

$R_h(t_e)$, which in this figure is anywhere to the left of ct_{max}, begins its journey propagating away from us, yet *stops* at $R_\gamma = R_h$ (black dot), and later reverses direction when R_h has increased faster than R_γ and overtaken it. These are the generic features of the null geodesic in any given FLRW cosmology. The behavior of R_γ, however, is heavily dependent on the equation-of-state, because the expansion dynamics directly affects $\dot{R}_h$.

7.5 PROPER SIZE OF THE VISIBLE UNIVERSE

Unlike the gravitational horizon in the Schwarzschild metric, which is also an event horizon, the gravitational horizon in cosmology generally grows as the Universe expands, exposing in our future previously unseen regions at larger proper distances. The proper size of our visible Universe today, at time t_0, hinges on the solution to Equation (7.27) for a given $R_h(t)$, starting at the Big Bang ($t = 0$) and ending at the present. Different assumptions concerning the equation-of-state in the cosmic fluid produce a variety of expansion scenarios from the solution to the Friedmann and Ray-chaudhuri equations (4.28, 4.29 and 4.31), so the proper size of the visible Universe depends on the cosmological model.

It is determined entirely by the greatest extent achieved in proper distance by those null geodesics that actually reach us at time t_0. Claims that photons may propagate without restriction back and forth across $R_{\rm h}$, thereby determining how far we can see, are therefore naively incorrect [150, 182, 152]. The critical point here is that only the null geodesics actually reaching us can determine the portion of the Universe we can see today. In other words, the gravitational horizon in cosmology is completely observer dependent: light rays that head off to infinity forever remain undetectable by us. We can formalize these general concepts as follows:

Theorem I: In a cosmology expanding monotonically with $\dot{H} \leq 0$ and $\dot{R}_{\rm h} \geq 0$ $\forall\ t \in [0, t_0]$, the proper size of the visible Universe today is always less than or equal to our gravitational horizon, i.e., $R_{\gamma, \max} \leq R_{\rm h}(t_0)$, where $R_{\gamma, \max} \equiv R_{\gamma}(t_{\max})$ (see Figure 7.3).

Proof: All null geodesics satisfy the initial boundary condition $R_{\gamma}(0) = 0$. Null geodesics that reach us must also satisfy the condition $R_{\gamma}(t) \to 0$ as $t \to t_0$. Therefore, $\exists$ a time $t_{\max} \in [0, t_0]$ at which R_{γ} has a turning point, i.e.,

$$\left. \frac{dR_{\gamma}}{dt} \right|_{t_{\max}} = 0 \,. \tag{7.30}$$

According to Equation (7.27), this means that

$$R_{\gamma}(t_{\max}) = R_{\rm h}(t_{\max}) \,. \tag{7.31}$$

But $\dot{R}_{\rm h} \geq 0\ \forall\, t \in [0, t_0]$, while

$$\frac{dR_{\gamma}}{dt} \leq 0 \quad \forall\, t \in [t_{\max}, t_0] \,. \tag{7.32}$$

Therefore $R_{\gamma, \max} = R_{\rm h}(t_{\max}) \leq R_{\rm h}(t_0)$.

To see how this works in practice, let us apply Equation (7.27) to two special cases. The first is the Milne universe (Section 5.1), in which $H(t) = 1/t$ (see Equation 5.2). As we shall see in Chapter 9, this functional form of the Hubble parameter also applies to a cosmological model in which $R_{\rm h}$ is always equal to ct, as the data now seem to suggest [159, 163]. In these cosmologies, the null geodesic equation becomes

$$\frac{dR_{\gamma}}{dt} = \frac{R_{\gamma}}{t} - c \tag{7.33}$$

which, together with the boundary conditions $R_{\gamma}(0) = 0$ and $R_{\gamma}(t_0) = 0$, has the simple solution

$$R_\gamma(t) = ct \ln\left(\frac{t_0}{t}\right) . \tag{7.34}$$

At the turning point, $dR_\gamma/dt = 0$, so $t_{max} = t_0/e$, and therefore

$$R_{\gamma,\,max}^{R_h=ct} = \frac{1}{e} R_h(t_0) \approx 0.37\, R_h(t_0) . \tag{7.35}$$

In the second example, de Sitter space (Section 5.2) has a constant Hubble parameter $H(t) = H_0$, so the gravitational radius $R_h^{\text{de Sitter}} = c/H_0$ never changes. In this case, the solution to Equation (7.27) is

$$R_{\gamma,\,max}^{\text{de Sitter}} = R_h(t_0) . \tag{7.36}$$

The standard model, ΛCDM (see Chapter 6), falls somewhere in between these two cases. In general, we find that $R_{\gamma,max}$ is typically $R_h(t_0)/2$ (except, of course, for de Sitter). The proper size of our visible Universe is therefore about half of our gravitational horizon today. This result is not so difficult to understand because, in all models other than de Sitter, there were no pre-existing sources a finite distance from any observer prior to the Big Bang at $t = 0$. Photons we detect today from the most distant sources could therefore be emitted only after the latter had reached their farthest detectable proper distance from us, which is about half of $R_h(t_0)$.

It has also been suggested that a Universe with phantom energy can violate this observability limit, because some null geodesics in such a cosmology extend into regions exceeding $R_h(t_0)$ [182, 152]. Phantom cosmologies have an equation-of-state $p < -\rho$ [183, 184]. But aside from the fact that phantom cosmologies allow the acausal transfer of energy—and are therefore unlikely to be relevant to the real Universe—this situation requires the existence of null geodesics that never reach us at t_0. They therefore cannot provide an indication of how far we can 'see.' This motivates a second important theorem on this topic:

Theorem II: In spite of the fact that a universe containing phantom energy (i.e., $p < -\rho$) may have a gravitational radius $R_h(t)$ that changes non-monotonically with time, none of the null geodesics reaching us today has ever exceeded our gravitational horizon.

Proof: Let $R_{h,i}^{max}$, $i = 1...N$, denote the N rank-ordered maxima of R_h on the interval $t \in [0, t_0]$, such that $R_{h,1}^{max} \geq R_{h,2}^{max} \geq ... \geq R_{h,N}^{max}$. In addition, let $t_{max} \in [0, t_0]$ be the time at which

$$R_h(t_{max}) = R_{h,1}^{max} . \tag{7.37}$$

Now suppose $R_\gamma(t_{max}) > R_{h,1}^{max}$. In that case, $dR_\gamma/dt \geq 0 \; \forall \; t \in [t_{max}, t_0]$, so $R_\gamma(t_0) \neq 0$, which violates the requirement that photons detected by us today follow null geodesics reaching us at time t_0. In the special case where the Universe is completely dominated by phantom energy, the horizon is always shrinking around the observer, so R_h has only one maximum (at $t_{max} = 0$), and light rays will reach us only if $R_\gamma < R_h$ at the Big Bang.

These two theorems show that no matter how $H(t)$ (and therefore R_h) evolve in time, none of the light we detect today could have originated from beyond our gravitational horizon $R_h(t_0)$. Except for the de Sitter cosmology, in which $R_h(t_0)$ is pre-existing and static (therefore leading to Equation 7.36), all other expanding universes have a visibility limit restricted to about half of the gravitational radius today, even less in some cases (see, e.g., Equation 7.35).

7.6 A COMPARISON OF THE GRAVITATIONAL, PARTICLE AND EVENT HORIZONS

We may now clearly understand the distinction between the gravitational (apparent) horizon R_h and the 'particle' and 'event' horizons in cosmology. The particle horizon, which is often used to establish the size of causally connected regions, is the maximum *comoving* distance covered by a particle from the Big Bang to cosmic time t. It is therefore given by the solution to Equation (7.25), which we shall call

$$r_p(t) = c \int_0^t \frac{dt'}{a(t')} \,. \tag{7.38}$$

The proper distance corresponding to this particle horizon is therefore

$$R_p(t) = a(t) c \int_0^t \frac{dt'}{a(t')} \,, \tag{7.39}$$

and if we now differentiate this equation with respect to t, we get

$$\dot{R}_p = c \left(\frac{R_p}{R_h} + 1 \right) \,. \tag{7.40}$$

To understand how the gravitational and particle horizons are related, we merely need to compare Equation (7.40) with (7.27). The differential equation for R_p represents the propagation of a photon (the null geodesic) *away* from the observer (hence the plus sign), so the solution in Equation (7.39) gives the maximum proper distance a particle could have traveled away from us up to time t.

This quantity is not the same as the maximum proper distance a photon could have traveled in reaching us at t_0, which instead is given by $R_{\gamma 0}(t_{max})$ in Figure 7.3. The most important difference between the two is that, whereas $R_{\gamma 0}(t_{max})$ is always less than $R_h(t_0)$, there is no limit to $R_p(t)$ because the right-hand side of Equation (7.40) is always greater than c. The proper particle horizon R_p can therefore increase past R_h without any problem, especially late in the ΛCDM expansion, when the cosmological constant starts to dominate the energy density, heralding a de Sitter expansion in which both H and R_h approach constant values.

Following our discussion in the previous section, we can now understand why R_h is much more relevant to the observations than R_p. We never again see the photons receding from us, reaching proper distances corresponding to R_p. Unlike R_h, which characterizes the size of the Universe that we can actually see—because the null geodesics that define it terminate at our location—R_p represents null geodesics that extend out to arbitrarily large distances, never again reversing direction to bring any information back to us.

The 'event' horizon represents something different yet again. It is defined to be the largest comoving distance from which light emitted now can ever reach us in the asymptotically infinite future. The corresponding proper distance at time t for this quantity is therefore

$$R_e(t) \equiv a(t)c \int_t^\infty \frac{dt'}{a(t')} \,, \tag{7.41}$$

and if we again differentiate this function with respect to t, we find that

$$\dot{R}_e = c \left(\frac{R_e}{R_h} - 1 \right) . \tag{7.42}$$

At least in form, this expression is identical to Equation (7.27) for R_γ. The physical meaning of R_e is thus similar to that of R_γ except, of course, that R_e represents our horizon for photons reaching us in our future, not today. The gravitational (apparent) radius is not necessarily an event horizon for this reason, because it characterizes only those photons we see that have been propagating solely within our past lightcone.

7.7 GAUGE TRANSFORMATIONS WITH R_H

The interval ds for the FLRW metric in Equation (4.19) has the same form in every cosmological model. The only dependence on the equation-of-state enters through the expansion factor $a(t)$. The remaining issue we need to address is how this form changes when we transform the coefficients from comoving coordinates (ct, r, θ, ϕ) to proper coordinates (ct, R, θ, ϕ). In particular, we seek to understand how R_h is modified from one cosmology to the next.

Our starting point is Equation (7.8) which, however, is still written in terms of the cosmic time t. Sometimes (as in de Sitter space) it is useful to 'complete' the transformation by introducing a new time coordinate T, such that

$$\Phi^{-1/2} \frac{c\,dT}{\eta(t,R)} \equiv \Phi^{1/2} c\,dt + \left(\frac{R}{R_h} \right) \Phi^{-1/2}\,dR \,, \tag{7.43}$$

where Φ is defined in Equations (7.7) and (7.13), and $\eta(t, R)$ is an integrating factor selected to ensure that dT is an *exact* differential [148]. In principle, there could be an infinite number of possible functions $\eta(t, R)$, but only one is dimensionless—ensuring that T has dimensions of time—while completely diagonalizing the metric. Once we make this substitution, the interval becomes

$$ds^2 = \Phi^{-1} \frac{c^2\, dT^2}{\eta^2(t,R)} - \Phi^{-1}\, dR^2 - R^2\, d\Omega^2 \ . \tag{7.44}$$

The guidance we get from requiring dT to be an exact differential is that the new time coordinate must satisfy the following condition:

$$\frac{\partial^2 T}{\partial R\, \partial t} = \frac{\partial^2 T}{\partial t\, \partial R} \ . \tag{7.45}$$

Thus, the function $\eta(t,R)$ must be a solution to the equation

$$\frac{\partial}{\partial R}\left[\Phi\, \eta(t,R)\right] = \frac{\partial}{\partial t}\left[\left(\frac{R}{cR_{\rm h}}\right)\eta(t,R)\right] \ . \tag{7.46}$$

This procedure completes the transformation, though the coordinate T has only limited applicability. It is usually physically impossible to integrate dT from $t = 0$ to the present, as we shall soon see.

7.7.1   de Sitter Space

Earlier in this chapter we pointed out that de Sitter himself published his new metric using a coordinate equivalent to the proper radius R. He was apparently not fully aware of its significance, however. Carrying out the above transformation for this spacetime is therefore a good starting point, providing a rich illustration of all the physics apparent in the FLRW metric written in this coordinate system. For this case,

$$a(t) = e^{H_0 t} \ , \tag{7.47}$$

and so the conventional form of the metric is written

$$ds^2 = c^2\, dt^2 - e^{2H_0 t}\left[dr^2 + r^2\, d\Omega^2\right] \ . \tag{7.48}$$

From Equation (7.3) we infer that

$$f(t) = \ln[a(t)] = H_0 t \ , \tag{7.49}$$

and therefore

$$\dot{f} = H_0 \ . \tag{7.50}$$

According to Equation (7.11), we thus determine that the gravitational radius in this metric is

$$R_{\rm h} = \frac{c}{H_0} \ , \tag{7.51}$$

identical to the Hubble radius identified in Equation (5.14).

de Sitter space is unique among cosmologies because its equation-of-state, $p = w\rho$, with $w = -1$, leads to a constant density ρ, as one can easily confirm from Equation (4.31). Therefore, the gravitational radius $R_{\rm h}$ is fixed in this model for all cosmic time t. This is to be contrasted with all other models with $w > -1$, for which

$\dot{R}_{\rm h} > 0$. The most important consequence of this distinction is that, whereas an observer may choose radii R that are always smaller than $R_{\rm h}$ in de Sitter space, he cannot physically do this when $\dot{R}_{\rm h} > 0$ because, for any given R, he would find that $R_{\rm h} < R$ when $t \to 0$.

As a result, ds is imaginary at early times for all but de Sitter space, so dT cannot be integrated starting at $t = 0$. But for the de Sitter metric, Equation (7.8) tells us that

$$\Phi^{1/2} \, c \, dT \equiv \Phi^{1/2} \left[c \, dt + \left(\frac{R}{R_{\rm h}} \right) \Phi^{-1} \, dR \right] , \tag{7.52}$$

and a comparison with Equation (7.43) immediately shows that the integrating factor $\eta(t, R)$ is here equal to Φ^{-1}. A straightforward integration of Equation (7.52) then gives

$$T(t, R) = t - \frac{1}{2H_0} \ln \Phi , \tag{7.53}$$

which is the coordinate transformation used previously in Equation (5.12) to cast the de Sitter metric into its static form. Let us see how this new time coordinate behaves when the radius is R instead of r. At the origin, $T = t$. Away from the observer, however, T includes an additional redshift effect due to time dilation. And as $R \to R_{\rm h}$, T diverges for a finite ds.

The de Sitter metric written in terms of T and R, shown in Equation (5.15), is one of the most important members of the class of static FLRW spacetimes. We repeat it here for completeness:

$$ds^2 = \Phi \, c^2 \, dT^2 - \Phi^{-1} \, dR^2 - R^2 \, d\Omega^2 . \tag{7.54}$$

This was the form originally published by de Sitter [70, 71].

7.7.2 A Cosmology with $\rho + 3p = 0$

This equation-of-state is highly relevant to the cosmology we shall explore in depth in Chapter 9, which appears to be highly motivated by our ever improving observations. A cosmic fluid with $p = -\rho/3$ produces zero acceleration (see Equation 4.29), so in this universe, the condition $R_{\rm h}(t) = ct$ is strictly maintained for all time.

It is straightforward to see that here

$$\dot{f} = \frac{1}{t} , \tag{7.55}$$

so that

$$ds^2 = \Phi \left[c \, dt + \left(\frac{R}{ct} \right) \Phi^{-1} \, dR \right]^2 - \Phi^{-1} dR^2 - R^2 \, d\Omega^2 , \tag{7.56}$$

with

$$\Phi = 1 - \left(\frac{R}{ct} \right)^2 . \tag{7.57}$$

As with de Sitter space, we confirm that the cosmic time dt diverges for a measurable line element when $R \to R_h$ though, of course, R_h is different in each cosmology.

Completing the exercise, let us also write the metric in terms of R and T. A dimensionless integrating factor that solves Equation (7.46) is

$$\eta(t, R) = \exp\left\{\frac{1}{2}\left(\frac{R}{ct}\right)^2\right\} . \tag{7.58}$$

This leads to the line element

$$ds^2 = e^{-(R/R_h)^2}\Phi^{-1} c^2 \, dT^2 - \Phi^{-1} \, dR^2 - R^2 \, d\Omega^2 . \tag{7.59}$$

Unlike the analogous situation with de Sitter, however, this form of the metric is not usable when $R_h < R$ and $dR = d\Omega = 0$, because in that case $ds^2 < 0$.

7.7.3 Radiation Dominated Universe

At least in the standard model, and possibly other scenarios, radiation dominated the equation-of-state (with $w = w_{\text{rad}} \equiv +1/3$) at the very beginning. The Friedmann Equation (4.28) may be solved for a spatially flat universe, yielding an expansion factor

$$a(t) = (2H_0 t)^{1/2} , \tag{7.60}$$

in which case,

$$f(t) = \ln[a(t)] = \frac{1}{2}\ln(2H_0 t) , \tag{7.61}$$

and

$$\dot{f} = \frac{1}{2t} . \tag{7.62}$$

During this expansion phase of the Universe, the gravitational radius was therefore $R_h = 2ct$. In this case, the proper size of the visible Universe (extending out to ct) never reaches R_h, and the metric is

$$ds^2 = \Phi\left[c\,dt + \left(\frac{R}{2ct}\right)\Phi^{-1}\,dR\right]^2 - \Phi^{-1}dR^2 - R^2\,d\Omega^2 , \tag{7.63}$$

with

$$\Phi = 1 - \left(\frac{R}{2ct}\right)^2 . \tag{7.64}$$

An observer making measurements at a fixed R, with $d\Omega^2 = 0$, sees a gravitationally-induced dilation of dt at progressively larger distances, but this effect never diverges within that portion of the Universe (i.e., $< ct_0$) that remains observable since the Big Bang.

From Equation (7.46), we see that the dimensionless integrating factor is

$$\eta(t, R) = 1 , \tag{7.65}$$

so the transformed metric is

$$ds^2 = \Phi^{-1} c^2 \, dT^2 - \Phi^{-1} \, dR^2 - R^2 \, d\Omega^2 \, . \tag{7.66}$$

This is another example of a universe in which R_h increases with time, so any fixed radius R will have been larger than the gravitational horizon near the beginning, making ds imaginary for $dR = d\Omega = 0$ back then.

7.7.4 Einstein-de Sitter Universe ($p = 0$)

The Einstein-de Sitter universe is spatially flat without a cosmological constant. With the additional assumption that matter is pressureless, so that $w \approx w_{\mathrm{matter}} \equiv 0$, the expansion factor grows according to

$$a(t) = \left(\frac{3}{2} H_0 t \right)^{2/3} . \tag{7.67}$$

Thus,

$$f(t) = \ln[a(t)] = \frac{2}{3} \ln \left(\frac{3}{2} H_0 t \right) , \tag{7.68}$$

and

$$\dot{f} = \frac{2}{3t} \, . \tag{7.69}$$

The gravitational radius in this cosmology is therefore $R_h = (3/2)ct$, and the metric may be written

$$ds^2 = \Phi \left[c\,dt + \left(\frac{2R}{3ct} \right) \Phi^{-1} \, dR \right]^2 - \Phi^{-1} dR^2 - R^2 \, d\Omega^2 \, , \tag{7.70}$$

with

$$\Phi = 1 - \left(\frac{2R}{3ct} \right)^2 . \tag{7.71}$$

As was the case for a radiation-dominated universe, the gravitational radius recedes from us faster than the speed of light. Dilation is present with increasing distance R, but this effect never becomes divergent.

The dimensionless integrating factor in this case is

$$\eta(t, R) = \sqrt{1 + \frac{1}{2} \left(\frac{R}{R_h} \right)^2} \, , \tag{7.72}$$

allowing us to write the metric using the coordinates R and T:

$$ds^2 = \Phi^{-1} \left[1 + \frac{1}{2} \left(\frac{R}{R_h} \right)^2 \right]^{-1} c^2 \, dT^2 - \Phi^{-1} \, dR^2 - R^2 \, d\Omega^2 \, . \tag{7.73}$$

7.7.5 The Standard (ΛCDM) Model

In Chapter 6, we learned that following the radiation-dominated era in the standard model, the expansion has been driven by a combination of matter and a cosmological constant. Given that $\rho_{\text{matter}} \propto a(t)^{-3}$, whereas $\rho_\Lambda = $ constant, the latter emerges more and more at later times. We therefore expect that the Universe should become more R_{h}-delimited as the cosmological constant takes over completely.

Putting $\rho = \rho_{\text{matter}} + \rho_\Lambda$ and $w = w_{\text{matter}} + w_\Lambda = -1$ into the Friedmann and Raychaudhuri equations (4.28) and (4.29), we infer that

$$a(t) = A \sinh^{2/3}\left(\frac{t}{t_\Lambda}\right) . \tag{7.74}$$

The time constant t_Λ subsumes the ratio $\rho_{\text{matter}}/\rho_\Lambda$, which is not known a priori. It is clear from this expression for the expansion factor that $a(t) \to$ constant $\times\, t^{2/3}$ when $t \ll t_\Lambda$, in agreement with Equation (7.67) for a matter-dominated universe, while $a(t) \to$ constant $\times \exp(2t/3t_\Lambda)$, as one would expect from Equation (7.47) for de Sitter space. This limit fixes the time constant in terms of the terminal expansion rate:

$$t_\Lambda \equiv \frac{2}{3H_\infty} , \tag{7.75}$$

where $H_\infty \equiv \lim_{t\to\infty} H(t)$ is the Hubble parameter during the cosmological-constant driven expansion.

We therefore find that

$$f(t) = \ln[a(t)] = \ln(A) + \frac{2}{3} \ln\left[\sinh(t/t_\Lambda)\right] , \tag{7.76}$$

and

$$\dot{f} = \frac{2}{3t_\Lambda \tanh(t/t_\Lambda)} = \frac{H_\infty}{\tanh(3tH_\infty/2)} . \tag{7.77}$$

In this case, the gravitational horizon is

$$R_{\text{h}} = \left(\frac{c}{H_\infty}\right) \tanh\left(\frac{3}{2}tH_\infty\right) , \tag{7.78}$$

so that

$$\dot{R}_{\text{h}} = \frac{3}{2}\left[1 - \tanh^2\left(\frac{3}{2}tH_\infty\right)\right] c . \tag{7.79}$$

The observer does not experience a divergent redshift with increasing R at early times, when $\dot{R}_{\text{h}} > c$, but she will start to see an observational limit at $\sim R_{\text{h}}$, after a 'transition' time t_{trans}, estimated from the constraint

$$ct_{\text{trans}} = R_{\text{h}}(t_{\text{trans}}) . \tag{7.80}$$

The solution to this equation (other than the trivial and irrelevant $t_{\text{trans}} = 0$) is $t_{\text{trans}} \sim 0.86/H_\infty$, not too different from t_Λ. This should not be too surprising, given that t_{trans} signals the point at which the Universe transitions from being matter-dominated to Λ-dominated. In our asymptotic future, the ΛCDM universe will become de Sitter, with $\dot{R}_{\text{h}} \to 0$ and R_{h} settling at the fixed value c/H_∞.

CHAPTER 8

Cosmological Redshift

T HE deeper appreciation we have developed for the gravitational horizon in the previous chapter will serve us well in resolving an interesting and important debate concerning the nature of cosmological redshift. Much of what we know about the expansion of the Universe is based on our observation of shifts in frequency ν of the light emitted at various distances. This effect is conventionally expressed in terms of the so-called redshift parameter

$$z \equiv \frac{\lambda_0 - \lambda_e}{\lambda_e}, \tag{8.1}$$

characterizing the fractional increase in wavelength between the emitter (labeled 'e') and observer (labeled '0'). Since $\lambda_0/\lambda_e = \nu_e/\nu_0$, the redshift equation may also be written

$$1 + z = \frac{\nu_e}{\nu_0}. \tag{8.2}$$

Vesto Melvin Slipher (1875–1969) (see Figure 8.1) is credited with having discovered the first evidence of a systematic redshift in the spectrum of radiation from distant sources, with an observing program at the Lowell Observatory 24-in. refractor in Flagstaff, Arizona. He observed 41 spiral nebulae from 1910 to the mid 1920's, reporting in 1922[1] that he had measured in 36 of them absorption lines shifted by up to $z \sim 0.006$. The other five, including the Andromeda galaxy, showed blue shifted lines, with $|z| \lesssim 0.001$.

This imbalance in the number of redshifted versus blueshifted sources argued against an initial interpretation at that time that the frequency shifts were merely due to the Doppler (or 'kinematic') effect associated with the motion of our solar system. The preponderance of redshifts from all directions in the sky gave credence to a new, alternative explanation—that these shifts were mostly due to a general recession of the nebulae away from Earth [186]. Within just a few more years, Wirtz [187] and Lundmark [188] had (separately) demonstrated that Slipher's redshifts were correlated with distance, implying that the farther the galaxies were from us, the

---

[1]These were listed in a table prepared for *The Mathematical Theory of Relativity* by Eddington (1924), p. 162 [185].

Figure 8.1 Vesto Slipher photographed while still a student (circa 1905), prior to joining the staff at Lowell Observatory in Flagstaff, Arizona. The 41 spiral nebulae he would observe over the next decade provided the first evidence that our Universe is expanding. (Courtesy AIP Emilio Segre Visual Archives)

faster they were receding. And Edwin Hubble's (Figure 8.2) announcement in 1929 that these velocities and distance satisfied a roughly linear relation [59], confirmed in most people's minds that this redshift must be due to a cosmological Doppler effect.

8.1 DEFINITION

But though a Doppler interpretation for z works reasonably well for nearby galaxies, it fails progressively more and more with increasing distance because, as we now understand it, Hubble's law is written in terms of the proper—not comoving—distance (see Equation 7.15). To see the impact of this distinction, consider an electromagnetic signal propagating radially towards us. The null condition for this ray gives

$$0 = c^2\, dt^2 - a(t)^2 \frac{dr^2}{1 - kr^2} \tag{8.3}$$

(for this discussion, it doesn't matter whether the Universe is flat or not, so we shall keep the spatial curvature constant k). Suppose the crest of a wave is emitted by a galaxy at comoving distance r_e, at time t_e. It will reach us at a later time t_0, given by the equation

$$f(r_e) = c \int_{t_e}^{t_0} \frac{dt}{a(t)}\,, \tag{8.4}$$

where

$$f(r_e) \equiv \int_0^{r_e} \frac{dr}{\sqrt{1 - kr^2}} = \begin{cases} \sin^{-1} r_e & k = +1 \\ r_e & k = 0 \\ \sinh^{-1} r_e & k = -1 \end{cases}. \tag{8.5}$$

Ignoring the peculiar motion of galaxies relative to the Hubble flow, we know that $f(r_e)$ is independent of time. Thus, if the next wave crest is emitted at r_e, at time $t_e + \delta t_e$, it will be detected by the observer at time $t_0 + \delta t_0$, given by the analogous relation to Equation (8.4):

$$f(r_e) = c \int_{t_e + \delta t_e}^{t_0 + \delta t_0} \frac{dt}{a(t)}\,. \tag{8.6}$$

If we now subtract Equation (8.4) from Equation (8.6), and use the fact that $a(t)$ changes very little during the increments in time δt_e and δt_0, we estimate that

$$\frac{\delta t_0}{a(t_0)} = \frac{\delta t_e}{a(t_e)}\,. \tag{8.7}$$

Therefore, since $\nu_e \propto 1/\delta t_e$ and $\nu_0 \propto 1/\delta t_0$, we see that

$$\frac{\nu_0}{\nu_e} = \frac{a(t_e)}{a(t_0)}\,. \tag{8.8}$$

Thus, one may calculate the cosmological redshift solely from the expansion factor, according to the very simple relation

Figure 8.2 Edwin Hubble at Caltech in 1931. His observations on the 100-inch Hooker telescope at Mount Wilson Observatory in the 1920's confirmed the view that 'nebulae' were external galaxies and supported the idea, based on Slipher's earlier observations, that their redshifts were due to a cosmological Doppler effect. Hubble discovered a linear relationship between their distance and radial velocities, which he published in 1929. This correlation became known as Hubble's law. (Courtesy of the Huntington Library, San Marino, California. HUB 1033/5, Edwin Powell Hubble papers)

$$1 + z = \frac{a(t_0)}{a(t_e)} \, . \tag{8.9}$$

The Weyl representation of proper distance, i.e., the product of a universal expansion factor $a(t)$ independent of position and a set of constant comoving coordinates, is often interpreted as evidence that space itself is dynamic, expanding with time. But this view is not accepted by everyone, chiefly because the distinction between a situation in which galaxies are fixed within an expanding space, and the alternative in which the galaxies move through a fixed space, is far more than merely semantic. Some question whether atoms can even be stable if space is expanding. At an even more fundamental level, energy cannot be conserved if the motion of particles is driven by the expanding vacuum. These are but two of the many issues that proponents of each side strongly contest in the long standing debate on this question, beautifully summarized in the work of Chodorowski [165, 166].

Underlying much of this discussion is in fact the origin of cosmological redshift itself [170, 168, 169]. It is the formulation in Equation (8.9), in particular, that seems to demonstrate that z is due to this stretching of space because it does not look like any of the classical sources of redshift we already know about, specifically, the Doppler (kinematic) effect and gravitational time dilation. This equation in fact suggests that the wavelength changes as a result of the universal expansion of space, which grows in proportion to $a(t)$.

In this chapter, we shall confront this issue with the new insights we have developed in Chapter 7, and answer this pivotal cosmological question. We seek to uncover whether the different formulation in Equation (8.9)—and therefore its interpretation—is really due to new physics, or whether it is merely a consequence of our choice of coordinates. In other words, is it possible to use another set of coordinates for which the cosmological redshift is then more like the 'traditional' *lapse function* in other applications of general relativity?

But resolving this important debate is quite difficult, as others have already discovered. Fortunately, we now have at our disposal the six static FLRW solutions (see Section 5.4) which, as we shall see, are crucial to finding the cosmological lapse function. In the following sections, we shall attempt to use these metrics with constant spacetime curvature to prove that z can be calculated—with equal validity—either from the conventional relation in Equation (8.9) involving the expansion factor $a(t)$, or from the well-known Doppler effect and gravitational time dilation more commonly found in other applications of general relativity. We shall do this by using the FLRW metric written both in terms of the original comoving coordinates (Equation 4.19), and its counterpart (Equation 7.12) expressed in terms of the coordinates (cT, R, θ, ϕ), for which the metric coefficients $g_{\mu\nu}$ are independent of time T.

8.2 REDSHIFT IN STATIC SPACETIMES

The motivation for the approach we shall follow here is that energy is conserved along a geodesic in any metric with constant spacetime curvature. A formal proof of this statement is based on the fact that, if a Killing vector field exists on a Riemannian manifold to preserve the metric, it generates a symmetry with its own unique conservation law [189]. The most useful aspect of Killing vectors is that their existence implies conserved quantities associated with the motion of free particles [190, 191]. In this context, if the metric admits a *timelike* Killing vector, it can be written in static form, i.e., a coordinate system can be found such that the metric coefficients are *time-independent* in that system. Most importantly, the conserved quantity associated with a timelike Killing vector is the energy.

If we know the gravitationally-induced time dilation at the source, then the redshifted energy of the photons is established relative to the observer at the time they are emitted. In a static spacetime, this energy is subsequently fixed as the radiation propagates along its null geodesic towards its ultimate detection at t_0, allowing the observer to separate the gravitational and Doppler components of redshift.

Our approach for finding the cosmological redshift as a lapse function therefore involves three essential steps. First, we identify the FLRW metrics that admit a timelike Killing vector field. These are simply the six special cases we considered in Section 5.4, which Florides (1980) proved were the only static FLRW cosmologies. In order to identify the gravitationally-induced time dilation at the source, we must use the transformed metric written in terms of the coordinates (cT, R, θ, ϕ), not Equation (4.19), in which the coefficients g_{ii} (with $i = 1, 2, 3$) are clearly time dependent. Second, we use the transformed metric to calculate the time dilation at the emitter's location relative to the proper time in a local free-falling frame. Finally, we include the kinematic contribution to the redshift by finding the *apparent* time dilation, which differs from its counterpart at the emitter's location because the relative arrival times of the photon's wave crests are altered by the motion of the source. Steps two and three are quite standard in relativity [69]. The first step we are introducing here has, to our knowledge, not been applied before [192].

It is important to stress, however, that our choice of static FLRW metrics has nothing to do with whether or not the spacetime curvature is zero. It is, in fact, not zero in four out of the six cases. This is a crucial point because the cosmological redshift is therefore not just a Doppler effect, as was surmised early on; it is typically a combination of both Doppler and gravitational effects, constituting a true lapse function like that in other applications of general relativity. By no means do static FLRW metrics simplify the redshift by eliminating one or more contributors. Gravitational effects are present even when the FLRW metric is time-independent, as is well known from the Schwarzschild and Kerr spacetimes.

8.2.1 Minkowski Spacetime

The metric in Equation (5.11) is already in static form. Quite trivially, we see from Equation (8.9) that $z = 0$ everywhere. And since there is no expansion or spacetime curvature, the Doppler and gravitational redshifts are also trivially zero in this case, so the lapse function is 1, which confirms that $z = 0$.

8.2.2 The Milne Universe

The static form of the metric in this model is given in Equation (5.6), written in terms of the coordinates (cT, R, θ, ϕ). To evaluate the time dilation in this system, we assume only radial motion, i.e., $d\theta = d\phi = 0$. In the comoving frame, the cosmic time t is also the proper time. Thus, for an interval growing solely with time, i.e., $ds^2 = c^2 dt^2$, we have from Equation (5.6)

$$\frac{dt}{dT} = \left[1 - \frac{1}{c^2}\left(\frac{dR}{dT}\right)^2\right]^{1/2}. \tag{8.10}$$

This time dilation is evaluated at the source when the light was emitted (at time t_e). It is not necessarily equal to the *apparent* time dilation, however. These two are equal only if the source was instantaneously at rest with respect to us at t_e. If the source was moving (as is the case here), successive wave fronts were separated by a time dT in Equation (8.10) but, during this interval, the proper distance measured (in the $R - T$ frame) to the source also increased by an amount $v_R\sqrt{g_{TT}}\, dT$, where

$$v_R \equiv \sqrt{\frac{g_{RR}}{g_{TT}}}\frac{dR}{dT} \tag{8.11}$$

is the proper distance per unit proper time measured along the line-of-sight.

Thus, as we typically find in such a situation, the ratio ν_0/ν_e is given by the expression

$$\frac{\nu_0}{\nu_e} = \left(1 + \frac{v_R}{c}\right)^{-1}\frac{dt}{dT}\bigg|_{T_e}. \tag{8.12}$$

Note that the quantities on the right-hand side of this equation formally must be evaluated at the time, T_e or, equivalently, t_e, when the light was emitted. The expansion velocity v_R at a fixed comoving distance χ is trivially constant for the Milne universe. This criterion will be much more important for the curved spacetimes we shall consider in the following subsections.

Let us now place the source within the Hubble flow. Then, with $dr = 0$, we find that

$$\frac{dR}{dT} = \frac{\partial R}{\partial t}\frac{dt}{dT} = c \tanh\chi. \tag{8.13}$$

Thus, Equations (8.10) and (8.13) imply that

$$\frac{dt}{dT} = \frac{1}{\cosh\chi}. \tag{8.14}$$

In the Milne cosmology $g_{RR} = g_{TT} = 1$, so the apparent frequency shift is

$$\frac{\nu_0}{\nu_e} = (1 + \tanh\chi)^{-1}\cosh^{-1}\chi$$

$$= e^{-\chi}. \tag{8.15}$$

Our procedure of using the lapse function to find z therefore suggests that the cosmological redshift should be

$$1 + z \equiv \frac{\nu_e}{\nu_0} = e^\chi . \tag{8.16}$$

But how does this compare with the conventional result using Equation (8.9)? According to the latter, the frequency shift should be given solely in terms of the universal expansion between the emission (t_e) and observation (t_0) times.

The light signal begins propagating away from the source at time t_e and travels along a null geodesic ($ds = 0$) until it reaches the observer at time t_0. Therefore, according to Equation (5.4) with $d\Omega = 0$, we see that

$$\int_0^\chi d\chi' = \int_{t_e}^{t_0} \frac{dt'}{t'} . \tag{8.17}$$

Note that the minus sign cancels in this expression because the photons are approaching us. That is,

$$\chi = \ln \left(\frac{t_0}{t_e} \right) . \tag{8.18}$$

The cosmological redshift we calculate from Equation (8.9) is therefore

$$\begin{aligned} 1 + z &= \frac{a(t_0)}{a(t_e)} = \frac{t_0}{t_e} \\ &= e^\chi , \end{aligned} \tag{8.19}$$

fully consistent with the result we derived in Equation (8.16) through a consideration of the time dilation between the two frames (Equation 8.10) and its subsequent modification as a result of the shift in arrival times (Equation 8.12). The cosmological redshift in the Milne universe may be calculated either from the expansion factor $a(t)$, or directly from the more conventional approach used in other applications of general relativity *that does not involve the assumption of an expanding space.*

8.2.3 de Sitter Space

This is the first of the six static FLRW metrics we shall encounter that has a constant—*though nonzero*—curvature. Whereas the Milne universe describes a flat universe with no gravitational acceleration, de Sitter space has $\rho \neq 0$. Objects within it not only recede from one another, but also accelerate under the influence of gravity. Thus, the inherent curvature in de Sitter's spacetime provides us with an important validation of our method, complementary to the Minkowski and Milne examples.

As we saw in Section 5.2, de Sitter space corresponds to a universe filled with a cosmological constant with an equation-of-state $p = -\rho$. The FLRW metric in this case is given in Equation (5.7). But to calculate the time dilation between a source at proper distance R and us, we must first rewrite the metric in its static form, which is given in Equation (5.15). The time dilation is then given as

$$\frac{dt}{dT} = \left[\Phi - \frac{1}{c^2}\Phi^{-1}\left(\frac{dR}{dT}\right)^2\right]^{1/2} , \qquad (8.20)$$

where we have again assumed that the source moves with the Hubble flow, and the function Φ is defined in Equation (5.13). Since r is then constant, we may also write

$$\frac{dR}{dT} = \frac{\partial R}{\partial t}\frac{dt}{dT} = \dot{a}r\frac{dt}{dT} = HR\frac{dt}{dT} . \qquad (8.21)$$

Equations (5.12), (8.20), and (8.21) therefore suggest that

$$\frac{dt}{dT} = \Phi . \qquad (8.22)$$

This time dilation includes *both* the effects of gravity and the effects due to the Hubble recession of the source. As before, however, this expression does not represent what is seen by the observer. To infer the shift in frequency of the light, we must first find the *apparent* time dilation, analogous to Equation (8.12).

Now, the proper velocity of the source along our line-of-sight is

$$v_R \equiv \sqrt{\frac{g_{RR}}{g_{TT}}}\frac{dR}{dT} . \qquad (8.23)$$

Neither g_{RR} nor g_{TT} are equal to 1 in de Sitter space. Thus, putting everything together, we see that

$$\frac{\nu_0}{\nu_e} = \left(1 + \frac{v_R}{c}\right)^{-1}\frac{dt}{dT}\bigg|_{T_e}$$
$$= \left(1 + \frac{R(T_e)}{R_{\rm h}}\right)^{-1}\Phi(T_e) , \qquad (8.24)$$

and therefore in de Sitter space

$$1 + z \equiv \frac{\nu_e}{\nu_0} = \left[1 - \frac{R(T_e)}{R_{\rm h}}\right]^{-1} . \qquad (8.25)$$

According to Equation (8.9), this redshift should be equal to $a(t_0)/a(t_e)$, so let us see if this is indeed the case.

Integrating along a null geodesic from t_e to t_0, we find that

$$\int_0^r dr' = c\int_{t_e}^{t_0}\frac{dt'}{\exp\left(Ht'\right)} , \qquad (8.26)$$

whose solution gives

$$r = \frac{c}{H}\left(e^{-Ht_e} - e^{-Ht_0}\right) . \qquad (8.27)$$

The proper distance at the time of emission is therefore

$$R(t_e) = a(t_e)r = \frac{c}{H}\left(1 - e^{-H(t_0-t_e)}\right) , \qquad (8.28)$$

and the redshift is thus

$$1 + z = \frac{a(t_0)}{a(t_e)} = e^{H(t_0-t_e)} = \left[1 - \frac{R(T_e)}{R_{\mathrm{h}}}\right]^{-1} , \qquad (8.29)$$

which agrees exactly with the result in Equation (8.25). Just as we concluded in the Milne universe, the cosmological redshift in de Sitter space may be calculated either from the expansion factor $a(t)$, or from the time dilation and frequency shift associated with motion of the source. The de Sitter spacetime is clearly more important than the previous two cases we considered because it represents a situation in which the redshift is due to *both* gravitational and kinematic effects. The bottom line is that, in all three cosmologies we have examined thus far, the *interpretation* of redshift as an expansion of space is dependent upon the coordinates we choose to calculate it.

8.2.4   The Lanczos Universe

The static form of the metric describing the gravitational field due to a rigidly rotating dust cylinder coupled to a cosmological constant is given in Equation (5.20). In terms of its metric coefficients, the time dilation is given by the expression

$$\frac{dt}{dT} = \left\{\left[1 - \left(\frac{R}{cb}\right)^2\right] - \frac{1}{c^2}\left[1 - \left(\frac{R}{cb}\right)^2\right]^{-1}\left(\frac{dR}{dT}\right)^2\right\}^{1/2} . \qquad (8.30)$$

But within the Hubble flow, we have $dr = 0$ (and therefore $dr/dT = 0$), so

$$\frac{dR}{dT} = cr\sinh(t/b)\frac{dt}{dT} . \qquad (8.31)$$

Thus, from Equations (8.30) and (8.31) we find that

$$\frac{dt}{dT} = \frac{1 - r^2\cosh^2(t/b)}{\sqrt{1 - r^2}} \qquad (8.32)$$

consistent also from the direct use of Equation (5.19). Thus, the ratio of the frequency of light actually measured by the observer to that emitted is given by Equation (8.12), but now with

$$v_R \equiv \sqrt{\frac{g_{RR}}{g_{TT}}}\frac{dR}{dT} = \frac{cr\sinh(t/b)}{\sqrt{1 - r^2}} . \qquad (8.33)$$

The ratio of frequencies is therefore

$$\frac{\nu_0}{\nu_e} = \frac{1 - r^2\cosh^2(t/b)}{\sqrt{1 - r^2} + r\sinh(t/b)}\bigg|_{T_e} , \qquad (8.34)$$

and the redshift in this cosmology is given by

$$1 + z = \frac{\nu_e}{\nu_0} = \left. \frac{\sqrt{1 - r^2} + r \sinh(t/b)}{1 - r^2 \cosh^2(t/b)} \right|_{T_e}. \tag{8.35}$$

Let us now compare this expression with the result we would have obtained from Equation (8.9). The geodesic equation describing the trajectory of a light signal from its emission at comoving distance r_e at time t_e to the observer at $r = 0$ and time t_0, derived from Equation (5.17) is

$$\int_0^{r_e} \frac{dr}{\sqrt{1 - r^2}} = \int_{t_e/b}^{t_0/b} \frac{du}{\cosh(u)}. \tag{8.36}$$

Its solution is

$$\sin^{-1}(r_e) = 2 \tan^{-1}\left(e^{t_0/b}\right) - 2 \tan^{-1}\left(e^{t_e/b}\right), \tag{8.37}$$

so that

$$r_e = 2 \sin\left[\tan^{-1}\left(e^{t_0/b}\right) - \tan^{-1}\left(e^{t_e/b}\right)\right] \cos\left[\tan^{-1}\left(e^{t_0/b}\right) - \tan^{-1}\left(e^{t_e/b}\right)\right]. \tag{8.38}$$

After some algebra, using the identities $\sin\left(\tan^{-1}[x]\right) = x\left(1 + x^2\right)^{-1/2}$ and $\cos\left(\tan^{-1}[x]\right) = \left(1 + x^2\right)^{-1/2}$, we find that

$$r_e = 2\left(e^{t_0/b} - e^{t_e/b}\right)\left(1 + e^{(t_0 + t_e)/b}\right)\left(1 + e^{2t_0/b}\right)^{-1}\left(1 + e^{2t_e/b}\right)^{-1}. \tag{8.39}$$

Further lengthy algebraic manipulations following the substitution of this expression into Equation (8.35) produce the final result:

$$1 + z = \frac{\cosh(t_0/b)}{\cosh(t_e/b)}, \tag{8.40}$$

in complete agreement with Equation (8.9) when the expansion factor is $a(t) = (cb)\cosh(t/b)$.

8.2.5 A Lanczos Universe with $k = -1$

The application of our procedure for finding the lapse function in this universe is very similar to the one we employed for the Lanczos model in Section 8.2.4. The time dilation here is also given by Equation (8.30) though, of course, R and T are now given in Equation (5.23). Thus, we now write

$$\frac{dR}{dT} = cr \cosh(t/b)\frac{dt}{dT}, \tag{8.41}$$

and

$$\frac{dt}{dT} = \frac{1 - r^2 \sinh^2(t/b)}{\sqrt{1 + r^2}}. \tag{8.42}$$

Then, the proper velocity is

$$v_R \equiv \sqrt{\frac{g_{RR}}{g_{TT}}} \frac{dR}{dT} = \frac{cr \cosh{(t/b)}}{\sqrt{1+r^2}} , \tag{8.43}$$

and the redshift may be written

$$1+z = \frac{\nu_e}{\nu_0} = \left. \frac{\sqrt{1+r^2} + r\cosh(t/b)}{1 - r^2 \sinh^2(t/b)} \right|_{T_e} . \tag{8.44}$$

We complete the comparison by calculating $r_e = r(t_e)$ from the geodesic equation

$$\int_0^{r_e} \frac{dr}{\sqrt{1+r^2}} = \int_{t_e/b}^{t_0/b} \frac{du}{\sinh(u)} , \tag{8.45}$$

whose solution is

$$\sinh^{-1}(r_e) = \ln{(\tanh{[t_0/2b]})} - \ln{(\tanh{[t_e/2b]})} . \tag{8.46}$$

The comoving distance to the emitter is therefore

$$r_e = \frac{1}{2} \left(\frac{\tanh(t_0/2b)}{\tanh(t_e/2b)} - \frac{\tanh(t_e/2b)}{\tanh(t_0/2b)} \right) . \tag{8.47}$$

Following the substitution of this expression into Equation (8.44), another lengthy algebraic manipulation produces the result

$$1+z = \frac{\sinh{(t_0/b)}}{\sinh{(r_e/b)}} , \tag{8.48}$$

which agrees completely with Equation (8.9) when we use the expansion factor appropriate for this metric (Equation 5.22).

8.2.6 Anti-de Sitter Space

In this sixth static FLRW metric, the proper velocity is given by the expression

$$v_R = \frac{cr \cos(t/b)}{\sqrt{1+r^2}} , \tag{8.49}$$

and the time dilation relation is

$$\frac{dt}{dT} = \frac{1 + r^2 \sin^2{(t/b)}}{\sqrt{1+r^2}} . \tag{8.50}$$

The redshift is therefore

$$1+z = \frac{\nu_e}{\nu_0} = \left. \frac{\sqrt{1+r^2} + r\cos(t/b)}{1 + r^2 \sin^2(t/b)} \right|_{T_e} . \tag{8.51}$$

The comoving distance r_e of the source at the time t_e the light was emitted may be found via the integral expression

$$\int_0^{r_e} \frac{dr}{\sqrt{1+r^2}} = \int_{t_e/b}^{t_0/b} \frac{du}{\sin(u)} \, , \tag{8.52}$$

with the solution

$$r_e = \frac{1}{2} \left(\frac{\tan(t_0/2b)}{\tan(t_e/2b)} - \frac{\tan(t_e/2b)}{\tan(t_0/2b)} \right) . \tag{8.53}$$

Thus, substituting this expression for r_e into Equation (8.51), we calculate the redshift to be

$$1 + z = \frac{\sin(t_0/b)}{\sin(t_e/b)} \, . \tag{8.54}$$

Once again, a comparison with Equation (8.9), in which the appropriate expansion factor for this metric is used (Equation 5.26), shows complete agreement.

8.3 EXPANSION OF SPACE

There are several very interesting conclusions we can draw from these six static FLRW spacetimes. These are the cosmological models with an equation-of-state that may or may not produce an expansion, sometimes accelerated, but always with the flexibility of allowing us to choose an appropriate set of coordinates such that the spacetime curvature is constant. This has been necessary in our procedure for determining the redshift because, in these static metrics, the gravitational redshift we calculate when the radiation was emitted remains constant along null geodesics. We have demonstrated in every case that, with this set of coordinates, the redshift has precisely the same contributions—Doppler and gravitational shifts—that we expect from the calculation of the lapse function in other applications of general relativity.

The interpretation of z as due to the 'stretching of light' in an expanding space is therefore coordinate dependent. In general relativity, an effect is physically real only if it is *independent* of the coordinate system. At least for the static FLRW metrics, we have therefore proven that the cosmological redshift is not due to a new type of time dilation, distinct from the more traditional, gravitational and kinematic ones. But this must be true in general, not solely for the static FLRW metrics, because Equation (8.9) for the redshift is universal; it is valid for any expansion factor $a(t)$, regardless of whether or not it produces a constant spacetime curvature.

It is true, however, that determining z for the FLRW spacetimes other than this group of six is complicated by the fact that the gravitational redshift calculated at the emission point does not remain constant as the radiation propagates towards the observer. Thus, in order to find the lapse function in such cases, one must not only calculate the kinematic shift due to the motion of the source, but also incorporate the change in energy of the radiation as the spacetime curvature evolves in time. Demonstrating that z is a lapse function even for the time-dependent FLRW metrics is therefore quite challenging, but it should be done. In this context, it may be helpful to consider the step taken by Mizony and Lachiéze-Rey [193], in which any FLRW metric may be transformed into a local static form which, interestingly, is equivalent to de Sitter in this limited domain. Following their approach, one may possibly be able to determine the change in gravitational redshift sequentially from one local static frame to the next.

Given that the formulation of cosmological redshift in terms of the expansion factor is identical in all cases suggests that our ability to calculate z using this simple equation is due to the high degree of symmetry in the FLRW metric. It allows us to bypass the more difficult task of finding z via the lapse function, which involves several additional steps. But this then raises a critical followup question. If there really were a third mechanism producing a redshift, beyond Doppler and gravity, why do we not see it manifested in the static FLRW spacetimes? Let us remember that FLRW spacetimes with constant spacetime curvature also satisfy Equation (8.9), just like all the others. And if Equation (8.9) is evidence that z is due to the stretching of light in an expanding space, this process should be happening regardless of whether or not the metric is static—and apparently it is not.

So our conclusion from the analysis in this chapter is that the evidence in favor of an expanding space is non-existent. There are several reasons why the distinction between an 'expanding space' and a 'fixed space' through which galaxies move is dynamically important. We have already discussed several of these earlier in this chapter. Another critical one has to do with whether or not signals can be transported superluminally across the Universe. It is not uncommon for us to read statements to the effect that light in cosmology can be carried across vast distances faster than c. Such claims are justified on the basis that the speed of light is restricted to c only in an inertial frame. If space is expanding, the argument goes, then light can be transported along with the expansion at arbitrarily high speeds, unrelated to c. But it is not at all difficult to understand that such notions arise from the improper use of coordinates.

In general relativity, the 'physical' velocity measured by an observer is the proper velocity (Equation 8.11), calculated using the proper distance and proper time. A null geodesic is defined by the condition $ds = 0$, so v_R for light is *always* equal to c, regardless of which coordinate system is being used, inertial or otherwise. The speed dR/dt, however, is *not* restricted to c. But this quantity is not the proper speed measured by a single observer using solely her rulers and clocks. As we have seen, the integrated proper radius $R = a(t)r$ is a community distance, assembled from the infinitesimal contributions of myriads of observers aligned between the endpoints. Our demonstration that cosmological redshift cannot be due to the stretching of space therefore affirms these conclusions by removing any possibility that light may be 'carried along' superluminally with the expansion.

III

Horizons

CHAPTER 9

Emergence of Zero Active Mass

T HOUGH the standard model has thus far successfully weathered many observational tests (Chapter 6), some tension with the data has started to emerge as the precision of the measurements continues to improve. Currently, a well-known example is the disagreement—at a level of confidence exceeding 4σ—between the value of H_0 measured using the local distance scale and that found by optimizing the parameters defining the *Planck* fluctuation spectrum [171, 194]. Nevertheless, its overall success thus far could not have prepared us for the recent emergence of a highly improbable coincidence seeking a physical explanation.

All the indicators today seem to be pointing to a gravitational radius, $R_h(t_0)$ (see Chapter 8), equal to ct_0—compelling evidence that the Universe's gravitational horizon is behaving like a *null* surface, more in line with what we would expect of an *event* horizon, rather than an apparent horizon. But the standard model does not require this. Indeed, even special initial conditions would not suffice to adequately account for such a 'coincidence,' for it somehow needs us to be participatory observers. As we shall see shortly, an $R_h(t) = ct$ constraint requires zero acceleration, $\ddot{a} = 0$, in Equation (4.29), implying that $\rho + 3p = 0$, i.e., the *zero active mass* condition in general relativity. In this chapter, we shall advance our discussion of the standard model beyond the empirical basis we explored in Chapter 6, and begin to assess how our current foundational view needs to be modified in order to properly account for the many exquisite, high-precision measurements leading us to this remarkable conclusion.

9.1   A MOST CURIOUS COINCIDENCE

As cosmological observations extend our view of the cosmos to greater redshifts, we probe physics ever closer to the Big Bang. But we are reminded of possible limits to our exploration from the influence of several horizons in general relativity, particularly the gravitational (or apparent) horizon we discussed in Chapter 7. This surface subdivides congruences of ingoing and outgoing null geodesics from a given compact region, restricting the light that can reach an observer as a function of time. Since the metric used to describe the cosmic spacetime is spherically symmetric, these are simply the ingoing and outgoing radial null geodesics from a two-sphere of symmetry.

The highly improbable coincidence we are discussing in this chapter originates from the fact that the gravitational horizon in cosmology has a time-dependent radius $R_h(t)$. It is not static, like an event horizon surrounding a Schwarzschild black hole. This radius may or may not turn into a true event horizon, depending on the equation-of-state in the cosmic fluid, which determines our future history via the cosmic expansion.

Yet the same optimization of free parameters in the concordance model that is starting to show some tension with the measured value of H_0 discussed above is also now revealing that the gravitational radius $R_h(t_0)$ today equals ct_0 within the measurement error. Given the inferred age $t_0 \sim 13.8$ Gyr of the Universe from *Planck*, we see that $ct_0 \sim 4,008$ Mpc. The tension with the measurement of H_0 provides some flexibility in the value of this quantity, but one may reasonably assume that $H_0 \sim 70$ km s^{-1} Mpc^{-1}. Thus, $R_h(t_0) \sim 4,286$ Mpc, so these two distances lie within a mere $\sim 6\%$ of each other. But t_0 is calculated from an integral over the Hubble parameter across the entire history of the Universe, while $R_h(t_0)$ depends on its value at just one time—today. As we shall see shortly, this strong similarity is thus highly improbable because these values could have differed from each other by many orders of magnitude.

To understand why the equality $R_h(t_0) = ct_0$ is unexpected and highly unlikely, let us return to the spherically-symmetric FLRW metric written in its standard form, using the comoving coordinates (ct, r, θ, ϕ) (Equation 4.19). Folding this metric through Einstein's field equations produces the Friedmann (4.28) and Raychaudhuri (4.29) equations. The range of possible values for $R_h = c/H$ is enormous due to the flexibility of the solution to Equation (4.28) with $k = 0$.

As we saw in Chapter 6, ΛCDM is an FLRW cosmology with an empirically motivated choice of components in the energy density ρ of the cosmic fluid. In the basic model, dark energy is assumed to be a cosmological constant Λ, so $\rho = \rho_m + \rho_r + \rho_\Lambda$, encompassing also the other contributions from matter and radiation. Radiation apparently dominated in the early Universe, while matter and (more recently) dark energy have dominated since then. The complexity in the solution for R_h arises because each of these constituents changes with $a(t)$ in its own particular way: $\rho_\Lambda =$ constant, $\rho_m \sim a^{-3}$ and $\rho_r \sim a^{-4}$.

It is not difficult to see from Equation (4.28) that

$$ct = R_h(t_0) \int_0^{a(t)} \frac{du}{(\Omega_m u^{-1} + \Omega_r u^{-2} + \Omega_\Lambda u^2)^{1/2}}, \qquad (9.1)$$

in terms of the critical density ρ_c (Equation 6.1) and the individual scaled densities $\Omega_i \equiv \rho_i(t_0)/\rho_c$ for each species "i". Using the conventional normalization $a(t_0) = 1$, we thus find that

$$\frac{R_h(t_0)}{ct_0} = \left\{ \int_0^1 du \left(\Omega_m u^{-1} + \Omega_r u^{-2} + \Omega_\Lambda u^2 \right)^{-1/2} \right\}^{-1}. \qquad (9.2)$$

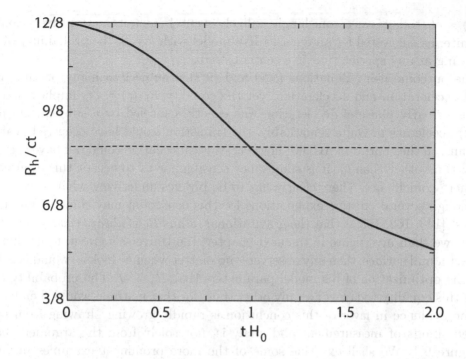

Figure 9.1 The ratio R_h/ct calculated from Equation (9.2) as a function of time, assuming the *Planck* (2015) optimized parameters, starting at the Big Bang ($t = 0$) and extending to twice the current age of the Universe, t_0. This ratio crosses the value 1 only once in the entire cosmic expansion history, and it must be happening now, just when we happen to be looking. (Adapted from ref. [195])

The right-hand side of this equation is clearly heavily dependent on the density ratios Ω_m, Ω_r and Ω_Λ. But we have no fundamental reason to expect any particular value for these quantities. The numbers we assign to them are entirely empirical, presumably set by unknown initial conditions at the time of the Big Bang. Since we don't have a theoretical basis for them, as far as we know in the context of ΛCDM, they could have been arbitrarily large or small.

But let us take a conservative approach, and ignore the evident randomness in H_0, Ω_m, Ω_r and Ω_Λ. Let us assume, for whatever reason, that there actually was a theoretical reason behind the measured values, and let us use the most recent *Planck* optimizations [29] for specificity: $H_0 = 67.4 \pm 0.5$ km s^{-1} Mpc^{-1}, $\Omega_m = 0.315 \pm 0.007$, $\Omega_r = (5.48 \pm 0.001) \times 10^{-5}$ and $\Omega_\Lambda = 1.0 - \Omega_m - \Omega_r$. Solving Equation (9.2) with these fixed parameters, we can calculate the ratio $R_h(t)/ct$ as a function of time and see how it varies before and after t_0. As shown in Figure 9.1, the answer to this question is very puzzling because R_h equals ct only once in the entire history of the Universe.

Even more troubling is the fact that this equality must be happening right now, at time t_0, just when we happen to be looking. This is why we mentioned earlier that, even with special initial conditions, the equality $R_h(t_0) = ct_0$ requires us to be participatory observers, since this is happening specifically and uniquely on our watch. There can be no doubt about the conclusion that witnessing the constraint

$R_h(t_0) = ct_0$ today is an astonishingly unlikely event. In fact, if the Universe's timeline is infinite, as suggested for an open FLRW model with $k = 0$, the probability of this happening at any specific time is essentially *zero*.

The outcome using Equations (9.1) and (9.2) is subject to many possible periods of acceleration and deceleration, yet the equality $R_h(t_0) = ct_0$ implies that the Universe's early phase of deceleration was exactly canceled by a subsequent (more recent) acceleration. Quite remarkably, the transition would have occurred near the midpoint. In the context of ΛCDM, however, this cancellation could not have occurred at any time other than t_0. It is simply not sensible for us to accept this eventuality as a mere coincidence. There clearly has to be physics underlying what we see.

Though several creative explanations for this constraint may come to mind, the simplest [148, 162, 163] is that the gravitational radius R_h is *always equal* to ct. If, for reasons we shall investigate in the next chapter, the Universe's gravitational horizon is in fact a null surface, then any observer—no matter when he looks—would conclude from the optimization of his model parameters that $R_h = ct$. The probability of us seeing this equality today—or at any other time, for that matter—would then be one.

The evidence in favor of this conclusion is rapidly growing, drawing from many different kinds of measurement and analysis, not solely from the argument based on Figure 9.1. We shall examine some of the more prominent examples in Chapter 12. Through the rest of this chapter, we shall consider how the timeline predicted by ΛCDM via Equation (9.2) has difficulties accounting for the apparently early formation of supermassive black holes and galaxies, seemingly requiring exotic modifications to the astrophysics we infer at low redshifts for the creation and evolution of these objects. In support of the above hypothesis, we shall find, instead, that a Universe expanding under the $R_h(t) = ct$ constraint has a timeline fully consistent with the standard formation and evolution scenario as we understand it today.

9.2   AGE OF THE UNIVERSE VERSUS REDSHIFT

When considering the formation and growth of high-redshift sources, it is helpful for us to understand the impact of the $R_h = ct$ constraint on the redshift-age relation characterizing the expansion history of the Universe. For the standard model, this expression may be obtained directly from Equation (9.2), using the conversion from $a(t)$ to z in Equation (8.9):

$$t_{\Lambda\text{CDM}}(z) = \frac{1}{H_0} \int_z^\infty \frac{dz'}{E(z')} , \tag{9.3}$$

where

$$E(z) \equiv (1 + z') \left\{ \Omega_m (1 + z')^3 + \Omega_r (1 + z')^4 + \Omega_\Lambda \right\}^{1/2} . \tag{9.4}$$

If $R_h(t) = ct$ for all t, however, then $H(t) = 1/t$ and $a(t) = (t/t_0)$ (assuming $k = 0$, which is strongly motivated by the observations; see Chapter 6). According to Equation (8.9), a Universe constrained by this condition would obey the simple redshift-age

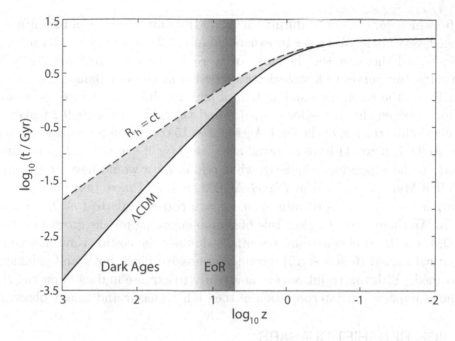

Figure 9.2 Age of the Universe versus redshift in the concordance model (solid curve), assuming the *Planck* (2015) optimization of model parameters in ΛCDM. For comparison, the redshift-age relation is also shown when the constraint $R_h = ct$ is assumed throughout its expansion history (dashed curve). For simplicity, the same Hubble parameter $H_0 = 67.4 \pm 0.5$ km s^{-1} Mpc^{-1} is adopted in both cases. The other assumed parameters are, $\Omega_m = 0.315 \pm 0.007$, $\Omega_r = (5.48 \pm 0.001) \times 10^{-5}$ and $\Omega_\Lambda = 1.0 - \Omega_m - \Omega_r$. In ΛCDM, the Dark Ages end around $z \sim 15$, i.e., $t \sim 300 - 400$ Myr, and are followed by the Epoch of Reionization (EoR), which lasts until $z \sim 6$, corresponding to $t \sim 940$ Myr.

relation

$$t_{R_h=ct}(z) = \frac{1}{H_0(1+z)}.$$ (9.5)

Figure 9.2 compares the functions $t_{\Lambda CDM}(z)$ and $t_{R_h=ct}(z)$, which assume the *Planck* (2015) parameter optimization (see Section 9.1 above). For simplicity, this figure also adopts the same value of H_0 when R_h equals ct.

The redshift range shown in this figure begins near the last scattering surface in the standard model ($z \sim 1100$), progresses through the dark ages down to $z \sim 15$, when Population III stars presumably formed, then across the Epoch of Reionization (EoR) from $z \sim 15$ to $z \sim 6$, and onwards to the present at $z \to 0$. Throughout this history, $t_{R_h=ct}(z)$ is greater than $t_{\Lambda CDM}(z)$, though the difference between them approaches zero at the present epoch, consistent with our discussion in Section 9.1. But if $t_{R_h=ct}(z)$ is always greater than $t_{\Lambda CDM}(z)$, isn't there an inconsistency as $z \to \infty$? Actually no, because an important reversal occurs during the hypothesized period of inflation in the standard model. As we saw in Chapter 6, this early phase must have expanded the Universe by at least 60 e-folds, dramatically altering the redshift (which goes as $\sim a^{-1}$ during this period) by many orders of magnitude while t changed very little. Were we to expand Figure 9.2 by another ~ 35 magnitudes in z, we would therefore see the curve of $t_{\Lambda\text{CDM}}(z)$ flatten significantly as $z \to \infty$, allowing the two curves to approach zero together at the Big Bang.

Insofar as the emergence and growth of high-redshift objects are concerned, the deviation between the curves for $t_{\Lambda\text{CDM}}(z)$ and $t_{R_{\text{h}}=ct}(z)$ in Figure 9.2 makes all the difference. While the end of the Dark Ages at $z \sim 15$ corresponds to an age $\sim 300-400$ Myr in ΛCDM, it would have occurred at ~ 940 Myr if the $R_{\text{h}} = ct$ constraint were imposed on the expansion. Similarly, while reionization would have been completed by $t \sim 980$ Myr (i.e., $z = 6$) in *Planck* ΛCDM, it would have taken another Gyr to do so with the $R_{\text{h}} = ct$ constraint, since $z = 6$ corresponds to $t \sim 2.1$ Gyr in this scenario. We therefore see a clear rule of thumb emerging for the growth of structure in ΛCDM: the $R_{\text{h}} = ct$ constraint essentially doubles the cosmic time elapsed during that critical period ($6 \lesssim z \lesssim 15$) when supermassive black holes and galaxies must have formed. With this result, we are now ready to explore in detail how the $R_{\text{h}} = ct$ hypothesis impacts our interpretation of the high-z quasar and galaxy observations.

9.3 HIGH-REDSHIFT QUASARS

Quasars are ideal objects for testing the redshift-age relation in cosmology. Their efficient extraction of energy from matter falling towards the black hole makes them the most powerful sources in the Universe. With their prodigious energy output, we can identify them well beyond $z \sim 6$, allowing us to study black-hole growth [196] and the formation of large-scale structure in the early Universe [197].

To fully utilize these unique objects for this test, we need to measure their mass. The reverberation technique for mapping their broad-lines has been indispensable for this purpose, yielding the radius of the line-emitting gas from the central ionizing source. Together with a measurement of the orbital speed of this plasma via its Döppler-broadened line width, these observations can yield the black-hole's mass [198] to within a factor ~ 3, even for quasars beyond $z \sim 6$, where masses $M > 10^9 \, M_\odot$ are surprisingly not uncommon.

The basic idea behind this technique is quite simple, though its execution requires relatively high precision measurements. Assuming that the observed physical state of the broad-emission line gas is set by the UV flux received from the central source with luminosity L_{UV}, its distance from the black hole is $R_{\text{BLR}} \sim L_{\text{UV}}^{1/2}$. And if these plasma concentrations are orbiting steadily about the center, their 'virialized' speed must be $v = (GM/R_{\text{BLR}})^{1/2}$. The latter may be measured from the Döppler-broadening of one of the lines, say MgII, indicated by its full-width at half maximum, FWHM(MgII). From these two measurements, we then have

$$M \sim \text{const} \times v^2 R_{\text{BLR}} = \text{const} \times [\text{FWHM(MgII)}]^2 \, L_{\text{UV}}^{1/2} , \qquad (9.6)$$

yielding the expression most commonly used in this type of work:

$$\log M = 6.86 + 2 \log \frac{\text{FWHM(MgII)}}{1,000 \text{ km s}^{-1}} + 0.5 \log \frac{L_{3000}}{10^{44} \text{ ergs s}^{-1}}. \qquad (9.7)$$

L_{3000} is the line luminosity at rest-frame 3,000 Å.

This scaling law [199] is based on several thousand high-quality spectra from the Sloan Digital Sky Survey (SDSS) third release (DR3) quasar sample [200]. It has also been applied very effectively in measuring quasar masses in the Canada-France High-z Quasar Survey (CFHQS) [201]. The F_{3000} flux density is currently measurable to an accuracy of about 10%, while the FWHM(MgII) is measurable to a corresponding accuracy of about 15%. The determination of the black-hole mass using these data thus yields a value accurate to approximately 0.3 dex [202].

To understand how high-z quasars evolved in time, we also need to know their total power output—a proxy for the rate at which they accreted matter from their environment—not just their line flux F_{3000}. Their monochromatic luminosity L_{3000} is only a fraction of the total power, so a 'bolometric' correction η must be used to infer its total luminosity, $L_{bol} \equiv \eta L_{3000}$. One must be careful in using such a simple, 'universal' relation, however, given that estimating L_{bol} from a single monochromatic luminosity can be quite uncertain for individual objects, since the quasar spectral energy distribution is somewhat diverse [203]. An additional complication is that the various fluxes used to infer these luminosities are in reality line-of-sight values, assuming that quasars emit isotropically. But this is not always completely correct. Such caveats notwithstanding, the inferred value of η ranges from about 5 to 6 across the board, representing a surprisingly small dispersion. The choice of wavelength 3,000 Å thus appears to be a robust choice for converting an observed monochromatic luminosity into the quasar's total power output.

From the mass M and bolometric luminosity L_{bol} of a given quasar, one may infer the rate at which it grew in time, based on the following straightforward physical argument. In the absence of any exotic mechanism that might enhance a black hole's accretion rate beyond what standard astrophysics tells us, its maximum rate of growth is the Eddington limit, defined from the maximum luminosity it can produce due to the outward radiation pressure acting on highly ionized infalling material [204]. This power depends on the composition of the gas, but for hydrogen plasma (the most abundant element) is

$$L_{Edd} \approx 1.3 \times 10^{38} \left(\frac{M}{M_\odot} \right) \ \text{ergs s}^{-1} \ , \tag{9.8}$$

in terms of the accretor's mass, M. To find the accretion rate $\dot{M}$ from L_{Edd}, we must then know the fractional efficiency ϵ with which rest mass energy is converted into radiation. This quantity is based on our theoretical understanding of how matter spirals inwards through the surrounding accretion disk towards the event horizon, converting gravitational energy into heat and radiation. For a Schwarzschild black hole described by the metric in Equation (3.19), the infalling matter converts a fraction $\epsilon \sim 6\%$ of its rest energy into these other forms, before its final plunge towards the event horizon from the disk's inner edge at $\sim 3r_S$. The disk's inner radius may be quite different for a Kerr black hole, however, lying closer to the event horizon for a prograde disk, and farther out for a retrograde disk (see Equation 3.41). In the former case, ϵ may be as high as 30% (or more) because frictional dissipation continues down to radii smaller than $\sim 3\,r_S$. Quasars evolve with a range of spins (see Equations 3.28 to 3.34), so the best one can do is to use a representative average of this efficiency factor. Most workers adopt the 'fiducial' value $\epsilon = 0.1$, which lies reasonably close to the Schwarzschild limit, but also allows for some influence due to the black hole's spin.

Under these conditions, we expect high-z quasars to experience a maximum accretion rate

$$\dot{M} = L_{\mathrm{Edd}}/\epsilon c^2 , \qquad (9.9)$$

implying that they grew their mass at a rate

$$\frac{dM}{dt} \approx 0.022 \left(\frac{M}{M_\odot}\right) M_\odot/\mathrm{Myr} . \qquad (9.10)$$

Solving this equation for the mass as a function of cosmic time then yields the so-called Salpeter relation [205],

$$M(t) = M_0 \exp\left(\frac{t - t_{\mathrm{seed}}}{45\ \mathrm{Myr}}\right) , \qquad (9.11)$$

where M_0 is the seed mass created at time t_{seed}. In principle, this relation tracks the black-hole mass $M(t)$ as a function of time for all $t \geq t_{\mathrm{seed}}$. Note, however, that this expression is valid only when the quasar is accreting at the Eddington rate (Equation 9.9). To see how it must be altered in cases where $L_{\mathrm{bol}} \neq L_{\mathrm{Edd}}$, it is useful to define the 'Eddington factor,'

$$\lambda \equiv \frac{L_{\mathrm{bol}}}{L_{\mathrm{Edd}}} . \qquad (9.12)$$

Then, if the quasar accretes with $\lambda \neq 1$, the timescale 45 Myr in Equation (9.11) correspondingly changes to $45/\lambda$ Myr. Intuitively, for a given mass M, a quasar producing a power smaller than L_{Edd}—and therefore accreting at a lower rate—takes longer to grow than the time implied by Equation (9.11).

As of today, we have reliable measurements [201] of M and L_{bol} (and therefore λ) for over 50 quasars at $z > 6$. Using Equation (9.11), we may devolve these objects backwards in time to infer when they started growing as a function of seed mass M_0. But there is still an important unknown to consider—how to deal with a possible time dependence of L_{bol}. Equation (9.8) is a function only of M, not time, but L_{bol} need not be equal to L_{Edd} during the entire black-hole growth—or even part of it, for that matter. The accretion may have been episodic, or it may have been concentrated in the early Universe, when plasma surrounding the black holes was more plentiful. As we shall see, however, such modifications to the accretion history can only strengthen our conclusion, not weaken it.

Fortunately, we now have several compelling observational clues that provide us with a fairly robust understanding of this process. To begin with, redshift 6 appears to have been an important threshold in the reionization of the Universe, as we learned in Chapter 6 and shall revisit shortly. Perhaps not surprisingly, this was also an important transition for black-hole growth.

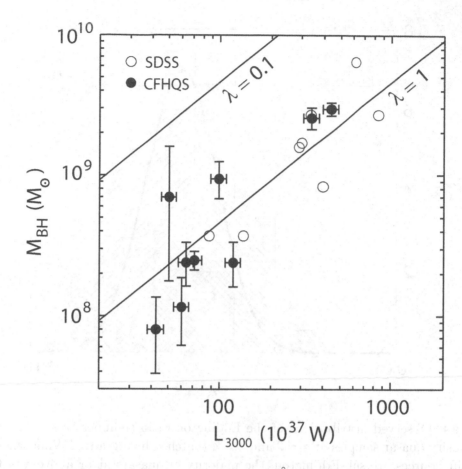

Figure 9.3 Virial black-hole mass versus the $3,000$ Å luminosity for the same quasars. The solid black lines correspond to accretion at 0.1 and 1.0 times the Eddington ratio λ. Quasars from both the SDSS and CFHQS samples are accreting at close to the Eddington limit across a wide range of masses. (Adapted from ref. [201])

As we can see in Figure 9.3, high-z quasar samples from the aforementioned SDSS (DR3) and CFHQS reveal a strong positive correlation between L_{3000} and the black-hole mass in the $z \sim 6$ sources, with M ranging between $\sim 10^8$ $M_\odot$ to $\sim 10^{10}$ $M_\odot$. The most telling feature in this diagram is the comparison between the quasar distribution and the lines of constant Eddington factor, one for $\lambda = 0.1$ and the other for $\lambda = 1$. There is no question that the $z \sim 6$ quasars cluster around $\lambda = 1$, showing that they are all accreting near the Eddington limit. There are no quasars at all with $\lambda = 0.1$, which is much more common at lower redshifts.

Previous studies of quasars up to $z = 4$ showed that the λ distribution in luminosity and redshift bins is typically lognormal, the most luminous among those at $2 < z < 3$ having a typical $\lambda = 0.25$ with a dispersion of 0.23 dex [202]. These quasars are represented by the histogram peaking to the left-hand side of Figure 9.4. By comparison, the histogram showing the λ distribution for the $z \sim 6$ quasars peaks to the right in this figure. The latter may be approximated by a lognormal with peak $\lambda = 1.07$ and a dispersion 0.28 dex. As we surmised from Figure 9.3, the typical

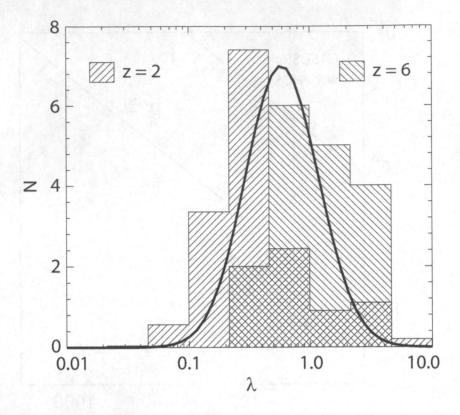

Figure 9.4 Observed distributions of the Eddington ratio (number versus λ) for the luminosity quasar samples at $z = 2$ and $z = 4$ (hatched histograms). While the lower redshift sources are sub-Eddington, the majority of quasars at or above $z \sim 6$ are accreting at close to the Eddington limit. The solid black curve shows the intrinsic distribution of the Eddington factor λ for quasars at $z \sim 6$ after accounting for the observational bias that high-luminosity sources are more likely to be found in flux-limited quasar samples than those with lower values of λ. (Adapted from ref. [201])

quasar at $z \sim 6$ is accreting right at the Eddington limit, with only a narrow λ distribution.

A more careful consideration of these data would take into account the observational bias arising from the fact that high-luminosity quasars are more likely to be found in flux-limited surveys than those with smaller values of λ. The 'intrinsic' distribution taking this effect into account is plotted as a solid curve in Figure 9.4, showing a small offset from the observed distribution (histogram). This intrinsic population is also lognormal with a peak at $\lambda = 0.6$ and a dispersion 0.30 dex.

All such surveys to date have shown that the distribution of Eddington factors gradually increases with redshift from ~ 0.25 at $z \sim 2$ up to ~ 1 at $z \sim 6$, and then saturates at that level. There is no evidence of super-Eddington accretion rates towards high redshifts. These population studies therefore suggest that, in the absence of mergers (see below) or other extraneous effects, typical high-z supermassive black holes grew at or below the Eddington rate throughout their history down to redshift

~ 6, and then settled into a more quiescent growth pattern towards the present time. As noted earlier, quasars may have gone through episodic cycles with lower accretion rates during this time, but such modifications to their overall evolution will only strengthen our result below.

We now consider the context within which supermassive black holes were created and how they could have grown towards $z \sim 6$. Such questions are necessarily woven into the history of the Dark Ages and the Epoch of Reionization that followed. As we saw in Chapter 6, the ΛCDM universe became transparent roughly 0.4 Myr after the Big Bang, ushering in a deepening darkness that would end only several hundred Myr later, when Population III stars started to form. Current observational constraints [206] suggest that the EoR lasted over the redshift range $6 \lesssim z \lesssim 15$ which, in the context of ΛCDM, corresponds to a cosmic time $t \sim 400 - 980$ Myr.

The intergalactic medium was neutral at the beginning of the EoR, but the temperature and ionized gas fraction increased as more and more ultraviolet sources formed, until essentially the whole Universe became ionized [207]. The high-z quasars themselves are some of the best probes of this reionization process. Their high-resolution spectra reveal a complete absence of structure bluewards of the quasar Lyman-α restframe emission, especially at $z > 6$, due to an increase in the Lyman-α optical depth as the ionized fraction decreases along the line-of-sight. The Universe was highly ionized at $z < 6$, but its neutral fraction increased with z, reaching ~ 1 by $z \sim 15$. Supporting evidence for this trend is provided by the Wilkinson Microwave Anisotropy Probe mission [106], whose measurements showed that the Universe was $\sim 50\%$ neutral at $z > 10$, and that reionization did not start before $z \sim 14$.

As we understand it today, it was the Population III stars, followed by (and partially overlapping with) Population II stars, that emitted ~ 13.6 eV photons and started the ionization process. Accreting supermassive black holes no doubt contributed as well, particularly towards the end of the EoR, but it is now recognized that their number was not sufficient for them to reionize the Universe on their own. From a theoretical perspective, we know that these earliest stars could not have started forming earlier than ~ 200 Myr after the Big Bang due to the inefficient cooling of the primordial gas. Another delay of at least ~ 100 Myr would have preceded the formation of Population II stars, while the hot gas ejected by Population III stars cooled and re-collapsed [208, 209]. This transition thus occurred towards the end of the Dark Ages, some $300 - 400$ Myr after the Big Bang in the ΛCDM universe. Based on these standard astrophysical arguments, we now believe that black-hole seeds created during supernova explosions of *evolved* Population II and III stars therefore could not have started their growth until the end of the Dark Ages.

The highest redshift quasar discovered to date [210], known as J1342+0928, is observed at $z = 7.54$. Like many of the other high-z quasars, this object is accreting right at the Eddington rate. In *Planck* ΛCDM, this redshift corresponds to a mere 690 Myr after the Big Bang (see Equation 9.3). Yet according to Equation (9.11), it should have taken ~ 820 Myr for it to grow to its observed mass, $M = 7.8^{+3.3}_{-1.9} \times 10^8 \ M_\odot$, even accreting at the maximum (Eddington-limited) rate, without any episodic downtime. So we are immediately confronted with a major problem: either the timeline in ΛCDM

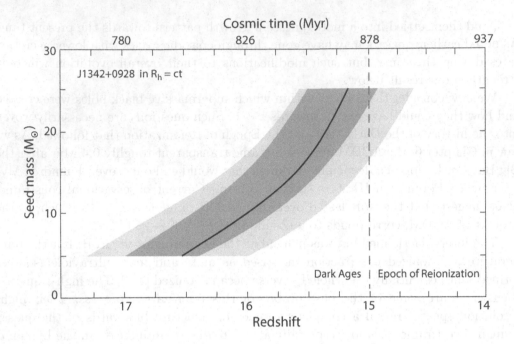

Figure 9.5 The calculated seed mass of quasar J1342+0928 versus redshift at the time it was created (solid black curve), assuming it reached its observed mass $M = 7.8^{+3.3}_{-1.9} \times 10^8 \ M_\odot$ at $z = 7.54$ via conventional Eddington-limited accretion. The redshift-age relation corresponds to the $R_h = ct$ constraint. The shaded regions represent the 1σ (dark) and 2σ (light) confidence regions based on the uncertainty in the measured mass. (Adapted from ref. [211])

is problematic, or our astrophysical understanding of how supermassive black holes formed and grew is incorrect.

Given the evidence presented earlier in this chapter (Section 9.1), let us first suppose that the expansion of the Universe was constrained by the $R_h = ct$ condition, and see how the corresponding timeline comports itself with the quasar data we have at hand. Figure 9.5 shows the seed mass M_0 versus the time t_{seed} (and corresponding redshift) at which it must have formed in order for it to grow into J1342+0928 at $z = 7.54$ (solid black curve), according to Equation (9.11). The cosmic time versus redshift on this plot is calculated from Equation (9.5). This figure also shows the 1σ (dark) and 2σ (light) confidence regions, estimated via error propagation from the uncertainty in M and, in addition, displays the demarcation between the Dark Ages (at $15 \lesssim z$) and the ensuing EoR ($6 \lesssim z \lesssim 15$). Quite remarkably, the ~ 820 Myr required to grow J1342+0928 via Eddington-limited accretion from its initial supernova-produced $\sim 10 \ M_\odot$ seed at $z \lesssim 16$ to its observed $7.8 \times 10^8 \ M_\odot$ mass at $z = 7.54$ coincides almost perfectly with the conditions we have described above, which are based on standard astrophysical principles. The mass of the seed, and the time/redshift at which it was created, are just as we would have expected from the Population III to Population II transition at the end of the Dark Ages. Moreover, the fact that we observe J1342+0928 to be accreting at the Eddington rate leaves

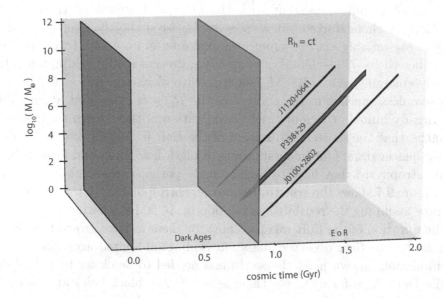

Figure 9.6 Growth of quasars J0100+2802, P338+29, and J1120+0641, versus cosmic time when the constraint $R_h = ct$ on the expansion in ΛCDM. For this calculation, there is only one free parameter, H_0, whose value is assumed to be the same as in ΛCDM for ease of comparison (see Figure 9.7). In this case, the EoR started at ~ 883 Myr and ended at ~ 2.0 Gyr ($6 < z < 15$). The 20 $M_\odot$ seeds would have formed at $t \sim 1.09$ Gyr (J0100+2802), ~ 1.06 Gyr (P338+29), and ~ 990 Myr (J1120+0641)—in each case, *shortly after the start of the EoR*. (Adapted from ref. [212])

no room to maneuver with possible Eddington factors $\lambda > 1$ in order to reduce the e-folding time of 45 Myr in the Salpeter relation (Equation 9.11). At least for the highest redshift quasar known to date, the timeline predicted with the $R_h = ct$ constraint fits the data very well.

But is J1342+0928 truly representative of all the high-z quasars, or is it anomalous in some way? The fact that we actually observe it to be accreting with an Eddington factor $\lambda \sim 1$ argues against it being special. The short answer to this question is no. At least for the 50 or so high-z quasars discovered to date, J1342+0928 is actually quite typical. To demonstrate how this category as a whole fits within the expansion history of the Universe, consider another three representative members of this set, whose growth trajectories are shown in Figure 9.6. The quasar J1120+0641 is noteworthy because it is ultraluminous, meaning that it is the most massive high-z quasar known today. Its estimated mass is $\sim 1.2 \times 10^{10}$ $M_\odot$, roughly 10 times greater than the rest of the sample [213]. Such a gargantuan object must have taken ~ 900 Myr to form, according to Equation (9.11), so time compression issues should emerge more prominently with this black hole than any of the others.

Yet from Figures 9.5 and 9.6, and every other high-z quasar studied thus far, we see that the imposition of the $R_h = ct$ constraint on the expansion of the Universe creates a timeline fully consistent with the birth and growth of these prominent objects. Each of them started out as a $\sim 10\ M_\odot$ seed shortly after the end of the Dark Ages, presumably via the supernova explosion of an evolved Population II or III star, followed by $700 - 900$ Myr of steady accretion at the Eddington rate, and then finally emerging as a $\sim 10^9\ M_\odot$ supermassive black hole at $z \sim 6$.

This seamless timeline in the context of the $R_h = ct$ constraint affirms the possibility, already hinted at by our brief consideration of the growth requirements for J1342+0928, that the shortened duration of the EoR in ΛCDM may be stressed by the high-z quasar data. Our understanding of black-hole birth and growth based on standard astrophysics may have to be modified—perhaps substantially—in this cosmology. Figure 9.7 shows the growth trajectories corresponding to those in Figure 9.6, though now assuming the redshift-age relation in ΛCDM (Equation 9.3). As we surmised, the duration of the EoR was too short for them to have grown from $\sim 10\ M_\odot$ seeds at $z \sim 15$ to their observed mass via Eddington-limited accretion. According to the simulations shown here, the seed mass needed to be closer to $10^5\ M_\odot$ at the end of the Dark Ages for us to see them as $\sim 10^9\ M_\odot$ black holes at $z \sim 6$. This is already very challenging, but the most serious drawback with this timeline is actually that they all would have had to start growing *before* the Big Bang to reach their observed endpoint. This makes no sense at all. The standard picture of black-hole birth and growth as we understand it from observations of the local Universe is thus not compatible with the timeline in ΛCDM.

Given the seriousness of this 'time compression' problem, it is hardly surprising to learn that several attempts have been made to modify the astrophysics in order to preserve the ΛCDM timeline. It would be fair to say, however, that all of them are 'exotic' in one way or another, having yet to find solid observational (or even theoretical) support. Mechanisms to resolve the mystery of how such ponderous objects could have assembled so quickly in ΛCDM have generally fallen into three broad categories: (1) an anomalously high accretion rate [214, 215, 216]; (2) the creation of enormously massive seeds in the early Universe [217, 218]; and (3) additional growth via mergers [219, 220, 221].

The first of these mechanisms is problematic because no compelling evidence has yet been found supporting such extreme conditions. Remember that the highest redshift quasar known to date, J1342+0928, is itself accreting right at the Eddington rate. And none of the other high-z supermassive black holes with a measured mass has a luminosity greatly exceeding the Eddington limit either (Figures 9.3 and 9.4). In principle, at least some of these objects—including J1342+0928—could have grown to their observed mass in only ~ 300 Myr if they had been accreting steadily throughout their growth at three times the Eddington rate. This modification would have allowed a $10\ M_\odot$ seed created at $z \sim 15$ to grow to $\sim 8 \times 10^8\ M_\odot$ by $z \sim 7.5$. But with so many supermassive black holes now detected at $z \sim 6$, we should have seen an abundance of them with $\lambda > 3$ at $z > 7$, if they existed. Unfortunately, none of the current surveys have ever seen such objects [222, 223, 224]. The measured accretion

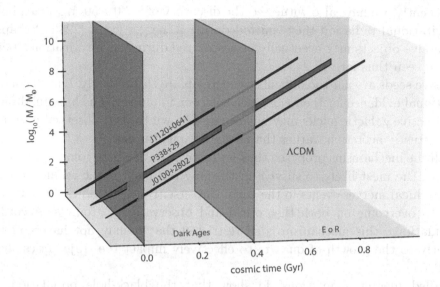

Figure 9.7 Growth of quasars J0100+2802, P338+29, and J1120 | 0641, versus cosmic time in the standard model, ΛCDM, assuming the *Planck* (2015) optimized parameters: $\Omega_m = 0.31$, $k = 0$, $w_\Lambda = -1$ and $H_0 = 67.3$ km s^{-1} Mpc^{-1}. For each quasar, the range in timelines corresponds to the range in seed mass, $M_{seed} = 5 - 20\ M_\odot$. The Dark Ages evolved into the Epoch of Reionization (EoR) at $t \sim 400$ Myr ($z \sim 15$), which then lasted until ~ 980 Myr ($z \sim 6$). For Eddington-limited accretion, these three black holes starting out with a seed mass $M_{seed} = 20\ M_\odot$ would have grown to $3.0 \times 10^5\ M_\odot$, $3.1 \times 10^5\ M_\odot$, and $5.4 \times 10^5\ M_\odot$, respectively, at the start of the EoR. But they would have been created ~ 33 Myr, ~ 35 Myr, and ~ 59 Myr *prior* to the Big Bang. (Adapted from ref. [212])

rates tend to be concentrated at, or below, the maximum Eddington rate, with λ trending even lower towards smaller redshifts.

The second proposal postulates that black hole seeds formed with a mass $\sim 10^5\ M_\odot$, which is even more difficult to confirm observationally than super-Eddington accretion rates. Such catastrophic events would have been too brief to be seen directly. The best we could hope for would be to find such 'intermediate-mass' black holes after they formed near us, so that we could detect their feeble emission. Unfortunately, the evidence is sparse and inconclusive. Several low-luminosity active galactic nuclei may be such candidates, including NGC 4395 which, at a distance of 4 Mpc, appears to harbor a $\sim 3.6 \times 10^5\ M_\odot$ black hole in its center [225].

Another category of object in nearby galaxies, known as 'Ultra-luminous X-ray sources,' may also be intermediate-mass black holes with $M \lesssim 1{,}000\ M_\odot$, but these masses are too small [226]. Some intermediate-mass black holes may be present in globular clusters, e.g., M31 G1, accounting for the large stellar velocities measured near their center. None of these candidates has stood up to followup scrutiny, however [227]. Recently, we have also witnessed the discovery of $\sim 30 - 50\ M_\odot$ black holes via the gravitational radiation they emitted during a merger event [228]. Perhaps even more massive objects may eventually be discovered during such transitions, but none have been seen thus far.

Massive seeds are not too difficult to contemplate theoretically, but no compelling observational evidence for their existence has been found yet. The handful designated as dwarf active galactic nuclei may have simply grown to their observed intermediate mass via steady accretion, rather than some incipient event.

Of all the mechanisms proposed thus far to mitigate the time compression problem in ΛCDM, the most likely to survive in the long run may be the reliance on growth via hierarchical merger events in the early Universe. But as of now, there are serious hurdles to overcome on both theoretical and observational grounds. According to the simulations, this mechanism is so restricted that it may not have contributed sufficiently to the growth of quasars to effectively impact the trajectories shown in Figure 9.7.

Detailed merger simulations do show that the black-hole population always evolves into a Gaussian distribution, irrespective of how the initial seeds form. The model can therefore be somewhat flexible. To be consistent with all of the available data, however, $\sim 100\ M_\odot$ seeds had to start forming earlier than $z \sim 40$, well before the EoR which, as we have seen, began at $z \sim 15$. More importantly, seed creation could not have extended past $z \sim 20 - 30$ to prevent the mass density in lower-mass (i.e., $10^5\ M_\odot - 10^7\ M_\odot$) black holes from being overproduced compared to what is actually observed [219]. This cutoff is quite serious because, without it, the lower-mass black holes would have been overproduced by a factor of $\sim 100 - 1,000$!

The merger history required to form supermassive black holes in the early Universe therefore does not sit comfortably with our current interpretation of Population III star-formation. It would be very difficult to see physically how the cooling time required to form the first generation of stars could be shortened substantially to allow black hole seeds to start forming by $z \sim 40$. It would be even more difficult to understand why these stars stopped forming at $z \sim 20 - 30$, well before the EoR even started. What this means, of course, is that almost certainly some mechanism other than Population III supernovae would be required to create these massive seeds well before the EoR. Theoretically, this would require new, unknown (and exotic) physics. Observationally, there is currently no evidence for such activity prior to $z \sim 15$. It all seems rather daunting to make the merger picture work given the already existing constraints.

The Population III supernova scenario for the formation of black holes that merge has been further hampered by evidence that the halo abundance at $z > 10$ was at least an order of magnitude smaller than previously thought [229]. Large (4 Mpc^3) high-resolution simulations of halo formation and the first generation of stars within them show that Population II and III stars formed coevally all the way to $z \sim 6$. The enhanced metal enrichment and molecule-dissociating Lyman-Werner photons responsible for the destruction of the coolants H_2 and HD required for matter condensation in the early Universe would have considerably altered the halo and Population III star formation rate.

When this new physics is introduced into the simulations, the calculated Population III star formation rate per comoving volume is found to be $\sim 10^{-4}$ $M_\odot$ yr^{-1} Mpc^{-3} at $z > 10$. But these same effects would have prolonged the period over which Population II and III stars formed and evolved coevally, thereby resulting in a higher stellar mass per unit volume at $z \sim 6$. The Population III star formation rate at $z \sim 6$ is found to be $\sim 10^{-5}$ $M_\odot$ yr^{-1} Mpc^{-3}, just an order of magnitude below its peak at $z \sim 10$. This is very interesting insofar as the star formation history is concerned, but such a net shift in the epoch when they formed reduces the density of Population III supernovae and black-hole seeds during the time ($z > 10$) when the frequency of mergers among these objects would have mattered most to black-hole growth.

The challenge of adjusting the astrophysics of star formation and black-hole growth to preserve the timeline predicted by ΛCDM stands in stark contrast with the effortless operation of fitting what we already know today about these processes within the context of a Universe expanding according to the $R_h = ct$ constraint. It is almost uncanny to see the time required to grow the high-z quasars within the EoR fitting so well compared to the observed end of the Dark Ages in this scenario. In the next section, we shall learn that the premature formation of galaxies provides corroborating evidence in support of these conclusions, arguing in favor of the $R_h = ct$ condition we discussed in Section 9.1 above.

9.4   HIGH-REDSHIFT GALAXIES

Our interest in probing the cosmic dawn is bolstered by the dramatic discovery of faint galaxies at redshifts well beyond the end of the Epoch of Reionization. Using gravitational lensing techniques with the WFC3/IR instrument on the *Hubble* Space Telescope, investigators have uncovered galaxies emerging as far back as $z \sim 10 - 12$, an unprecedented leap towards the earliest moments of structure formation in the Universe [230, 231, 232, 233]. These very early galaxies contributed to the reionization of the intergalactic medium, and may even have dominated this process at certain times.

But finding galaxies at $z \sim 10$ was a complete surprise for reasons not unlike the discovery of supermassive black holes at $z \sim 6$. The issue is again the time compression problem, perhaps even more severe here than the one we considered in Section 9.3 above. According to the simulations we have already discussed, Population III stars could form at a rate of only one per halo because the UV radiation from a single ~ 100 $M_\odot$ first generation star could destroy all of the molecular hydrogen in the parent condensation. These earliest stellar structures were therefore not what we would call galaxies.

By the time the hot gas expelled by Population III stars had cooled and recondensed, at least 300 Myr had elapsed since the Big Bang. It was this metal enriched plasma, we believe, that hosted the second generation stars assembled into the first galaxies. The most detailed theoretical simulations reveal that these would have been atomic cooling halos—condensations with mass $\sim 10^8 \; M_\odot$ and virial temperatures $\sim 10^4$ K, cooling via the emission of atomic lines [234]. The harrowing question, however, is how could an assembly of stars with mass $10^8 - 10^9 \; M_\odot$ possibly have formed and aggregated into galaxies by $z \sim 10 - 12$, a mere 200 Myr later?

Theoretical work on the formation of galaxies is quite varied among the different groups working on this problem, but their conclusions tend to confirm each other's results because their calculations rely on a common set of physical principles. We shall discuss the premature formation of high-z galaxies in the context of one of these approaches [235] due to its completeness and straightforward application.

An examination of the duty cycle and history of condensation in high-z galaxies simulated with this approach shows that the star formation in individual objects may have occurred in bursts. Averaged over a prolonged period, however, their evolution follows an exponential growth with a characteristic timescale ranging from $t_c = 70$ Myr to 200 Myr, for galaxy masses $M_* \sim 10^6 \; M_\odot$ to $\sim 10^{10} \; M_\odot$. One therefore infers from these hydrodynamic calculations that the stellar mass in a typical galaxy must have grown at a rate

$$\frac{dM_g}{dt} = K \, \exp\left(\frac{t - t_*}{t_c}\right) , \qquad (9.13)$$

where $M_* \equiv M_g(t_*)$, and $K \approx 1 \; M_\odot \; \mathrm{yr}^{-1}$. With Equation (9.13), we can see right away how the time compression problem emerges for these high-z galaxies. For example, solving this equation for a $10^8 \; M_\odot$ galaxy at $z \sim 10$, where $t \sim 550$ Myr in ΛCDM, we find that its mass at $t \sim 300$ Myr would have been $\sim 8 \times 10^6 \; M_\odot$. Clearly, this initial condition would be inconsistent with the transition from Population III to Population II stars described above.

The two key quantities at the core of this discussion are the inferred mass, M_*, and star-formation rate, K, of these high-z galaxies. Perhaps the problem arises artificially due to incorrect inputs into the simulations. For example, an actual rate at which these galaxies form stars ten times greater than that implied by the theoretical modeling would largely mitigate the time compression problem.

But the observations broadly confirm the calculations. Several groups have utilized an assortment of techniques, including gravitational lensing by foreground clusters of galaxies and spectral energy distribution fitting techniques on objects selected from a deep multi-band near-infrared stack. The Cluster Lensing and Supernova survey with *Hubble* (CLASH) [236] has discovered several $z > 8.5$ candidates, three [237] at $z \sim 9 - 10$ and a multiply-imaged source [232] at $z = 10.7$. And one of the deepest searches of star-forming galaxies to date, based on the Hubble Ultra Deep Field, has yielded 7 candidates [231] at $z > 8.5$. All of the 11 or so high-z galaxies discovered thus far have measured star-formation rates in the range $\sim 1 - 4 \; M_\odot \; \mathrm{yr}^{-1}$, fully consistent with the simulations. Their masses, M_*, fall within the range $\sim 1 - 10 \times 10^8 \; M_\odot$. Below, we shall highlight two representative members of this set: MACS0647-JD at $z = 10.7$, with $M_* = 1 - 10 \times 10^8 \; M_\odot$ and a star-formation rate $\sim 1 - 4 \; M_\odot \; \mathrm{yr}^{-1}$; and UDF12-3947-8076 at $z = 8.6$, with $M_* \sim 10^9 \; M_\odot$, growing at a rate of $\sim 3 \; M_\odot \; \mathrm{yr}^{-1}$.

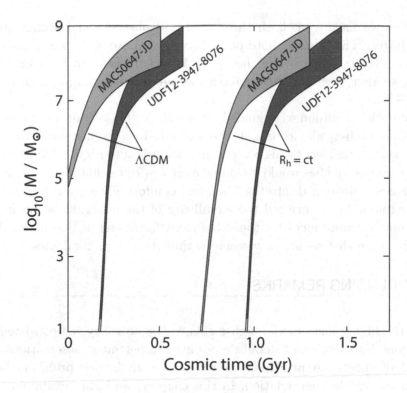

Figure 9.8 Growth of two representative high-z galaxies, MACS0647-JD at $z = 10.7$, and UDF12-3947-8076 at $z = 8.6$, as a function of cosmic time t. The other high-z galaxies lie between these two redshifts. The galaxy growth rate is based on the observed star-formation rate at $z \sim 9 - 12$, supported by the theoretical simulations discussed in the text. The shaded region for each galaxy includes the uncertainty (of about a factor ~ 10) in the inferred galaxy mass. The EoR would have started at ~ 400 Myr for ΛCDM, and at ~ 883 Myr for an expansion constrained by $R_h = ct$. (Adapted from ref. [238])

The evolutionary history of these two galaxies may be traced by solving Equation (9.13), yielding their mass as a function of time $t \leq t_*$. Figure 9.8 shows their trajectory in the mass vs. time plane, for both the timeline in ΛCDM (Equation 9.3) and that corresponding to a Universe expanding under the $R_h = ct$ constraint (Equation 9.5). As a reminder, the EoR would have begun at $t \sim 400$ Myr in ΛCDM, and at ~ 883 Myr with $R_h = ct$. To be ultra-conservative, this plot also takes into account a possibly large uncertainty in the inferred mass, M_*, showing how the trajectory would vary if M_* were to change by a factor 10.

It is quite clear from this figure that the time compression problem with high-z galaxies in ΛCDM may be even worse than the one with quasars. Whereas one could envisage new physics, albeit exotic, to enhance the growth rate of individual black holes, it is much more challenging to do this with an assembly of $\sim 10^8$ individual stars. Perhaps the star-formation rate is wrong. Certainly, the time compression would be largely mitigated if these galaxies grew at a rate $\sim 20\ M_\odot\ \text{yr}^{-1}$ instead of $\sim 2\ M_\odot$

yr^{-1}. But as we have seen, both theoretical and observational arguments tend to favor the latter. The other remote possibility is that we are simply measuring the wrong mass. To fix the problem, however, M_* would need to be 4 or 5 orders of magnitude smaller, but then these galaxies would be far too faint for us to see them at $z \sim 10 - 12$.

If the cosmic expansion were guided by the $R_h = ct$ constraint, however, the evolution of these high-z galaxies would fit seamlessly into the history of the early Universe, just as we found for the high-z quasars. According to Figure 9.8, we see that the growth of these assemblies would have begun at $t \sim 700 - 900$ Myr, around the time the Dark Ages transitioned into the EoR, and comfortably after the first generation stars had exploded and enriched the metallicity of the intergalactic medium, creating the 'right' circumstances for the subsequent formation of Population II stars—presumably those that we are now seeing within these high-z galaxies.

9.5 CONCLUDING REMARKS

Together, the high-z quasars and high-z galaxies paint a very self-consistent picture. We have gone into significant detail describing the pertinent observations and their theoretical interpretation, principally because these are the best probes we have today concerning the redshift-age relation. In this chapter, we began examining the most puzzling coincidence in cosmology, having to do with the current age of the Universe in relation to its gravitational 'size,' and then surmised that the most likely physical reason for its existence is that the gravitational horizon is null. Not surprisingly, this inference impacts our theoretical understanding of the data in very serious ways. In this chapter, we considered the most direct connection between the constraint $R_h = ct$ and the cosmic expansion, i.e., its implied timeline. Our foray into the temporal evolution of the best studied high-redshift sources suggests that these data strongly favor the evolutionary history implied by $R_h = ct$ over that predicted by *Planck* ΛCDM.

But there are many other kinds of measurement in cosmology, some having to do with the redshift-dependent rate of expansion, others based on integrated distances. The development of 'standard candles' and 'standard rulers' in recent decades has enriched our database, stretching our view from the nearby Universe all the way to the last scattering surface at $z \sim 1081$. Our exploration of the $R_h = ct$ constraint has therefore just begun. In Chapter 12, we shall greatly broaden our range of tests of this hypothesis by incorporating into our analysis several other prominent and influential cosmological measurements.

CHAPTER 10

Theoretical Basis for Zero Active Mass

BY setting $g_{tt} = 1$ in the FLRW metric (Equation 4.4), we complied with the requirements of the Cosmological Principle (Section 1.2), which incorporates isotropy and homogeneity as essential symmetries of the cosmic spacetime. But we also highlighted the fact that this assumption was made prior to folding the FLRW metric coefficients through Einstein's equations and subjecting them to the constraints imposed by our choice of stress-energy tensor $T^{\mu\nu}$ (Equation 4.27). The equation-of-state in the cosmic fluid determines whether or not the cosmic expansion is accelerated, so we should at least work through a self-consistency check to prove that our choice of unitary lapse function is physically correct under all circumstances. This step is now imperative, given our discussion in Chapter 9, which demonstrates quite compellingly that the Universe is apparently evolving according to the $R_h = ct$ constraint. This feature did not emerge from our arguments in Chapter 4, but there are good reasons to suspect that the issue with $g_{tt} = 1$ may have something to do with a null gravitational horizon. If $R_h = ct$, then the expansion factor $a(t) = (t/t_0)$ is linear in time, implying zero acceleration and no time dilation. Understanding whether or not this suspected relationship is correct will be the primary focus of this chapter.

10.1 THE LAPSE FUNCTION

Given its high degree of symmetry, based on the presumed isotropy and homogeneity of the Universe, the FLRW metric is a special, well-studied member of the general class of spherically-symmetric spacetimes describing systems undergoing gravitational collapse or expansion [180, 239, 173, 240]. For various historical reasons, however, FLRW was developed using a different approach than that of the others. More precisely, whereas the equations governing gravitational collapse are derived from Einstein's Equations (2.167) using a general form of the metric coefficients, the Friedmann Equations (4.28), (4.29) and (4.31) describing the cosmic expansion were obtained after all the symmetries underlying the cosmological principle were introduced to greatly simplify the metric (Equation 4.19) prior to its insertion into the field equations—regardless of which equation-of-state (in terms of the total energy density ρ and pressure p) is subsequently adopted for the stress-energy tensor $T^{\mu\nu}$.

This difference has an enormous impact on how the Universe evolves, because g_{tt} is then independent of whether or not the cosmic expansion is accelerating. We must be wary of this relaxation, however, given that Einstein already realized from his early work that the new gravity theory correctly reduces to Newton's law in the weak-field limit only if $g_{tt} = 1 + 2\Phi/c^2$, where Φ is the gravitational potential. Thus, if $g_{tt} = 1$, then the observer sees no acceleration and no time dilation.

Scant attention is paid to this issue, mainly because conventional wisdom has it that g_{tt} in FLRW can—at most—be a function only of time, not space, for consistency with the cosmological principle, which is undoubtedly true. It is therefore believed that, if necessary, one can change the gauge $dt \to dt' \equiv \sqrt{g_{tt}}\, dt$ to again eliminate any coordinate dependence from the lapse function. But one should be suspicious of this procedure because, at the very heart of it, a change in gauge in general relativity is really a transformation of the coordinates, i.e., a change in the frames of reference. The FLRW metric was constructed to describe the spacetime in the comoving frame, with coordinates (ct, r, θ, ϕ), but if one must carry out a gauge transformation to again reduce g_{tt} to 1 given a choice of $T^{\mu\nu}$, then where does the gauge transformation $dt \to dt' \equiv \sqrt{g_{tt}}\, dt$ take us? In this section, we shall use the Local Flatness Theorem (Equations 2.127–2.131) in general relativity to formally answer this question, i.e., to test whether the choice of lapse function $g_{tt} = 1$ is in fact valid for any equation-of-state, or whether the symmetries it formally represents restrict its validity to a particular subset of possible expansion scenarios.

In this analysis, it will be easier for us to use a Cartesian system, so let $x^\mu = (ct, x, y, z)$ be the coordinates in the comoving frame, $g_{\mu\nu}$ the metric coefficients, and $\Gamma^\lambda_{\ \mu\nu}$ the corresponding Christoffel symbols. As we discussed in Chapter 6, it is safe for us to assume spatial flatness, given that most of the observations today appear to be supporting this initial condition in the cosmic expansion. Thus, the relevant FLRW metric may be written simply as

$$ds^2 = L(t)c^2 dt^2 - a(t)^2 \left(dx^2 + dy^2 + dz^2 \right) , \qquad (10.1)$$

where $a(t)$ is the usual expansion factor, and the metric coefficients are

$$g_{\mu\nu} = \begin{pmatrix} L & 0 & 0 & 0 \\ 0 & -a^2 & 0 & 0 \\ 0 & 0 & -a^2 & 0 \\ 0 & 0 & 0 & -a^2 \end{pmatrix} . \qquad (10.2)$$

To keep the calculations as general as possible, we have introduced the lapse function $g_{tt} \equiv L(t)$, allowing us to examine the constraints it must satisfy in order to solve Equation (10.4) below. As noted earlier, this simple functional form of L is consistent with the assumption of isotropy and homogeneity in the cosmological principle, which permits the lapse function to depend only on time, not space. It is straightforward to see that the corresponding non-zero Christoffel symbols in the comoving frame are therefore

$$\Gamma^0{}_{00} = \frac{1}{L}\frac{\partial L}{\partial ct}$$

$$\Gamma^0{}_{ii} = \frac{1}{Lc}a\dot{a}$$

$$\Gamma^i{}_{i0} = \Gamma^i{}_{0i} = \frac{1}{c}\frac{\dot{a}}{a}\,. \tag{10.3}$$

According to the Local Flatness Theorem [69, 49], there exists a locally inertial coordinate frame $\xi^\mu(X)$ in the neighborhood of any spacetime point X^μ in x^μ. The ξ^μ frame is in free fall at X^μ, so any non-inertial effects in the coordinate system x^μ—such as a time dilation—can be measured absolutely relative to ξ^μ. These effects are absent only if the comoving frame itself were in free fall, in which case it would be inertial to begin with.

For convenience, and without loss of generality, let us place X^μ at the common origin of x^μ and ξ^μ. Given the evident symmetry, it is also safe to assume that ξ^0 may depend solely on t and the radius $r = \sqrt{x^2 + y^2 + z^2}$, while ξ^i may be a function only of t and x^i (with $i = 1, 2$ or 3). The equations satisfied by ξ^μ are (see Section 2.6)

$$\frac{\partial^2 \xi^\alpha}{\partial x^\mu \partial x^\nu} = \Gamma^\lambda{}_{\mu\nu}\frac{\partial \xi^\alpha}{\partial x^\lambda}\,. \tag{10.4}$$

The $\alpha = \mu = \nu = 0$ component in this expression thus yields the following relation that must be satisfied by the lapse function $L(t)$ in terms of $a(t)$:

$$\xi^0 = f(r)\int^{ct} L(t')\,d(ct') + g(r)\,, \tag{10.5}$$

where $f(r)$ and $g(r)$ are functions yet to be determined. The $\alpha = \mu = 0$, $\nu = i$ component gives

$$\frac{\partial}{\partial x^i}\left(\frac{\partial \xi^0}{\partial ct}\right) = \frac{1}{c}\frac{\dot{a}}{a}\left[\frac{x^i}{r}\frac{df}{dr}\int^{ct} L(t')\,d(ct') + \frac{x^i}{r}\frac{dg}{dr}\right]\,. \tag{10.6}$$

Evidently, these two expressions show that $dg/dx^i = 0$, and we may simply choose $g(r) = 0$, given that it represents a constant temporal translation. Most importantly, we find from Equations (10.5) and (10.6) that

$$\int^{ct} L(t')\,d(ct') = cL(t)\frac{a}{\dot{a}}\,. \tag{10.7}$$

The choice of lapse function $L(t) = 1$ is therefore consistent only with the expansion factor

$$a(t) \equiv \left(\frac{t}{t_0}\right)\,, \tag{10.8}$$

written in conventional form for a spatially flat metric using the current age of the Universe, t_0.

Of course, Equation (10.4) is also trivially satisfied with $\xi^0 = ct$ and $L(t) = 1$ when $a = $ constant, in which case the spacetime curvature is identically zero, i.e., $\Gamma^\alpha_{\mu\nu} = 0$. So the complete statement should be that, relative to a local inertial frame ξ^μ at $\vec{x} = 0$, the lapse function in FLRW may be set equal to 1 only for the very special cases (1) $a = $ constant (i.e., Minkowski space) and (2) $a(t) = (t/t_0)$. For any other expansion factor representing acceleration, the time t in the comoving frame must necessarily be dilated relative to the time ξ^0 in the inertial frame.

If $L(t) = 1$, and only then, the time coordinate (Equation 10.5) in the local inertial frame may be written $\xi^0 = f(r)ct$, in terms of a function $f(r)$ that we may now identify. Equation (10.4) provides us with one more relevant component, corresponding to $\alpha = 0$, $\mu = \nu = i$:

$$\frac{(x^i)^2}{r^2}\frac{d^2 f}{dr^2} + \frac{(x^j)^2 + (x^k)^2}{r^3}\frac{df}{dr} = \frac{1}{(ct_0)^2}f \,, \tag{10.9}$$

where $i \neq j \neq k$. The solution to this equation, correct to second order in r, is

$$f(r) = 1 + \frac{1}{2}\left(\frac{r}{ct_0}\right)^2 \,. \tag{10.10}$$

Recalling that ξ^μ are the coordinates in a *locally* inertial frame in the vicinity of $\vec{x} = 0$, we see that $\xi^0 \to ct$, as one would expect.

Let us next consider the spatial coordinates. According to Equation (10.4),

$$\xi^i = h(x^i)\int^{ct} L(t')\,d(ct') \,, \tag{10.11}$$

in which the function $h(x^i)$ will be derived below. Thus, if $L(t) = 1$, we have

$$\xi^i = h(x^i)ct \,. \tag{10.12}$$

These coordinates must also satisfy the equation

$$\frac{\partial^2 \xi^i}{\partial ct\, \partial x^i} = \frac{1}{c}\frac{\dot{a}}{a}\frac{\partial \xi^i}{\partial x^i} \tag{10.13}$$

(with no implied summation over 'i'), so that

$$\frac{dh(x^i)}{dx^i} = \frac{1}{c}\frac{\dot{a}}{a}\left(\frac{dh(x^i)}{dx^i}ct\right) \,, \tag{10.14}$$

which means that $a(t)$ must again satisfy the equation

$$\frac{1}{c}\frac{\dot{a}}{a}ct = 1 \,, \tag{10.15}$$

reproducing the result in Equation (10.8).

Our final result comes from the $\mu = \nu = 1$ component in Equation (10.4), from which we get

$$h(x^i) = h_0 e^{x^i/ct_0} \,, \tag{10.16}$$

where h_0 is an integration constant. With a redefinition of the spatial scale,

$$\chi^i \equiv h_0 \, ct_0 \, e^{x^i/ct_0} \,, \tag{10.17}$$

we may formally write

$$\xi^i = a(t)\chi^i \,. \tag{10.18}$$

The coordinates ξ^μ thus reveal that the choice of lapse function $L(t) = 1$ in the comoving frame necessarily constrains the expansion factor to be uniquely $a(t) = (t/t_0)$ (or $a = $ constant for Minkowski), and that the locally inertial frame is then simply the Hubble flow with $\xi^\mu = (ct, a\chi^1, a\chi^2, a\chi^3)$.

The conclusion drawn from Equations (10.5)–(10.8) is the most important result discussed in this book. The Local Flatness Theorem is foundational to general relativity, for it formalizes the concept of equivalence that allows one to account for spacetime curvature with reference to a local inertial frame. Unlike velocity, which is relative, acceleration is absolute. The time measured by an accelerated observer must be dilated relative to that in a free-falling frame.

We have become accustomed to ignoring this effect in FLRW, regardless of whether or not the expansion is accelerated, with the belief that one may always redefine the time coordinate in order to cast the metric in standard form with $g_{tt} = 1$. This presumed flexibility is the reason we can find a multitude of solutions for $a(t)$, providing a wide assortment of FLRW metrics (Chapter 5). As it turns out, however, the imposition of the lapse function $g_{tt} = 1$ on the FLRW spacetime is inconsistent with the existence of a local inertial frame and the strict relationship (Equation 10.4) that must be respected between its coordinates and those of the comoving frame. When viewed properly in this fashion, the expansion factor $a(t)$ must be highly constrained in order for it to produce the lapse function $L(t) = 1$. In retrospect, this should have been obvious: only an equation-of-state yielding a non-accelerating Universe is permitted by the Local Flatness Theorem when the spacetime is assumed to be FLRW.

We are entitled to ask, therefore, why the standard model (ΛCDM; Chapter 6), with an empirically inferred blend of baryonic and cold dark matter, radiation and a cosmological constant, seems to be consistent with $L(t) = 1$, even though it produces a complicated redshift-dependent equation-of-state and $\ddot{a} \neq 0$? Chapter 9 provided a partial answer to this question. In spite of the apparent flexibility afforded the expansion dynamics by its highly parametrized formulation, ΛCDM ironically adheres to the constraint $R_{\rm h} = ct$, equivalent to $a(t) = (t/t_0)$, when we view its evolution averaged over a Hubble time (i.e., the quantity $t_{\Lambda\rm CDM}$ at $z = 0$ in Equation 9.3). Even though ΛCDM's changing constituents could have produced all manner of acceleration or deceleration, the observations are telling us that the Hubble radius today, $R_{\rm h} \equiv c/H_0$, in terms of the Hubble constant H_0, equals ct_0 within the measurement error (Section 9.1). But this equality is precisely what we would have measured if $a(t) = (t/t_0)$ all along.

Given the impact of this argument, doesn't it make sense to simply impose the constraint $a(t) = (t/t_0)$ or, equivalently, $R_{\rm h} = ct$, on ΛCDM and test whether it is supported by the data? We have actually already started doing this, specifically in

Sections 9.3 and 9.4, where we tested the predicted redshift-age relation with and without this condition. Insofar as the timeline is concerned, there is no question that the observations strongly favor the zero active mass condition, $\rho + 3p = 0$, which gives $R_h = ct$ (via Equation 4.29), thereby supporting the theoretical result we have derived in this section.

Many other kinds of observation are available, and much of the material in Chapters 11 and 12 is focused on directly addressing this question. As we shall see, the data overwhelmingly support the idea that the cosmic spacetime is FLRW, driven by an expansion with $a(t) = (t/t_0)$. In Chapter 12, we shall advance this discussion even further, demonstrating that—as a largely empirical model—ΛCDM is really just an approximation to a more formal cosmology constrained by the zero active mass condition, the basis for $R_h = ct$. And we shall also show that several yet unresolved issues plaguing the standard model, notably its temperature and electroweak horizon problems, completely disappear when we impose the zero active mass condition as the equation-of-state in ΛCDM.

10.2   THE COSMOLOGICAL ORIGIN OF $E = mc^2$

An FLRW universe constrained by the zero active mass equation-of-state, $\rho + 3p = 0$, should leave a tell-tale signature on fundamental physics, given that it uniquely has a gravitational radius R_h receding at lightspeed from the observer. The causally-connected region in this cosmos is well defined and easily interpreted, thereby providing an accessible link to other branches of physics. For example, a clear implication of a Universe sub-divided into two well-understood regions, one causally-connected and surrounding the observer, the other a disconnected exterior domain, is that particle interactions are always strictly limited to a known, finite volume. Our result in the previous section leaves no ambiguity about how big this region is—we know it precisely as a function of cosmic time t. Here, we begin to explore the impact of this null gravitational horizon on local physics by calculating the binding energy of a mass m due to its gravitational attraction to that segment of the Universe with which it is causally connected. The outcome of this calculation will be surprisingly auspicious.

In Chapter 7, we learned how to re-write the FLRW metric (Equation 4.19) in terms of the proper distance $R(t) = a(t)r$ instead of the comoving radius r, thereby showing its explicit dependence on the gravitational radius $R_h = c/H$. This expression (Equation 7.12), is based on the coordinate system (ct, R, θ, ϕ) and contains the factor Φ defined in Equation (7.13). It is specifically this function of R and R_h that determines all of the features we associate with time dilation and length contraction as the observer approaches the apparent horizon.

Let us now define the 4-momentum of a particle

$$p^\mu \equiv (E/c, p^R, p^\theta, p^\phi), \tag{10.19}$$

where E is its energy, and p^i are the usual spatial components, and consider the invariant contraction $p^\mu p_\mu$. For the metric coefficients in Equation (7.12), one has

$$\Phi \left[1 + \left(\frac{R}{R_h} \right) \Phi^{-1} \frac{\dot{R}}{c} \right]^2 \left(\frac{E}{c} \right)^2 - \Phi^{-1} \left(m\dot{R} \right)^2 = \Xi^2 , \tag{10.20}$$

where Ξ is a scalar constant yet to be determined, and we have assumed purely radial motion with $p^\theta = p^\phi = 0$ and

$$p^R = m\dot{R} , \tag{10.21}$$

in terms of the particle's rest mass m. We point out that no additional factor, such as a time dilation, is necessary in Equation (10.21) because the cosmic time t, used in the derivative, is also the local proper time at every spacetime point in the cosmic fluid. Later in this section, we shall confirm that the contraction of p^μ with itself, based on the definitions in Equations (10.19) and (10.21), is a scalar constant in the spacetime described by Equation (7.12). Equation (10.20) thus yields the particle's energy E in terms of its momentum $m\dot{R}$ everywhere in the medium, starting from the observer's location at the origin ($R = 0$) all the way to the gravitational horizon at R_h.

We shall write Equation (10.20) in a somewhat more conventional form

$$E^2 = \frac{(c\Xi)^2 \Phi + (mc)^2 \dot{R}^2}{\left[\Phi + (R/R_h) \left(\dot{R}/c \right) \right]^2} , \tag{10.22}$$

and first consider what happens at the horizon, where $R = R_h$ and $\dot{R} = c$, while $\Phi = 0$. Clearly,

$$E(R_h) = mc^2 . \tag{10.23}$$

But notice that this value comes—not from Ξ, which we might have assumed ab initio, but rather—from the momentum transitioning to its relativistic limit, $p^R \to mc$, while the contribution from Ξ itself gets redshifted away completely as a result of $\Phi \to 0$ when $R \to R_h$. This is a remarkable result because it tells us that the particle's escape energy as it nears the gravitational horizon is what we have become accustomed to calling its rest-mass energy mc^2. Of course, the emphasis here is on the phrase *escape energy* because this value of E is *entirely* due to p^R at R_h.

To examine how this energy would change away from the apparent horizon, assume that it has no peculiar velocity at $R < R_h$, so that

$$m\dot{R} = mc \left(\frac{R}{R_h} \right) . \tag{10.24}$$

The general expression for the total energy is therefore

$$E^2 = (mc^2)^2 \left[1 - \left(\frac{R}{R_h} \right)^2 \right] \left(\frac{\Xi}{mc} \right)^2 + (mc^2)^2 \left(\frac{R}{R_h} \right)^2 . \tag{10.25}$$

A quick inspection of Equation (7.12) shows that, for the zero active mass equation-of-state, i.e., the FLRW spacetime with $R_h = ct$, the metric coefficients g_{tt} and g_{RR} are time-independent because both $R(t)$ and R_h are proportional to t. But energy is conserved along a particle geodesic (Equation 10.24) when the spacetime metric is independent of time (Secction 8.2). The fact that $\ddot{a} = 0$ when $\rho + 3p = 0$ (Equation 4.29), means that the particle cannot accelerate and gain or lose energy to the background. Therefore, E in Equation (10.25) must be constant within the framework of $R_h = ct$. But according to this energy conservation equation, E can be constant only when $\Xi = mc$, in which case the equation

$$E = mc^2 \tag{10.26}$$

must hold everywhere and at all times.

This result is just as remarkable as Equation (10.23), for it tells us that the *total* energy E can remain constant even though p^R ranges from 0 at the origin to its maximum value mc at R_h. Thus, the particle's *binding energy* mc^2 at the origin, which has no kinetic contribution at all, is gradually converted into kinetic energy as R approaches R_h, and becomes entirely kinetic when $R = R_h$. Throughout this range, the total energy remains fixed and equal to mc^2.

Moreover, none of these results are affected by the time-dependence of R_h. Even though R_h increases with time, E is always constant and p^R depends only on the *ratio* R/R_h. So the value $E = mc^2$ and its transition from binding to kinetic energy (via Equations 10.24 and 10.25) remain valid forever. This is a crucial point, because rest-mass energy today would therefore be identical to its value moments after the Big Bang, as long as inertial mass has not changed. In this context, we must remind ourselves of the clear distinction we are making between the origin of inertia, i.e., rest mass, m, and the nature of rest-mass energy, mc^2. Our current understanding is that the Higgs mechanism, with its $SU(2)$ internal symmetry group, endows inertia to elementary particles that couple to the Higgs field [128, 129]. Why inertia is associated with an energy mc^2 is a different issue. We are finding in this chapter that an FLRW universe expanding under the $R_h = ct$ constraint uniquely allows us to identify rest-mass energy as the gravitational binding energy of an inertial mass m within its causally-connected spacetime.

It is appropriate for us to identify the quantity $E = mc^2$ as a gravitational binding energy because, according to us, this is how much energy the particle would need to free itself from its gravitational coupling to the Universe within R_h. The Universe exterior to R_h is not causally connected to us and therefore does not participate in this gravitational interaction. We have found in this section that E is gradually converted into kinetic energy (in the form of p^R), reaching its *escape* value $p^R c = (mc)c$ at our gravitational horizon. In this sense, mc^2 is literally the binding energy required to climb out of the gravitational potential well.

Mathematical consistency with these concepts is ensured by the invariance of the contracted 4-momentum vector, $p^\mu p_\mu$, which we must now prove. To do this, we recall the discussion concerning the selected worldlines with proper speed $\dot{R} \equiv dR/dt$ preceding Equation (7.12), and simplify the procedure by assuming zero peculiar motion everywhere, i.e., $\dot{r} = 0$. Thus, $\dot{R} = \dot{a}r = HR$, where H is the Hubble constant

$H \equiv \dot{a}/a$. And since $R_{\rm h} = ct$, it is trivial to see that $\dot{R}_{\rm h} = c$. The geodesics in an isotropic and homogeneous Universe are radial, so the four-velocity $U^\mu \equiv dX^\mu/d\tau = dX^\mu/dt$, where $X^\mu = (ct, R, \theta, \phi)$, may be written as

$$U^\mu = (c, \dot{R}, 0, 0) . \tag{10.27}$$

The 4-momentum is simply $p^\mu \equiv mU^\mu$. To see whether its contraction $p^\mu p_\mu$ changes with time, we evaluate the derivative

$$\frac{d}{dt}\left(p^\mu p_\mu\right) = m^2 \frac{dU^\mu}{dt} U_\mu + m^2 U^\mu \frac{dU_\mu}{dt} . \tag{10.28}$$

With the four-velocity in Equation (10.27), and its covariant analogue

$$U_\mu \equiv g_{\mu\nu} U^\nu , \tag{10.29}$$

in which only the metric coefficients

$$g_{tt} \equiv \Phi\left[1 + \left(\frac{R}{R_{\rm h}}\right)\Phi^{-1}\frac{\dot{R}}{c}\right]^2 \tag{10.30}$$

and

$$g_{RR} \equiv -\Phi^{-1} \tag{10.31}$$

are non-zero, Equation (10.28) becomes

$$\frac{d}{dt}\left(p^\mu p_\mu\right) = 2m^2\dot{R}\left(-\Phi^{-1}\right)\ddot{R} - m^2\dot{R}^2\frac{d}{dt}\Phi^{-1} +$$
$$m^2 c^2 \frac{d}{dt}\left\{\Phi\left[1 + \left(\frac{R}{R_{\rm h}}\right)\Phi^{-1}\frac{\dot{R}}{c}\right]^2\right\} . \tag{10.32}$$

The right-hand side of this equation may be evaluated using

$$\frac{d}{dt}\Phi = \frac{d}{dt}\Phi^{-1} = 0 , \tag{10.33}$$

and

$$\frac{d}{dt}\left[1 + \left(\frac{R}{R_{\rm h}}\right)\Phi^{-1}\frac{\dot{R}}{c}\right]^2 = 2\Phi^{-2}\frac{R}{R_{\rm h}}\frac{\ddot{R}}{c} , \tag{10.34}$$

which allow us to see that

$$\frac{d}{dt}\left(p^\mu p_\mu\right) = -m^2\Phi^{-1}\frac{d}{dt}\dot{R}^2 + 2m^2 c^2\Phi^{-1}\frac{R}{R_{\rm h}}\frac{\ddot{R}}{c}$$
$$= -m^2\Phi^{-1}\frac{d}{dt}\dot{R}^2 + m^2\Phi^{-1}\frac{d}{dt}\dot{R}^2$$
$$= 0 . \tag{10.35}$$

The contraction of the 4-momentum p^μ is thus clearly a scalar constant in this space-time.

We may therefore fully rely on the invariance of $p^\mu p_\mu$ to tell us exactly how the energy E is changing in terms of the particle's momentum p^R, as we saw in Equation (10.22). Our physical description of the binding energy informs our understanding of what is happening, but it is ultimately the invariance of the scalar Ξ that yields the dependence of p^R on R. We actually do not have to calculate E from the gravitational interaction itself. This is already being done for us via the function $\Phi(R)$ in the metric coefficients, since it represents the redshift effects due to the gravitational attraction of the particle to the rest of the Universe within $R_{\rm h}$.

Of course, this begs the question of why a particle's inertia is proportional (or even equal) to its gravitational mass. This is a more subtle question than what we are attempting to answer here. As is well known, this is the basis for the Principle of Equivalence in general relativity (Section 2.6). And as long as this principle is realized in nature, we can use inertia to characterize the strength of the gravitational interaction with the surrounding medium, so it is indeed legitimate to ask what the gravitational binding energy is in terms of m. This is the reason we can interpret mc^2 as a gravitational binding energy in the first place. If inertia and gravitational mass were unrelated, there would be no physical reason to suppose that the rest energy associated with m should have anything to do with gravity.

It may also be helpful to pause for a moment and consider the role played by the observer in this discussion. From our perspective at the origin of the coordinates (ct, R, θ, ϕ), the Universe is not static. Every particle moves away from us at the local Hubble speed, $\dot{R}$, increasing steadily towards c at $R = R_{\rm h}$. We thus see a cosmic fluid with a total energy commensurate with its momentum p^R. Thus, if the nature of rest energy mc^2 were independent of the recessional velocity, the Hubble flow should be getting progressively more energetic as $R \to R_{\rm h}$. But this is not confirmed by the invariance of $p^\mu p_\mu$. When we view the Universe using these coordinates, we therefore conclude that the quantity mc^2 represents a blend of stored and kinetic energy, transitioning completely to $p^R c$ at the gravitational horizon.

If we were instead to view the Universe from within the comoving frame, the cosmic fluid would always be at rest (neglecting inconsequential peculiar motions). We would then see only the energy $E = mc^2$ corresponding to $p^r = 0$. But there is a different free-falling frame at each new location, so the rest energy would then be measured by different observers at different spacetime points. This switching from one observer to the next would replace the variation of p^R with R seen in the accelerated frame.

The approach we have followed in deriving our result overlaps somewhat with a method commonly used in other contexts, such as the calculation of the mass-energy of 'cosmological black holes.' Real black holes do not exist in a static, flat spacetime environment. They are necessarily embedded within an expanding FLRW metric [241, 242, 243, 244, 245]. Interestingly, modeling extended bodies in a curved spacetime introduces various degrees of coupling between their mass-energy and the geometry of the Universe, notably its apparent horizon (Chapter 7), which affects their dynamics and their own horizon. This topic does not directly refer to the nature of rest-mass energy per se, but the relationship between the enclosed energy of cosmological black holes and the presumed background metric arises from the same gravitational interaction restricted to a causally connected region that we have used to calculate the binding energy of a fundamental particle within the Universe's gravitational horizon. Several issues concerning how to properly define the mass-energy of a cosmological black hole still remain unresolved (see also the discussion in Section 7.2), but the arguments to merge the Kerr and FLRW metrics follow the same physical principles that we have used in this chapter.

The identification of rest-mass energy with the binding energy inside our gravitational horizon is quite compelling. The fact that only an FLRW metric with the zero active mass equation-of-state, producing a null gravitational horizon, uniquely allows us to draw this conclusion is another important confirmation—along with the observational evidence concerning the redshift-age relation in Chapter 9—of the theoretical argument we developed in Section 10.1. There are many more observations to consider, however, and in Chapter 12 we shall begin the process of methodically testing the $R_{\mathrm{h}} = ct$ hypothesis step by step using all of the accessible cosmological probes.

CHAPTER 11

The Horizon Problems

T HE emerging zero active mass (i.e., $R_h = ct$) constraint on the cosmic expansion brings into question the relevance of inflation in the early Universe. The standard model today relies heavily on the viability of a highly accelerated expansion shortly following the Big Bang to remove several undesirable features related to the size of the Universe and its large-scale structure. Indeed, experience over the past four decades has shown that an inflationary paradigm may be the best option for resolving both the temperature horizon problem and the origin of a primordial fluctuation spectrum.

Inflation both sustains the viability of ΛCDM and provides some of its most exciting possibilities going forward (Chapter 6), particularly if an improved understanding of the inflaton field can lead to meaningful extensions to the standard model of particle physics. Much has been invested in the inflationary concept, and expectations are high that observational breakthroughs are just around the corner. The CMB and its anisotropies are viewed as the most fertile ground for providing us with at least some of these notable discoveries.

And yet a recent result based on the *Planck* observations has created a new challenge to the basic slow-roll concept (Section 6.6). The measurement of a non-zero cutoff k_{min} in the fluctuation spectrum $\mathcal{P}_\Theta^{obs}(k)$ (Equation 6.153) directly impacts the viability of inflation to simultaneously solve the temperature horizon problem and account for the formation of structure in the early Universe. Moreover, as we navigate our way through these complex issues, a second—perhaps even more serious—horizon problem emerges. Whereas the idea of inflation was motivated by a possible phase transition at the GUT scale ($\sim 10^{15}$ GeV), we are now quite certain that the Universe must have passed through a subsequent electroweak phase transition at 159.5 GeV, approximately 10^{-11} seconds after the Big Bang. Inflation could not have fixed this problem. In fact, there is currently no established mechanism for resolving this issue in the context of ΛCDM. Given the emergence of $R_h = ct$, however, we shall also explore the possibility that *both* horizon problems may find a satisfactory resolution in a ΛCDM cosmology modified by this constraint.

11.1 INFLATION: A STATUS UPDATE

The *Wilkinson* Microwave Anisotropy Probe (WMAP) [106, 28] and *Planck* [171] have confirmed the existence of several large-scale anomalies first seen in the CMB by the Cosmic Background Explorer (COBE) [26, 27]. Over the past three decades, at least one of these—the lack of any significant correlation at angles $> 60°$—has emerged as a particularly difficult issue for inflation. Its possible impact on the inflationary paradigm has spawned a vigorous debate concerning the physical reality of this phenomenon and whether it may simply be due to some unrecognized observational systematic effect.

Initial counter arguments to a physical explanation invoked the idea that large-angle correlations may be absent due to cosmic variance [246, 247]. Estimated probabilities for the missing correlations are typically $\lesssim 0.24\%$, however, so this possibility is excluded at better than 3σ.[1] The anomaly could also arise from inaccuracies in the foreground subtraction, or some unrecognized instrumental systematics, but no firm conclusion one way or the other has yet been reached.

The largest angular scales probe different physics than the anisotropies seen at separations smaller than $2°$. The latter are highly consistent with the predictions of ΛCDM (Section 6.4). Conventional wisdom has it that fluctuations on larger scales reflect predominantly the growth of overdense regions due to metric perturbations in the expanding medium, a process known as the Sachs-Wolfe effect [118]. This is the principal reason why the absence of large-angle correlations in the high-fidelity CMB maps poses a very serious challenge to inflation and the internal self-consistency of ΛCDM, because it cannot be due to local physical effects.

Another possible issue related to the lack of large-angle correlations is the unexpectedly weak power seen by both WMAP and *Planck* in the low-ℓ multipoles (see Equation 11.5 below) compared to the higher ℓ's [251]. These two anomalies may or may not be related, and arguments have been made supporting both sides of this possibility. If they are unrelated, however, their existence reduces the probability of cosmic variance even further, exacerbating the tension with standard inflationary cosmology [252]. Clearly, we should not downplay the power deficit at large angular scales, which manifests itself in many ways, not just via these two signatures. Its impact on the interpretation of the CMB fluctuations is extensive. The broader issues associated with the low angular power continue to be studied by the *Planck* Collaboration, and will feature prominently in many followup experiments. In this chapter, we shall focus our attention specifically on the interpretation of the angular correlation function, which can now be measured with unprecedented accuracy using the latest *Planck* data release.

Some of the earliest attempts at addressing the weak power at large angles included the idea that inflation might have evolved through an early *fast-roll* stage prior to entering its more conventional slow-roll phase (Sections 6.6 and 11.2). Such a two-step process would have produced a characteristic scale when the transition mode

---

[1]A sample of the many publications discussing this topic includes refs. [248, 249, 250].

(called k_c below) crossed the Hubble horizon.[2] But the data used in this exploratory work were not precise enough to establish the existence of such a scale at better than $\sim 2\sigma$. As we shall see shortly, this situation has improved dramatically over the past two decades with the *Planck* observations.

An example of these modifications was the functional form [255]

$$\mathcal{P}_\Theta(k) = A_s k^{n_s-1}[1 - e^{-(k/k_c)^\alpha}] \tag{11.1}$$

for the scalar power spectrum, based on the original ansatz in Equation (6.153), i.e., $\mathcal{P}_\Theta(k) \sim A_s k^{n_s-1}$, where k_c is a characteristic wavenumber, and $\alpha = 1.8$. A Bayesian model comparison using the WMAP data available at that time favored an attenuated power spectrum with $k_c \approx 0.0005$ Mpc^{-1}. In the standard model, we see the last scattering surface at $z_{\rm cmb} \sim 1081$, implying an expansion factor $a(z_{\rm cmb}) = 1/(1 + z_{\rm cmb}) \approx 0.000925$ in a spatially flat Universe. The mode k_c would therefore correspond to a physical fluctuation size $\lambda_{\rm max} \sim 10$ Mpc (Equation 6.144). Given the relatively low precision of WMAP compared to *Planck*, however, this particular analysis showed that a truncated model is favored over the usual power-law distribution at only the $\sim 1.4\sigma$ confidence level.

The most recent analysis [256] of the angular correlation function in the CMB has made full use of the unprecedentedly accurate maps produced by *Planck*, focusing exclusively on the fluctuation spectrum at angles $\gg 1°$. The motivation for this simplified approach is that the strongest evidence of a non-zero $k_{\rm min}$ in the *Planck* data is likely to come from the large-angle fluctuations, for which the expansion dynamics is most important. An accurate measurement of this scale can severely test the inflationary paradigm, and perhaps even point to new physics. As we shall see in Section 11.2 below, a non-zero $k_{\rm min}$ would affect both the calculation of the angular correlation function and the number of e-folds during inflation.

The basic assumption required for the introduction of a $k_{\rm min}$ is that quantum fluctuations were seeded in the early Universe with a well-defined power spectrum $\mathcal{P}_\Theta(k)$, which then evolved via linear growth towards $t_{\rm cmb}$, though with $k_{\rm min} \neq 0$. This cutoff could have been produced, e.g., by an early transition from fast to slow-roll evolution, or might even have been generic to non-inflationary scenarios. The conventional approach until now has been to assume $k_{\rm min} = 0$ and grow the fluctuations to all observable scales during the rapid inflationary expansion.

As we saw in Section 6.4, the microwave temperature $T(\hat{\mathbf{n}})$ in every direction $\hat{\mathbf{n}}$ is assumed to be a Gaussian random field in the plane of the sky. By convention, we write it as an expansion in spherical harmonics $Y_{lm}(\hat{\mathbf{n}})$ (Equation 6.63), whose random coefficients a_{lm} have zero mean. The angular correlation function coupling directions $\hat{\mathbf{n}}_1$ and $\hat{\mathbf{n}}_2$ then depends only on $\cos\theta \equiv \hat{\mathbf{n}}_1 \cdot \hat{\mathbf{n}}_2$, which may be expanded in terms of Legendre polynomials:

$$C(\cos\theta) \equiv \langle T(\hat{\mathbf{n}}_1)T(\hat{\mathbf{n}}_2) \rangle = \frac{1}{4\pi}\sum_\ell (2\ell + 1)C_\ell P_\ell(\cos\theta) . \tag{11.2}$$

---

[2]Several of the more prominent publications contributing to this discussion include refs. [253, 254].

As discussed in Section 6.4, the coefficients are statistically independent, so they must satisfy a constraint that we repeat here for convenience:

$$\langle a_{\ell m}^* a_{\ell' m'} \rangle \propto \delta_{\ell\ell'} \, \delta_{mm'} \; . \tag{11.3}$$

And as we saw earlier in Chapter 6, statistical isotropy ensures that the constant of proportionality depends only on ℓ,

$$\langle a_{\ell m}^* a_{\ell' m'} \rangle = C_\ell \, \delta_{\ell\ell'} \, \delta_{mm'} \; , \tag{11.4}$$

in which the expansion coefficient

$$C_\ell = \frac{1}{2\ell + 1} \sum_m |a_{\ell m}|^2 \tag{11.5}$$

in Equation (11.2) represents the *angular power* of multipole ℓ. If the data included a full-sky coverage, the angular correlation function $C(\cos\theta)$ would provide a means, complementary to the angular power spectrum of the C_ℓ's itself, of analyzing the CMB anisotropies (Section 6.4). This is because $C(\cos\theta)$ contains the same information as the C_ℓ's, though it describes the anisotropies in a more readily understandable fashion.

A complete theoretical calculation of the angular correlation function involves a high degree of complexity, due in part to the many physical processes that contribute to C_ℓ. As we have seen, some of these physical influences act preferentially at large angles (i.e., $\theta \gg 1°$), while others—such as baryon acoustic oscillations (BAO)—are dominant on smaller scales. Of particular interest to our discussion of the large-angle correlations is the aforementioned Sachs-Wolfe effect, representing metric perturbations due to scalar fluctuations in the matter field. It relates the anisotropies observed in the CMB temperature today to inhomogeneities of the metric-fluctuation amplitude on the surface of last scattering. Figure 11.1 shows the range ($\ell \lesssim 30$) over which Sachs-Wolfe is dominant, including the 2013 *Planck* power spectrum, along with two theoretical fits that we shall discuss shortly. The vertical axis on this plot shows the quantity D_ℓ, defined as

$$D_\ell \equiv \ell(\ell + 1)C_\ell/2\pi \; . \tag{11.6}$$

The well-known dichotomy between the effects at large and small angles allows us to greatly simplify the calculation of C_ℓ for the purpose of finding $C(\cos\theta)$ on large scales, whose calculation would otherwise take considerable computational resources. Assuming a streamlined power-law fluctuation spectrum

$$\mathcal{P}_\Theta(k) = \begin{cases} A_s k^{n_s} & (k \geq k_{\min}) \\ 0 & (k < k_{\min}) \end{cases} \; , \tag{11.7}$$

and incorporating only the Sachs-Wolfe effect, we find that the angular power in Equation (11.5) reduces to the simplified integral expression

$$C_\ell = N \int_{k_{\min}}^{\infty} k^{n-2} \, j_\ell^2(kc\,\Delta\tau_{\text{cmb}}) \, dk \; , \tag{11.8}$$

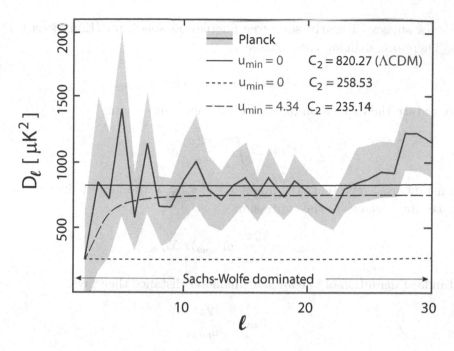

Figure 11.1 Power spectrum (black) estimated with the *NLIC* method (Planck Collaboration 2014). The 1σ Fisher errors are indicated by the grey shaded area. The best fit ΛCDM model produced by *Planck* (with $u_{\min} = 0$ and $C_2 = 820.27$), based solely on the Sachs-Wolfe effect, is represented by the thin black line. The long-dash curve with $u_{\min} \neq 0$ represents solely the Sachs-Wolfe effect, and is optimized via the best fit to $C(\cos\theta)$ in Figure 11.2. The short-dash curve, also based solely on Sachs-Wolfe, is similarly optimized for $u_{\min} = 0$. The Sachs-Wolfe effect dominates at large angles $\gg 2°$, corresponding to $\ell < 30$. Local physical effects, such as baryon acoustic oscillations, dominate for $\ell > 30$. (Adapted from ref. [256])

where j_ℓ is the spherical Bessel function of order ℓ, and $r_{\mathrm{cmb}} \equiv c\Delta\tau_{\mathrm{cmb}}$ is the comoving radius of the last scattering surface written in terms of the conformal time difference between t_0 and t_{cmb}. This expression is derived from the complete integral formalism [257, 258] encompassing all of the relevant effects at all scales, but is only valid when we calculate the power at large angles, where the Sachs-Wolfe influence dominates over everything else.

In Equation (11.8), N is the normalization constant that must be set directly from the data. Its value is typically found by optimizing the fit to the power spectrum shown in Figure 11.1, but may also be found from a best fit to the angular correlation function itself. The most influential factor from Equation (11.8) that relates to the missing correlations at large angles is $k_{\min}$. This makes sense, of course, because a non-zero value of $k_{\min}$ maps into a maximum wavelength ($\lambda_{\max} = 2\pi/k_{\min}$) of the fluctuations.

To keep this analysis as simple and transparent as possible, allowing us to avoid excessive numerical calculations that do not materially change our results, we shall use the approximation $n_s = 1$ in Equation (11.8), justified by the fact that both WMAP and *Planck* suggest a nearly scale-free fluctuation spectrum (Equations 6.153 and 6.154). Therefore, defining the variable

$$u_{\min} \equiv k_{\min} c\Delta\tau_{\text{cmb}} , \tag{11.9}$$

we may reduce the expression for C_ℓ to the even simpler form

$$C_\ell = N \int_{u_{\min}}^{\infty} \frac{j_\ell^2(u)}{u} \, du . \tag{11.10}$$

The constant $u_{\min}$ represents the angular size $\theta_{\max}$ of the largest fluctuation on the last scattering surface via the expression

$$u_{\min} = \frac{2\pi}{\lambda_{\max}} \, a(t_{\text{cmb}}) \, c\Delta\tau_{\text{cmb}} . \tag{11.11}$$

The standard definition of the angular-diameter distance then gives

$$\theta_{\max} = \frac{2\pi}{u_{\min}} . \tag{11.12}$$

Figure 11.2 shows the angular correlation function measured by *Planck* (thick solid curve) and the corresponding 1σ error bars (grey). The calculated angular power C_ℓ for multipoles $2 \le \ell \le 30$ in the conventional ΛCDM model (thin solid line) is compared to the observed power spectrum in Figure 11.1, optimized to fit the *Planck* 2013 data (thick solid lines). This procedure results in a very poor fit (long dashed curve) to the angular correlation function in Figure 11.2, as will be quantified shortly. For comparison, Figure 11.1 also shows the calculated angular-power spectrum with $u_{\min} = 0$ (thin dashed line), though with a normalization obtained from the optimized angular correlation function in Figure 11.2 (reflected in the value of C_2). In the end, neither of these approaches produces a reasonable fit to the angular correlation function. However, Figure 11.2 also shows the dramatic improvement of the fit when $u_{\min}$ is allowed to differ from zero (short dashed curve). Its optimized value is $u_{\min} = 4.34 \pm 0.50$. A zero cutoff, corresponding to the traditional ΛCDM prediction (long dashed curve) in Figure 11.2, is therefore excluded by the *Planck* measurements at better than 8σ.

We have known since the early days of WMAP that the angular correlation function predicted by inflationary ΛCDM is in tension with the measured $C(\cos\theta)$, but the comparison between the calculated angular correlation functions and the data in Figure 11.2 is quite dramatic nonetheless. Although we already had a strong inkling of this result, the robust nature of this outcome is still quite surprising. The *Planck* result for $k_{\min}$ confirms the earlier analysis by WMAP based on Equation (11.1), but does it with better precision and a much higher degree of confidence.

The cutoff $u_{\min} = 4.34$ corresponds to a maximum fluctuation size $\theta_{\max} \approx 83°$ in the plane of the sky (Equation 11.12). At a redshift $z_{\text{cmb}} = 1081$ in flat ΛCDM with the *Planck* optimized parameters, the implied value of $\lambda_{\max}$ is ~ 20 Mpc, roughly twice the scale inferred earlier from k_c. The difference may be attributed to the fact that these characteristic modes are not defined in exactly the same way. Moreover,

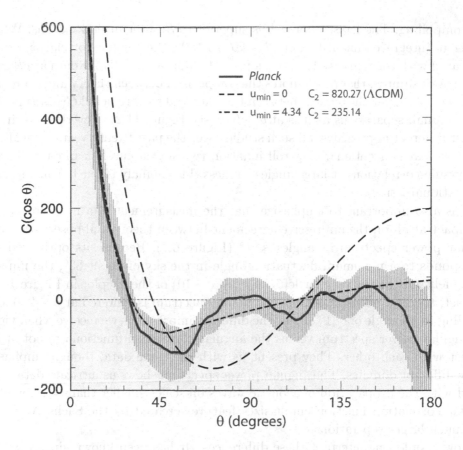

Figure 11.2 The angular correlation function (dark solid curve, with 1σ errors in grey) measured by *Planck* (Planck Collaboration 2014). Also shown are: (long dash) the prediction of standard inflationary ΛCDM, with an assumed $u_{\min} = 0$ and a value of C_2 optimized with the power spectrum in Figure 11.1; and (short dash) a cosmology (inflationary or otherwise) with a truncated fluctuation spectrum, characterized by an optimized lower limit $u_{\min} = 4.34 \pm 0.50$ and amplitude $C_2 = 235.14$. (Adapted from ref. [256])

the measurement of $k_{\min}$ has benefited greatly from the higher precision of the *Planck* data.

This measurement of $k_{\min}$ is a culmination of two-decades of effort to understand the missing large-angle correlations in the CMB. During this time, the anomaly has become more prominent as the quality of the observations has improved. At first, this large-angle deficiency was characterized using the so-called $S_{1/2}$ statistic [28], basically an integral of $C(\cos\theta)^2$ from $\cos(\theta) = -1$ to $\cos(\theta) = 1/2$. The likelihood of the anomaly being just a statistical outlier due to cosmic variance was estimated using Monte-Carlo simulations to build a distribution of $S_{1/2}$ values. The probability ($\lesssim 0.24\%$) quoted earlier for conventional ΛCDM was obtained using this comparison.

The introduction of a cutoff in $\mathcal{P}_\Theta(k)$ has moved us beyond merely quantifying the anomaly, attempting instead to account for its existence with a physical explanation. The analysis of the angular correlation function has thus become much more rigorous and compelling. The latest approach is superior to $S_{1/2}$ for several reasons. Whereas $S_{1/2}$ is an integrated quantity from $\theta \sim 60°$ to $180°$, the angular correlation function calculated with the truncated fluctuation spectrum is used to fit the data at *all angles*. Notice, for example, that $S_{1/2}$ ignores the comparison between theory and observation at angles $\lesssim 60°$, yet the tension between the data and the predicted $C(\cos\theta)$ is just as large at small separations as it is at $\theta > 60°$ (see Figure 11.2). Nevertheless, in spite of their different approaches, all such studies over the past twenty years have affirmed the view that conventional slow-roll inflation (without a cutoff) cannot account for the missing correlations at large angles—unless this anomaly is due to some unknown observational issue.

It is also important to emphasize that the measurement of a non-zero k_{min} has no impact at all on the impressive agreement between theory and observation of the angular power spectrum at angles $\lesssim 1°$ (Figure 6.7). Because its optimized value corresponds to a maximum fluctuation angle in the sky of about $83°$, the impact of k_{min} is felt only at the very far-left (i.e., $\ell \sim 1-10$) of the D_ℓ plot in Figure 11.1. In contrast, the first acoustic peak in the power spectrum is centered at $\ell \sim 200$, corresponding to an angle of $\sim 1°$. Thus, the different impressions we receive when viewing the angular power spectrum versus the angular correlation function are not at all in conflict with each other. They present us with the same data, though emphasizing vastly different domains. The angular power spectrum shows us intricate details associated with the propagation of acoustic waves on scales smaller than $\sim 1°$, while the angular correlation function emphasizes features created by the Sachs-Wolfe effect over much larger separations.

How should one interpret these differences? It has been known since the 1990's that BAO features in the angular power spectrum depend only weakly on the cosmology [122], essentially because the propagation of sound waves over relatively small distances is affected mostly by local physical effects, rather than the expansion dynamics farther out. This may be understood in the context of the Birkhoff-Jebsen theorem (Section 7.2), which explains how the influence of the large-scale expansion cancels out locally when the Universe is isotropic. The situation for the angular correlation function changes, however, because large-angle features directly probe the competition between dilution from the Hubble expansion and growth due to gravity. The angular correlation function is therefore understandably much more sensitive to any imperfections of the inflationary model.

11.2   VIABILITY OF SLOW-ROLL INFLATION

Given an inflaton potential, $V(\phi)$, and the assumption that a minimum wavenumber corresponds to the *first* mode crossing the Hubble horizon (Section 6.7), the measurement of k_{min} is critically important to understanding the dynamics of the early Universe because it signals a precise cosmic time, t_{start}, at which slow-roll inflation would have started.

Without this constraint, slow-roll inflation would have stretched all fluctuations beyond the horizon, producing a $\mathcal{P}_\Theta(k)$ with $k_{min} = 0$ and strong CMB correlations at all angles. The *measured* wavenumber cutoff is instead

$$k_{\min} = \frac{4.34 \pm 0.50}{r_{\mathrm{cmb}}} , \tag{11.13}$$

where r_{cmb} is the comoving distance to redshift $z_{\mathrm{cmb}} = 1081$, corresponding to the last scattering surface (LSS) in flat ΛCDM. For the latest *Planck* parameters [29] ($H_0 = 66.99 \pm 0.92$ km s^{-1} Mpc^{-1}, $\Omega_{\mathrm{m}} = 0.321 \pm 0.013$, $\Omega_\Lambda = 0.679 \pm 0.013$, and $\Omega_{\mathrm{r}} = 9.3 \times 10^{-5}$), we may use the Friedmann Equation (4.28) in the form

$$H^2(a) = H_0^2 \left(\frac{\Omega_{\mathrm{m}}}{a^3} + \frac{\Omega_{\mathrm{r}}}{a^4} + \Omega_\Lambda \right) , \tag{11.14}$$

together with Equation (11.19) below, to find that $r_{\mathrm{cmb}} \approx 13{,}804$ Mpc, implying a minimum wavenumber

$$k_{\min} = (3.14 \pm 0.36) \times 10^{-4} \text{ Mpc}^{-1} . \tag{11.15}$$

As usual, the scaled densities $\Omega_i \equiv \rho_i(t_0)/\rho_c$ for each species "i" (Equation 6.3) are written in terms of the critical density ρ_c (Equation 6.1).

As we saw in Section 6.7, mode k would have crossed the Hubble horizon at time t_c, defined by the equality [133]

$$\frac{\lambda_k(t_c)}{2\pi} = \frac{c}{H_c} , \tag{11.16}$$

in terms of the wavelength $\lambda_k(t_c) = 2\pi a(t_c)/k$, the expansion factor $a(t_c)$ in the FLRW spacetime (Equation 4.19), and the Hubble constant H_c at that instant. The presence of a cutoff $k_{\min}$ therefore implies that basic slow-roll inflation is subject to the initial condition

$$a(t_{\mathrm{start}})H_{\mathrm{start}} = 94.3 \pm 10.9 \text{ km s}^{-1} \text{ Mpc}^{-1} . \tag{11.17}$$

We shall use this observational constraint to critically examine whether slow-roll inflationary models can still solve the temperature horizon problem in light of the $k_{\min}$ measurement. We shall begin with pure slow-roll inflation on its own, and then add some flexibility by allowing a kinetic-dominated (KD) or radiation-dominated (RD) phase to precede the slow-roll expansion. The introduction of such a phase prior to a flattening of the inflaton potential (Figure 6.12) was one of the motivations for attempting to mitigate the large-angle anomaly in the CMB with the hypothesized scalar fluctuation spectrum in Equation (11.1).

The simplest assumption we can make is a pure exponential (i.e., de Sitter) expansion. Today we see a CMB temperature equilibrated across the sky, which means that photons had enough time to travel over a comoving distance $\geq 2r_{\mathrm{cmb}}$ prior to decoupling from the matter. The minimal condition required to solve the temperature horizon problem is therefore

$$r_{\mathrm{preCMB}} = 2r_{\mathrm{cmb}} \equiv 2c \int_{t_{\mathrm{cmb}}}^{t_0} \frac{dt}{a(t)} . \tag{11.18}$$

Rewriting r_{cmb} in terms of $H = \dot{a}/a$, we have

$$r_{\text{cmb}} = c \int_{a_{\text{cmb}}}^{a_0} \frac{da}{a^2 H} ,$$ (11.19)

where $H(a)$ is a function of a, and a_0 is the expansion factor today, which may be normalized to 1 if the cosmology is spatially flat.

The comoving distance r_{preCMB} is calculated from the start of inflation ($a_{\text{start}} \equiv a[t_{\text{start}}]$) to decoupling, but its main contribution comes from the expansion up to a_{end}, after which the inflaton field becomes sub-dominant. To a good approximation then, we may put

$$r_{\text{preCMB}} \approx c \int_{a_{\text{start}}}^{a_{\text{end}}} \frac{da}{a^2 H} .$$ (11.20)

But $H(a)$ $(= H_{\text{start}})$ is constant during the de Sitter expansion, so the integration is trivial:

$$r_{\text{preCMB}} = \frac{c}{H_{\text{start}}} \left(\frac{1}{a_{\text{start}}} - \frac{1}{a_{\text{end}}} \right) .$$ (11.21)

In addition, $a_{\text{start}} \ll a_{\text{end}}$, so

$$r_{\text{preCMB}} = \frac{c}{H_{\text{start}} a_{\text{start}}} .$$ (11.22)

The constraint in Equation (11.17) thus suggests that r_{preCMB} is only $\approx 3,181$ Mpc, significantly smaller than the comoving distance $2r_{\text{cmb}} \approx 27,608$ Mpc required to overcome the temperature horizon problem. This factor 9 disparity therefore rules out pure exponential inflation.

With slow-roll conditions, however, the inflationary $H(a)$ would not have been exactly constant. It is easy to see from Equation (6.88) that $H(a) \leq H_{\text{start}}$ for all $a \geq a_{\text{start}}$ when the small parameter ϵ is monotonic. One should therefore expect r_{preCMB} to be bigger than the value in Equation (11.22) if the initial condition in Equation (11.17) remains the same.

To calculate the difference, we start by defining the new variable

$$\beta(a) \equiv \frac{1}{Ha^2} ,$$ (11.23)

i.e., the integrand in Equation (11.20). This function is bounded by the constraints shown schematically in Figure 11.3. The measured cutoff k_{min} corresponds to the thin solid curve labeled β_{cutoff}, where

$$\beta_{\text{cutoff}} \equiv \frac{1}{(H_{\text{start}} a_{\text{start}}) a} \propto \frac{1}{a} .$$ (11.24)

Any inflationary expansion must begin at a_{start} somewhere on this curve. In the de Sitter example we just considered (thin dashed curve), H is constant, so inflation

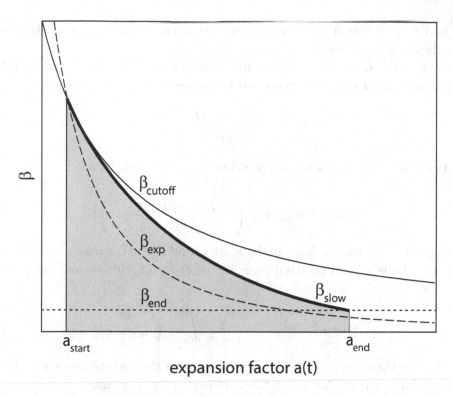

Figure 11.3 Example of a permitted $\beta(a)$ trajectory versus the expansion factor a for slow-roll inflationary models (thick solid curve). The other curves correspond to: $\beta_{\text{cutoff}} = (H_{\text{start}} a_{\text{start}})^{-1} a^{-1} \propto 1/a$ (thin solid); $\beta_{\text{exp}} = (H_{\text{start}})^{-1} a^{-2} \propto 1/a^2$ (thin long dash); and $\beta_{\text{end}} = (H_{\text{end}} a_{\text{end}}^2)^{-1} = $ constant (thin short dash). The grey shaded area shows the dominant contribution to the integral for r_{preCMB}, which should be compared to the comoving distance r_{cmb} to decoupling. (Adapted from ref. [259])

would have started at the intersection of the thin-solid and thin-dashed curves, and evolved according to

$$\beta_{\text{exp}} \propto \frac{1}{a^2} \tag{11.25}$$

after that.

In the standard model, the Universe was also dominated by radiation at the end of inflation, so

$$H_{\text{end}}^2 = H_0^2 \left(\frac{\Omega_{\text{r}}}{a_{\text{end}}^4} \right) . \tag{11.26}$$

For de Sitter inflation with $H(a) = H_{\text{start}} = H_{\text{end}}$, the condition in Equation (11.26) is represented by the horizontal short-dashed line near the bottom of the figure. On this curve, we have

$$\beta_{\text{end}} \equiv \frac{1}{H_{\text{end}} a_{\text{end}}^2} = \text{constant} . \tag{11.27}$$

A typical slow-roll inflationary model with a slowly varying H would have followed the trajectory labeled $\beta_{\text{slow}}(a)$, shown as a thick solid curve, sandwiched between

β_{cutoff} and β_{exp}. It would never have been able to cross the β_{cutoff} boundary because H was always smaller than its starting value H_{start}.

Next, it is not difficult to show from Equations (6.104), (6.109) and (6.120) that the small parameter ϵ may be expressed in the form

$$\epsilon \equiv \frac{m_P^2}{4\pi} \left(\frac{1}{H} \frac{\partial H}{\partial \phi} \right)^2, \tag{11.28}$$

where m_P is the usual Planck mass (Section 6.6). A straightforward integration then gives

$$H(\phi) = H_{\text{start}} \exp \left(-\int_{\phi_{\text{start}}}^{\phi} \sqrt{\frac{4\pi \epsilon(\phi)}{m_P^2}}\, d\phi \right), \tag{11.29}$$

where ϕ_{start} is the inflaton field at the start of inflation. We may also calculate the number of e-folds (Equation 6.90) from the start of inflation to any given ϕ, which we write as

$$N(\phi_{\text{start}}, \phi) \equiv \ln \left(\frac{a}{a_{\text{start}}} \right) = \int_{\phi_{\text{start}}}^{\phi} \sqrt{\frac{4\pi}{m_P^2\, \epsilon(\phi)}}\, d\phi. \tag{11.30}$$

To check that these expressions are consistent with the requirements we introduced in Section 6.6, we note that $\epsilon = 0$ when H is strictly constant, and it is small if H changes slowly. Slow-roll inflation must therefore end when $\epsilon \to 1$, since the slow-roll approximation breaks down when $\epsilon > 1$.

Strictly speaking, true slow-roll inflationary models actually have $\epsilon \ll 1$ during most of the inflationary phase, and approach $\epsilon = 1$ only at the end, when $\partial H / \partial \phi$ becomes very large and the inflaton field somehow decays into standard model particles. To represent such models, we define another parameter $0 < \varepsilon < 1$ such that $\epsilon^2 \leq \varepsilon$ during most of the inflationary expansion, possibly exceeding this limit only at the very end. The integration of Equation (11.28) now gives

$$H(\phi) > H_{\text{start}} \exp \left(-\int_{\phi_{\text{start}}}^{\phi} \sqrt{\frac{4\pi\varepsilon}{m_P^2 \epsilon(\phi)}}\, d\phi \right), \tag{11.31}$$

which may be simplified to

$$H(\phi) > H_{\text{start}} \exp(-\sqrt{\varepsilon}N) = H_{\text{start}} \left(\frac{a}{a_{\text{start}}} \right)^{-\sqrt{\varepsilon}}. \tag{11.32}$$

Assuming that $a_{\text{end}} \gg a_{\text{start}}$, we therefore find that

$$r_{\text{preCMB}}(\epsilon^2 < \varepsilon) < r_{\text{preCMB}}(\epsilon^2 = \varepsilon)$$

$$\equiv \frac{1}{(1 - \sqrt{\varepsilon})\, H_{\text{start}}\, a_{\text{start}}}. \tag{11.33}$$

Thus, a solution to the temperature horizon problem with the minimal condition $r_{\text{preCMB}} = 2r_{\text{cmb}}$ requires $\sqrt{\varepsilon} > 0.875$.

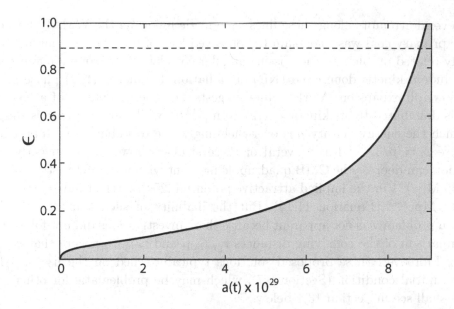

Figure 11.4 The run of ϵ as a function of $a(t)$ for the 'small-field' inflaton potential in Equation (11.34). The dashed line indicates the value required for the model to agree with the *Planck* measurement of k_{min}, while simultaneously also solving the temperature horizon problem. (Adapted from ref. [259])

This is problematic because an ϵ this large is not at all consistent with conventional slow-roll inflationary models. In fact, inflaton potentials with $\epsilon \sim 1$ during the whole expansion have already been well studied and strongly rejected on observational grounds [260]. They either produce an inflationary phase that is too short to solve the temperature horizon problem, or they predict an extremely red spectral index ($n_s \ll 1$) in $\mathcal{P}_\Theta(k)$ substantially different from the observed value 0.9649 ± 0.0042 (see Section 6.7).

As a concrete example, consider the so-called 'small-field' model with an inflaton potential

$$V(\phi) = V_0 \left[1 - (\phi/\mu)^p\right] , \tag{11.34}$$

where $p = 2$ and $\epsilon(a_{\text{end}}) = 1$. The scale μ does not affect $\epsilon(a)$, so we can ignore its specific value. Such potentials also have higher-order terms, but these become important only towards the end of inflation. The numerical solution for ϵ as a function of a, based on the *Planck* optimized parameter values, is plotted in Figure 11.4. As one can see, ϵ does eventually exceed 0.875, but only during the very brief interval $8.5 \times 10^{-29} \lesssim a \lesssim a_{\text{end}} = 8.9 \times 10^{-29}$. It is far too small everywhere else to generate a comoving distance r_{preCMB} exceeding $2r_{\text{rec}}$.

Pure slow-roll inflation therefore cannot simultaneously satisfy the initial condition in Equation 11.17, associated with the cutoff k_{min} in $\mathcal{P}_\Theta(k)$, and expand the Universe sufficiently to overcome the temperature horizon problem. Some modification to the basic premise is required, and this may be as simple as adding a fast-roll phase prior to the main event.

Several attempts along these lines were made following the WMAP data release. Their primary goal was to explain the observed lack of power on the largest scales, a closely related problem to the missing angular correlations. A common approach was to include a kinetic-dominated (KD) or radiation-dominated (RD) phase preceding the slow-roll expansion. As the name suggests, the energy density of a KD inflaton field is dominated by its kinetic energy term, $\dot{\phi}^2/2$, while an RD phase is simply one in which the energy density ρ is overwhelmingly due to radiation with an equation-of-state $p = \rho/3$. At least several of these studies showed that an early fast-roll inflation can depress the CMB quadrupole moment with a characteristic scale $k_{KD} \sim (3,759 \text{ Mpc})^{-1}$ in the implied attractive potential [254], interestingly close to $k_{min} = (3,442 \text{ Mpc})^{-1}$ (Equation 11.15). But the inability of such a model to solve the horizon problem was not apparent because these investigations did not follow up with a comparison of the comoving distances r_{preCMB} and r_{cmb} to ensure that $r_{preCMB} \geq 2r_{cmb}$. Moreover, these proposed solutions typically relied on the use of a Bunch-Davies initial condition (Section 6.7), which may be problematic for other reasons, as we shall see in Section 13.1 below.

This general approach of modifying how inflation started was also pursued by the Planck Collaboration [29], which considered the impact of a cutoff on the angular power spectrum, though not the angular correlation function itself. They also ignored the impact such a cutoff would have on r_{preCMB} versus r_{cmb}. Several other studies have modified the inflaton potential in order to reduce the low-multipole power compared to the ΛCDM prediction, though again without simultaneously considering whether such corrections would still allow inflation to fix the horizon problem [261, 262, 263, 264].

It should now be clear that in order to understand whether or not inflation can be confirmed by the data, we must be able to fix *both* problems simultaneously, with the same inflaton field. Starting with the basic slow-roll picture we have been describing in this chapter, let us now augment the inflaton potential by adding a KD or RD phase to the inflationary expansion, and see whether this two-step process can overcome the deficiency of slow-roll inflation on its own, and mitigate the inconsistency between r_{preCMB} and r_{cmb}, even when the missing large-angle correlations require a cutoff k_{min} in $\mathcal{P}_\Theta(k)$.

A scalar-field dominated Universe satisfies the dynamical Equations (6.100) and (6.101), from which one can show that

$$\dot{H} = -\frac{4\pi}{m_P^2}\dot{\phi}^2 , \tag{11.35}$$

and

$$\dot{\phi} = -\frac{m_P^2}{4\pi}\frac{\partial H}{\partial \phi} . \tag{11.36}$$

During a KD phase, Equation (6.101) reduces to

$$H(\phi)^2 \approx \frac{8\pi}{6m_P^2}\dot{\phi}^2 . \tag{11.37}$$

Thus, solving for H, we find that

$$H(\phi) = H_{\text{start}}\, e^{\frac{2\sqrt{3\pi}}{m_{\text{P}}}(\phi - \phi_{\text{start}})} \tag{11.38}$$

where, as always, the subscript 'start' denotes the beginning of the slow-roll phase. In addition,

$$\frac{\partial H}{\partial \phi} = \frac{2\sqrt{3\pi}}{m_{\text{P}}} H_{\text{start}}\, e^{\frac{2\sqrt{3\pi}}{m_{\text{P}}}(\phi - \phi_{\text{start}})} . \tag{11.39}$$

Assuming that Equation (6.150) is at least approximately valid during the KD expansion, it is not difficult to show from Equations (11.37)–(11.39) that the fluctuation spectrum produced by this phase is $\mathcal{P}_\Theta(k) \propto k^3$. This represents an important modification to the predicted $C(\cos\theta)$ because the angular power C_ℓ of each multipole ℓ (Equation 11.10), from which the angular correlation in Equation (11.2) is calculated, depends on the entire fluctuation spectrum. Modifications to $\mathcal{P}_\Theta(k)$ produced during the KD or RD phase can alter $C(\cos\theta)$ from that predicted by pure slow-roll conditions. Before attempting to optimize the transition mode—which we shall call k_{start} to distinguish it from k_{min}—it is therefore necessary to first assess these changes quantitatively.

Using the modified fluctuation spectrum

$$\mathcal{P}_\Theta(k) = \begin{cases} A_s(k/k_*)^{n_s-1} & \text{if}\, k \geq k_{\text{start}} \\ A_s(k_{\text{start}}/k_*)^{n_s-4}(k/k_0)^3 & \text{if}\, k < k_{\text{start}} \end{cases} , \tag{11.40}$$

where k_* is the pivot scale, we find that Equation (11.10) must now be written

$$C_\ell = N \int_0^{u_{\text{start}}} \left(\frac{u}{u_{\text{start}}}\right)^3 \frac{j_\ell^2(u)}{u}\, du + N \int_{u_{\text{start}}}^\infty \frac{j_\ell^2(u)}{u}\, du . \tag{11.41}$$

The normalization constant N encompasses A_s and several other factors. The angular correlation function $C(\cos\theta)$ is then recalculated using Equation (11.2), and its fit to the *Planck* data (Figure 11.2) is optimized by varying the transition wavenumber k_{start}. This procedure produces the best-fit value

$$k_{\text{start}} = (4.12 \pm 0.47) \times 10^{-4}\ \text{Mpc}^{-1} , \tag{11.42}$$

somewhat larger than k_{min} (Equation 11.15), but not dissimilar to it.

Figure 11.5 shows the optimized angular correlation functions for $\mathcal{P}_\Theta(k)$ with a hard cutoff k_{min} (solid curve), and $\mathcal{P}_\Theta(k)$ given in Equation (11.40) with the transition mode k_{start} (dashed curve). The differences between these two theoretical curves are tiny compared to the 1σ error bars (Figure 11.2), so they are indistinguishable. At this level of accuracy, either one of them would be a satisfactory fit to the *Planck* data. So the central issue is whether the addition of a KD phase to the inflationary expansion can alleviate the tension between r_{preCMB} and r_{cmb}.

With the inclusion of the KD phase, the initial condition in Equation (11.17) must be modified in order to reflect the value of k_{start} rather than k_{min}. We now have

$$a(t_{\text{start}})H_{\text{start}}\,|_{k_{\text{start}}} = 123.7 \pm 14.3\ \text{km s}^{-1}\ \text{Mpc}^{-1} , \tag{11.43}$$

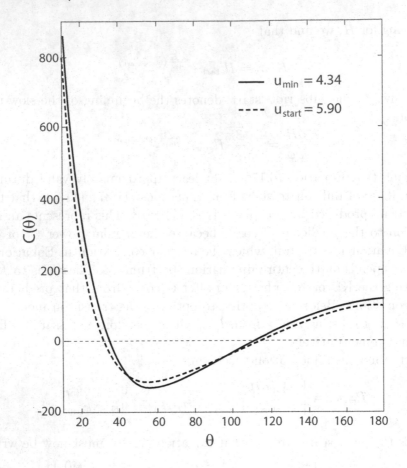

Figure 11.5 A comparison of the best-fit angular correlation functions for $\mathcal{P}_\Theta(k)$ with a cutoff $k_{\min}$ (solid curve), corresponding to the short-dashed curve in Figure 11.2; and for $\mathcal{P}_\Theta(k)$ in Equation (11.40), with an optimized transition scale $k_{\text{start}} = 4.12 \times 10^{-4}$ Mpc^{-1} (dashed curve). (Adapted from ref. [259])

which actually produces a *smaller* r_{preCMB} when used with Equation (11.22). But there are additional contributions to the comoving distance from the expansion of the Universe prior to the slow-roll phase.

From Equation (11.36), we see that

$$\frac{dt}{d\phi} = -\sqrt{\frac{\pi}{3}} \frac{2}{m_\text{P} H_{\text{start}}} e^{\frac{2\sqrt{3\pi}}{m_\text{P}}(\phi_{\text{start}}-\phi)} , \qquad (11.44)$$

whose integration gives the time as a function of ϕ:

$$t(\phi) = \frac{1}{3H(0)} e^{\frac{-2\sqrt{3\pi}}{m_\text{P}}\phi} = \frac{1}{3H} . \qquad (11.45)$$

A further integration then yields the expansion factor,

$$a(t) = \mu_{\text{KD}} \, t^{1/3} , \qquad (11.46)$$

where μ_{KD} is a constant whose specific value is irrelevant for this argument. The comoving distance traversed by a photon during the KD phase, starting at a_{init}, is thus

$$r_{KD} = \frac{3c}{2\mu_{KD}^3} \left(a_{start}^2 - a_{init}^2 \right) . \tag{11.47}$$

As long as the KD expansion began right after the Big Bang, we may assume that $a_{init} \ll a_{start}$, and Equations (11.38) and (11.39) then lead to the final result

$$r_{KD} \approx \frac{c}{2a_{start}H_{start}} . \tag{11.48}$$

When added to the comoving distance traveled by a photon during slow-roll, this KD contribution increases r_{preCMB} by about 50%, but when coupled to the change in $a(t_{start})H_{start}$ (Equation 11.43 versus 11.17), the overall effect is an increase by a mere $\sim 14\%$. So instead of $\approx 3,181$ Mpc, the combination of slow-roll inflation preceded by a KD phase would have resulted in $r_{preCMB} \approx 3,626$ Mpc, still a factor 8 too small to fix the temperature horizon problem.

The only remaining remedy is to include an RD phase prior to the inflationary expansion. If the Universe was dominated by radiation from the Big Bang to the onset of inflation, the Hubble constant during that time would have been

$$H = \mu_{RD}a^{-2} , \tag{11.49}$$

where μ_{RD} is another constant of no particular consequence. Thus, $a(t) \propto t^{1/2}$, and the comoving distance traveled by a photon prior to the slow-roll phase would have been

$$r_{RD} = c \int_0^{t_{start}} \frac{dt}{a} = \frac{ca_{start}}{\mu_{RD}} . \tag{11.50}$$

Thus, combining this with Equation (11.49), we have

$$r_{RD} = \frac{c}{a_{start}H_{start}} . \tag{11.51}$$

The addition of an RD expansion preceding the slow-roll phase may double the comoving distance traveled by a photon prior to the end of the inflation, but this too is insufficient to solve the temperature horizon problem, which requires the comoving distance to be at least 5–10 times bigger.

The conclusion from this analysis is straightforward and robust. After two decades of growing concerns with the absence of large-angle correlations and missing low-ℓ multipole power in the CMB anisotropies, the most recent *Planck* observations have improved the measurement accuracy to the point where we can now confidently interpret these anomalies as being due to a hard cutoff in the primordial fluctuation spectrum, $\mathcal{P}_\Theta(k)$. The mode k_{min} tells us the time at which slow-roll inflation could have started, which directly impacts how much the Universe could have expanded during the quasi de Sitter phase. But none of the slow-roll inflaton potentials studied thus far, irrespective of whether an earlier KD or RD phase is included, could have produced sufficient stretching of the cosmos to fix the temperature horizon problem unless the small parameter ϵ was close to one throughout the inflationary acceleration. Such a scenario, however, predicts an extremely red spectral index completely at odds with the measured value. There does not appear to be any way of simultaneously resolving both the large-angle anomalies and the horizon problem with the same inflaton potential.

11.3   THE TEMPERATURE HORIZON PROBLEM

This outcome is quite troubling for basic ΛCDM cosmology because it creates a gap in our understanding of how the Universe could have come into existence at the time of the Big Bang. The situation is actually much worse than this, because we shall discover in the next section that the standard model suffers from a second, better understood, horizon problem associated with the aforementioned electroweak phase transition at $t_{ew} \sim 10^{-11}$. Based on the quantum mechanics we have today, it is not at all clear how even special initial conditions might have ensured a uniform Higgs vacuum expectation value throughout the Universe, including all of the causally disconnected regions. So the growing tension between the expectations of inflation and the ever improving observations is not necessarily unique, though this paradigm has been with us for four decades and a great deal of hope still rests on the promise it offers for explaining, e.g., how the primordial fluctuation spectrum might have been produced.[3]

Section 6.5 left no doubt that a temperature horizon problem exists today. But we also learned in that chapter that this issue is not endemic to all FLRW cosmologies. Had the inflationary paradigm worked as expected, it would have strongly supported the expansion history predicted by ΛCDM. The growing challenges it now faces suggest that we ought to consider an alternative approach in which the expansion history was somewhat different. Several rather strong arguments have been made in Chapters 9 and 10 (see also Chapter 12) that the Universe appears to have evolved subject to the constraint $R_h = ct$. Indeed, the use of the Local Flatness Theorem in general relativity leaves very little room for FLRW to function without it (Section 10.1). This would apply to the dynamics in the very early Universe as well, so can we demonstrate whether or not the $R_h = ct$ condition can mitigate the temperature horizon problem (and later also the electroweak horizon problem) without any additional, anomalous acceleration? The answer is yes.

Let us continue with where we left off at the end of Section 6.5. The null geodesics and proper distances provided by the $R_h = ct$ constraint, analogous to those illustrated for ΛCDM in Figure 6.10, are shown in Figure 11.6, again from the perspective of an observer in patch C (see Figure 6.9). The principal difference between this situation and ΛCDM is that now $a(t) \propto t$ for all cosmic time t. Thus, without having

---

[3]We shall see in Section 13.1, however, that inflation is not necessarily required to create a primordial fluctuation spectrum matching the features seen in the CMB. It appears that a non-inflationary scalar field can also do this, as long as the quantum fluctuations emerged into the semi-classical Universe at roughly the Planck scale.

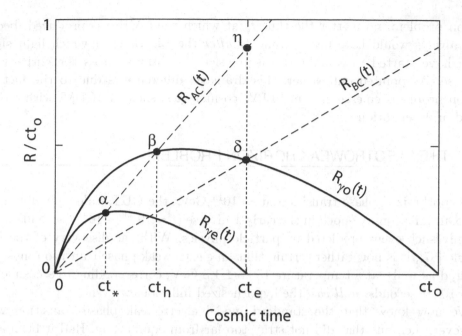

Figure 11.6 Null geodesics (solid curves) seen by an observer in patch C of Figure 6.9, for a Universe expanding under the $R_h = ct$ constraint. All the symbols have the same meaning as those of Figure 6.10. (Adapted from ref. [127])

to repeat the steps we took in deriving Equations (6.81) and (6.83), we have

$$R_{AC}(t_e) = ct_e \ln(t_e/t_*), \qquad (11.52)$$

and

$$R_{BC}(t_e) = ct_e \ln(t_0/t_e). \qquad (11.53)$$

Recall that t_* is the time at which observer A emitted a signal in order to communicate with C at or before the emission time t_e. We found in the case of ΛCDM that even putting $t_* = 0$ was not sufficient to ensure the equilibration of the CMB temperature at A and C if t_e was as early as $\approx$ 380,000 years. The CMB should have been produced at least 2.7 Gyr after the Big Bang in order for us to see a homogeneous temperature across the sky.

The situation is very different if the expansion of the Universe was guided by the $R_h = ct$ constraint at early times, as we can see in Figure 11.6. We seek a time t_* such that A and C would have been causally connected at t_e, with $R_{AC}(t_e) = 2R_{BC}(t_e)$. That is, here on Earth, we would be seeing light emitted by A and C at time t_e from opposite sides of the sky. Clearly, this condition is met when

$$t_* = t_e \left(\frac{t_e}{t_0}\right)^2. \qquad (11.54)$$

The time t_* is small—*but not zero*. Thus, if the Universe expanded subject to the $R_h = ct$ constraint throughout its early history, there would never have been a temperature horizon problem, no matter the time t_e at which the CMB was produced, because there always would have been a time t_* *after* the Big Bang at which light signals would have started to equilibrate the physical conditions across the entire visible Universe. As pointed out earlier, this dramatic difference is due to the fact that horizon problems emerge only in FLRW cosmologies, such as ΛCDM, with an early period of deceleration.

11.4   THE ELECTROWEAK HORIZON PROBLEM

The hypothesized phase transition at $\sim 10^{15}$ GeV (the GUT scale), possibly giving rise to an inflationary epoch in the early Universe ($t \sim 10^{-36} - 10^{-33}$ seconds), is not the only such event predicted by particle physics. With the discovery of the Higgs particle [87], it is now rather certain that an electroweak phase transition must have occurred as well—at a temperature 159.5 ± 1.5 GeV, corresponding to a cosmic time $t_{ew} \sim 10^{-11}$ seconds, *well past* the hypothesized inflationary event.

We now know that the standard model electroweak phase transition was a 'crossover,' i.e., one that did not stray too far from equilibrium. Had it been a first order (i.e., discontinuous) transition, it could have provided a viable explanation for the origin of baryon asymmetry, a property in particle physics responsible for the matter left over after matter-antimatter annihilations ended at very early times (Section 6.3). But this situation is not entirely clear because many extensions to the standard model of particle physics predict additional Higgs fields, possibly reopening the viability of a first-order phase transition at the electroweak scale [265]. Nevertheless, the crucial function of the electroweak phase transition was to generate fermionic mass, thereby separating the electric and weak forces, a process analogous to the symmetry breaking that may have separated the strong and electroweak forces (see discussion in Section 6.6).

A third phase transition must have occurred at roughly $100 - 150$ MeV after the Universe cooled down further (Section 6.2). In ΛCDM, this would have happened at $t \sim 10^{-6}$ seconds, from the transformation of quarks behaving like free particles in a quark-gluon plasma into the confined states we call baryons and mesons in the hadronic phase.

Our primary interest here, though, is the electroweak phase transition, because the properties of the Higgs field inevitably lead to another horizon problem, analogous to the CMB temperature horizon problem we have been discussing in this chapter. The issue now is the vacuum expectation value (the 'vev') of the Higgs field, which all of our cosmological measurements show is universal—even on scales exceeding the size of causally-connected regions. This problem is actually worse than the temperature horizon problem because inflation—if it occurred at all—would have ended well before the electroweak era, and could not have affected it in any significant way.

The electroweak horizon problem has become much more prominent in recent years, though the earliest signs of its potential impact on cosmology actually surfaced gradually, in various guises, over the past half century. But in spite of this relatively long history, we have yet to find a well-accepted solution for it. When the problem first emerged, the idea was to search for domain walls that might have been created in the expanding cosmos due to the localized transitions in the early Universe [266, 267]. As the observations improved, however, the existence of such topological defects has become less likely because they would have produced easily measurable signatures in the CMB temperature distribution [268, 269, 270]. It has become increasingly clear that the most likely solution to the electroweak horizon problem is to simply avoid it altogether or, at least, to suppress it somehow. Suggestions have been made that a second, late-time weak-scale, inflation must have occurred [271, 272, 273, 274, 275]. But none of these proposals has had any impact on our interpretations of the CMB data thus far.

What we have seen in this chapter raises the possibility that the electroweak horizon problem could be resolved with a modified expansion history of the Universe. Just as the $R_h = ct$ constraint completely eliminated the temperature horizon problem, the same concept can remove this one as well, without the introduction of highly speculative and poorly motivated extensions to the standard model of particle physics. We remind ourselves that the critical factor introduced by the $R_h = ct$ condition is that the Universe would have expanded without any deceleration, thus avoiding the principal cause of horizon incompatibilities.

With the discovery of the Higgs particle, the Higgs mechanism [128, 129] has been confirmed to a high level of confidence. There are two essential ingredients in this picture: (1) the 'turning on of the Higgs field,' which signals the time at which a non-zero vev was acquired; and (2) the size of the coupling constants characterizing the strength of the interaction with the Higgs field. In the early Universe, the electroweak symmetry was broken (Section 6.6) when the particle energies dropped below the point where their momentum per unit energy started to be affected by the relative size of their inertial masses. Symmetry was present at asymptotically high temperatures because the ratio p/E (with $E^2 = p^2 c^2 + m^2 c^4$) was virtually identical for all bosons, regardless of how their inertial mass was generated. But the Higgs mechanism broke the symmetry when the temperature dropped to a specific value—corresponding to a particular redshift, or cosmic time.

Nonetheless, a critical factor is still missing in this theory: what sets the Higgs vev? There is no known theoretical constraint on this essential property of the Higgs field. The critical temperature at which this happened could have been quite different from the one we have inferred. But whatever conditions established the vev, we can be sure that it was set uniformly throughout a causally-connected spacetime region. The electroweak horizon problem arises from the observation that the same vev has emerged everywhere, even at distances exceeding the observer's causal horizon (Figure 11.7).

One way of establishing how the vev might have been set is to estimate its likelihood function using anthropic constraints on the existence of atoms [277]. Nuclei and atoms would not exist if the light-quark and electron masses were even modestly different from the values we have measured in the laboratory [278, 279, 280]. These anthropically permitted bounds therefore provide constraints on the accessible Higgs vev distribution because the fermionic masses are proportional to it. The Higgs vev distribution function is shaped by allowed variations in the cosmology, though always

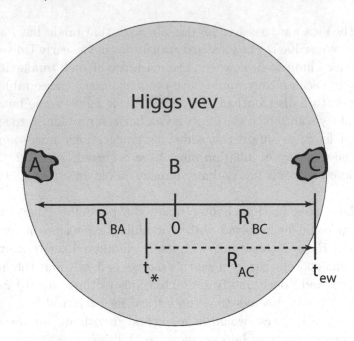

Figure 11.7 Observer B becomes causally connected with two opposite patches A and C, a proper distance $R_{BA}(t) = R_{BC}(t)$ away, within a region of uniform Higgs vev. Patch A connects with C via a null geodesic starting at time t_*, a proper distance $R_{AC}(t_{ew})$ away, which reaches the latter at the electroweak phase transition time t_{ew}. (Adapted from ref. [276])

with the requirement that nuclei and atoms must appear. At a deeper philosophical level, this assumes that a range of vev's does in fact exist, based on some unknown variable in the fundamental theory. As far as we can tell from particle physics today, the Higgs vev may extend all the way up to the Planck scale, many orders of magnitude beyond the current electroweak energy. Incidentally, the disparity in energy between the electroweak and Planck scales creates its own puzzle [281], known as the 'hierarchy problem.'

Within such an anthropic framework, there is no physical reason to expect a Higgs vev equal to its measured value, commonly referred to as v_0. The anthropic arguments have shown that its likelihood function has a median value $2.25v_0$ and a 2σ range extending from $0.10v_0$ to $11.7v_0$. Nuclear and atomic properties could therefore have varied by at least one to two orders of magnitude from one causally-disconnected region in the Universe to another, but no such variation has ever been seen.

Demonstrating how the horizon problem created by this null result may be remedied follows very similar steps to those we took in Section 11.3 when considering the CMB temperature across the sky. In fact, Figure 11.7 is analogous to 6.2, except that our concern now is the matched vev in patches A and C, rather than the temperature of the radiation emitted by these two regions. The key question is whether a light signal could have been emitted by one of these patches (A) at some $t_* \geq 0$, to reach the other (C) by the time the Higgs vev was established.

Inserting the *Planck* optimized parameters (Section 11.2) into the Hubble constant in Equation (11.14), it is easy to show that a backward integration starting at, say, the time t_{cmb} on the last scattering surface, yields the expansion factor $a(t)$ at any $t < t_{cmb}$:

$$t_{cmb} - t = \int_a^{a_{cmb}} \frac{da}{a\, H(a)} \,. \tag{11.55}$$

At the electroweak phase transition (i.e., $t_{ew} = 10^{-11}$ seconds), this integration yields an expansion factor

$$a(t_{ew}) \approx 1.93 \times 10^{-4} \tag{11.56}$$

(based on the normalization $a[t_0] = 1$ today), with a corresponding Hubble (i.e., gravitational) radius

$$R_h(t_{ew}) \equiv \frac{c}{H(t_{ew})} \approx 0.016 \ \text{Mpc} \,. \tag{11.57}$$

Recalling our discussion in Section 7.5, solutions to the null geodesic in Equation (7.27) show that the proper size of the visible Universe at any given time t is approximately $R_h(t)/2$. Thus, the causally-connected region at t_{ew} would have had a proper size

$$R_{ew}(t_{ew}) \lesssim R_h(t_{ew})/2 \approx 0.008 \ \text{Mpc} \,. \tag{11.58}$$

And shifting forward to today, this scale would have expanded to

$$R_{ew}(t_0) \sim \frac{a(t_0)}{a(t_{ew})} R_{ew}(t_{ew}) \sim 41.5 \ \text{Mpc} \,. \tag{11.59}$$

This is the proper size of the largest region we should expect to see with a uniform Higgs vev today. But the proper size of our visible Universe right now is $\lesssim R_h(t_0)/2 \approx 2,212$ Mpc—*more than* 50 *times bigger*. Thus, assuming our understanding of the electroweak phase transition is correct, we should be observing one to two order-of-magnitude variations of the fermionic and atomic properties at multiple locations across the Universe. This is absolutely not the case. Moreover, it would be very difficult to see how the electroweak phase transition could have produced the baryon asymmetry, given that we should then be seeing many pockets of antimatter at proper distances exceeding $R_{ew}(t_0)$. Again, there is simply no evidence of such exotic domains.

The conflict between our understanding of the electroweak phase transition, and the predicted expansion history in ΛCDM, is not easily resolvable because we have no viable mechanism for inducing an inflation-like expansion at t_{ew}. On the other hand, the inclusion of the $R_h = ct$ constraint in ΛCDM would modify the dynamics somewhat and prevent the electroweak horizon problem from emerging in the first place, as it did with the CMB temperature.

The geodesics we need to understand this are shown in Figure 11.8, which is clearly very similar to Figure 11.6. The only difference between them is the physical meaning of t_e versus t_{ew}, which should be obvious from the context of this discussion. The critical issue is whether there exists a time $t_* \geq 0$ such that the diametrically

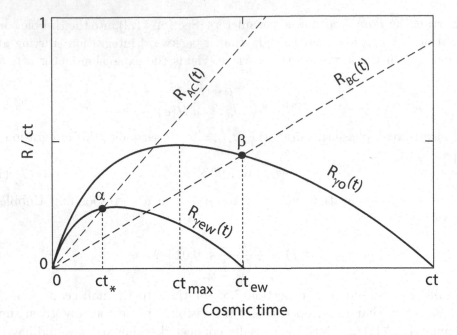

Figure 11.8 Null geodesics seen by the observer in patch C (see Figure 11.7) in a Universe expanding under the $R_h = ct$ constraint. The null geodesic labeled $R_{\gamma\text{ew}}(t)$ arrives at C at the electroweak phase transition time t_{ew}, carrying the light signal emitted at time $t < t_{\text{ew}}$ by any source a proper distance $R_{\text{src}}(t) = R_{\gamma\text{ew}}(t)$ away. Correspondingly, the null geodesic labeled $R_{\gamma 0}(t)$ arrives at C at time $t > t_{\text{ew}}$. The label 't_{max}' denotes the time at which the null geodesic $R_{\gamma 0}(t)$ attains its maximum proper distance. The proper distance from A to C and from B to C is represented by the dashed lines $R_{\text{AC}}(t)$ and $R_{\text{BC}}(t)$, respectively. The other labels and symbols are as defined previously. (Adapted from ref. [276])

opposite patches A and C within the Higgs vev were causally connected at t_{ew}, with a proper separation $R_{\text{AC}}(t_{\text{ew}})$ that grew to fill the whole visible Universe today. And the required condition, analogous to Equation (11.54), is simply

$$t_* = t_{\text{ew}} \left(\frac{t_{\text{ew}}}{t_0} \right)^2 . \tag{11.60}$$

No matter when the electroweak phase transition took place, there was always a time t_* following the Big Bang when patches throughout the visible Universe back then could have started exchanging light signals and remained causally connected up to the point where the Higgs vev was established. Very importantly, this causally-connected region would have subsequently expanded to fill all of our visible Universe today, ensuring that a single, uniform vev is manifested everywhere. Our Universe would thus have a homogeneous structure based on just a single Higgs field whose properties were set at the electroweak phase transition.

The electroweak horizon problem is not as well known as its GUT partner, probably because the latter is viewed as a more integral component of the standard model.

For example, the conventional wisdom today is that all of the large-scale structure probably originated with quantum fluctuations in the inflaton field. But the foundational theory underlying the Higgs mechanism for separating the electric and weak forces is well established, more so than the hypothesized inflaton field which lies beyond the standard model of particle physics. A conflict between the electroweak phase transition and the expansion history in ΛCDM should therefore be a reliable indicator that there may be something wrong—or, more likely, incomplete—with the cosmological framework.

In this chapter, we have highlighted the strong similarity between the two outstanding horizon problems in cosmology. Both remain problematic insofar as matching the data are concerned, including the CMB temperature, whose uniformity is heavily reliant on inflation. In a way, the resolution we have considered here will complement our discussion in Section 12.9, where we shall demonstrate that several kinds of data are pointing to ΛCDM as a largely empirical approximation to a more formal FLRW cosmology based on the $R_{\rm h} = ct$ constraint. From a purely theoretical perspective, this condition seems to be unavoidable (Section 10.1). And the horizon problems now appear to be leaving us with no other choice.

CHAPTER 12

Observing Zero Active Mass

O UR iconic founders of general relativity and cosmology could not possibly have imagined a day when the breadth and quality of cosmological observations would rival those of many, even most, other branches of physics. Today, we observe the expansion of the Universe both near ($z \sim 0$) and far ($z > 10$), we measure luminosity distances out to the edge of the quasar population at $z \sim 7$, and actually trace a unique and remarkable property of the angular-diameter distance—the maximum at $z \sim 1.7$ and its subsequent reduction back to zero as we peer ever closer to the Big Bang. We already sensed how exquisite the cosmological observations can be when we studied the redshift-age relation of supermassive black holes and galaxies in Chapter 9. And we have just witnessed in Chapter 11 the effectiveness of analyzing the CMB anisotropies extracted from the high-quality *Planck* data.

But this is only the beginning. In this chapter, we shall broaden the coverage of observations that inform our understanding of the cosmic spacetime, encompassing several measurable signatures, always with a focus on refining the equation-of-state of the cosmic fluid. Very few would even consider alternatives to the FLRW metric anymore, yet we have learned that the application of this very specialized spacetime seemingly requires an equally specialized zero active mass constraint, $\rho + 3p = 0$. The central theme of this chapter is to gauge how strongly the wide assortment of observations now at our disposal supports this condition over the unconstrained blend of matter, radiation and cosmological constant that drive the expansion in ΛCDM.

12.1 STATISTICAL METHODS USED IN MODEL SELECTION

A possible complication with this type of work, not usually encountered when the primary goal is merely to optimize a model's free parameters by adjusting its fit to the data, arises when the cosmologies being compared are formulated differently, often with unequal numbers of unknowns. In such situations, the statistical techniques used to identify the model favored by the observations can be as varied as the data themselves, though two particular implementations appear to have become quite common in cosmology now. Before addressing the actual observations, we should therefore introduce these techniques and briefly describe how they work.

12.1.1 Information Criteria

The evidence for and against competing models may be estimated using the Akaike Information Criterion (AIC), which will be featured in several applications below [282, 283, 284]. Viewed as an enhanced 'goodness of fit' criterion, the AIC extends the more familiar χ^2 statistic by taking account of the number of free parameters in each model. Like the other information criteria we shall introduce shortly, it favors the more *parsimonious* models, i.e., those with fewer parameters compared to the others, unless the latter provide a substantially better fit to the data. The inclusion of the number of free parameters reduces the possibility of overfitting. Optimizing over a greater number of parameters allows one to simply fit the 'noise,' thereby providing an advantage to those competing models with a larger set of unknowns.

The AIC ranks two or more competing models and provides a numerical measure of confidence that each model is the best [285, 286]. The confidences are analogous to likelihoods or posterior probabilities in traditional statistical inference, though not exactly because, unlike traditional inference methods, the AIC can be applied to models that are not 'nested.' When models are nested, one is a specialization of the other, and comparing them is straightforward: after each model is fit to the data, one computes the χ^2 per degree of freedom, and uses the comparison to decide which is a better fit. An alternative approach is to calculate (say, by applying an F-test) the likelihood of the null hypothesis that the simpler model is a better approximation to the 'true' one. The AIC generalizes this procedure by allowing one to compare a pair of models, neither of which is a specialization of the other. In the data analysis described below, this approach will be used to compare the basic ΛCDM model with a version that includes the $R_{\rm h} = ct$ constraint. As we shall see, the latter cosmology *always* has fewer free parameters that need to be optimized.

A typical application of the AIC begins with the following regression. Let the values $z_1, \ldots, z_n$ of an independent variable have the measured values $h_1, \ldots, h_n$ of a dependent one, with known (normally distributed) error bars $\pm\sigma_1, \ldots, \pm\sigma_n$. Now let model $\mathcal{M}$ predict values $\hat{h}_1, \ldots, \hat{h}_n$, computed from a formula $\hat{h}_i = \hat{h}_i(\vec{\beta})$ that involves a parameter vector $\vec{\beta}$ comprising k unknown parameters, i.e., $\vec{\beta} = (\beta_1, \ldots, \beta_k)$. In other words, the data model $\mathcal{M}$ is statistical and of the form

$$h_i = \hat{h}_i(\vec{\beta}) + \sigma_i Z_i , \tag{12.1}$$

where $Z_1, \ldots, Z_n$ are independent standard normal random variables. If this were linear regression, then $\hat{h}_i(\vec{\beta})$ would be $\sum_{j=1}^{k} X_{ij}\beta_j$ for known coefficients X_{ij}; typically, $X_{ij} = \hat{h}^{(j)}(z_i)$ for known functions $\hat{h}^{(1)}, \ldots, \hat{h}^{(k)}$ of z.

For model $\mathcal{M}$, the χ^2 goodness of its fit to the data is a (weighted) sum of squared errors given by

$$\chi^2 = \sum_{i=1}^{n} \frac{[h_i - \hat{h}_i(\vec{\beta})]^2}{\sigma_i^2} , \tag{12.2}$$

and the reduced χ^2 (i.e., the χ^2 per degree of freedom) is then

$$\chi^2_{\rm dof} = \frac{\chi^2}{(n-k)} . \tag{12.3}$$

We assume that the number of data points exceeds the number of free parameters, i.e., $n > k$. The parameters $(\beta_1, \ldots, \beta_k)$ are optimized to minimize the χ^2, yielding the best fit to the data. The definition of the AIC for the resulting fitted model is then

$$\text{AIC} = \chi^2 + 2k . \tag{12.4}$$

When there are two or more competing models, $\mathcal{M}_1, \ldots, \mathcal{M}_N$, and they have been separately fitted, the most likely to be nearest to the 'truth' is assessed to be the one with the least resulting AIC. The 'truth' in this case is the unknown model $\mathcal{M}_*$ that generated the data. To get a more quantitative ranking of the models, one calculates the unnormalized likelihood that $\mathcal{M}_\alpha$ is closest to the truth using the 'Akaike weight' $\exp(-\text{AIC}_\alpha/2)$, where AIC_α is assumed to come from model $\mathcal{M}_\alpha$. Informally, $\mathcal{M}_\alpha$ then has likelihood

$$\mathcal{L}(\mathcal{M}_\alpha) = \frac{e^{-\text{AIC}_\alpha/2}}{e^{-\text{AIC}_1/2} + \cdots + e^{-\text{AIC}_N/2}} \tag{12.5}$$

of being the best choice. Of course, the 2's could be omitted by redefining AIC, but the normalization implicit in Equation (12.4) is traditional. When we work with a pair of models $\mathcal{M}_1, \mathcal{M}_2$, the difference $\text{AIC}_2 - \text{AIC}_1$ determines the extent to which $\mathcal{M}_1$ is favored over $\mathcal{M}_2$.

The $2k$ term in Equation (12.4) clearly disfavors models with too many free parameters. These models can still be preferred by the observations, however, if they do a much better job of fitting the data. The choice of proportionality constant (i.e., 2) is based on an argument from information theory that has close ties to statistical mechanics. From this standpoint, any two statistical models of the data set $(h_1, \ldots, h_n)$, say, the 'true' model $\mathcal{M}_*$ and another model $\mathcal{M}$, may be thought of as probability density functions (PDF's) on $\mathbb{R}^n$, which we may denote $f_*(h_1, \ldots, h_n)$ and $f(h_1, \ldots, h_n)$, respectively. In information theory the discrepancy of the PDF f from f_* (which is a measure of distance) is given by the Kullback–Leibler formula

$$D(\mathcal{M}_* \| \mathcal{M}) = \int_{\mathbb{R}^n} dh_1 \ldots dh_n \, f_*(h) \ln \frac{f_*(h)}{f(h)} \geq 0 , \tag{12.6}$$

using the argument h to stand for the entire data set $[h_1, \ldots, h_n]$. The best model $\mathcal{M}$ from a set of candidate models is the one with the minimum $D(\mathcal{M}_* \| \mathcal{M})$, though this cannot be done literally when $\mathcal{M}_*$ is not known a priori. Nevertheless, $\mathcal{M}$ is special when it is a parametrized model optimized to minimize χ^2. In that case, it turns out that the AIC of the fitted model $\mathcal{M}$ is a good approximation to $2D(\mathcal{M}_* \| \mathcal{M})$, up to an ignorable additive constant.

The quantity AIC/2 is therefore an unbiased estimator of the distance $D(\mathcal{M}_* \| \mathcal{M})$. This statement is exact for linear regression, and correct to leading order for non-linear regression. The fitted model $\mathcal{M}$ depends on the data set, however, so both $D(\mathcal{M}_* \| \mathcal{M})$ and AIC/2 are random variables. In probabilistic language, the lack of bias means that they have the same expectation.

The extent to which the fitted AIC is an *accurate* estimate of $2D(\mathcal{M}_* \| \mathcal{M})$ has been investigated theoretically [287], and its variability has been studied empirically.

An example relevant to the data we shall consider shortly is the repeated comparison of ΛCDM to other cosmological models using data sets generated by a bootstrap method [284]. Such investigations show that the AIC is increasingly accurate when n is large. But quite generally, the magnitude of the difference $\Delta = \text{AIC}_2 - \text{AIC}_1$ provides a numerical assessment of the evidence that model 1 is preferred over model 2, which may be expressed quantitatively as follows: if $\Delta \lesssim 2$, the evidence is weak; if $\Delta \approx 3$ or 4, it is mildly strong; and if $\Delta > 5$, it is quite strong.

As noted earlier, however, one may also weight each candidate model in a Boltzmann-like fashion using its Akaike weight in Equation (12.5). In this way, the likelihood $\mathcal{L}(\mathcal{M}_\alpha)$ of model $\mathcal{M}_\alpha$, which depends on the differences between AIC_α and the AIC's of the other model(s), mirrors the posterior probability in statistical inference, even though it is not being computed by a Bayesian procedure.

One sometimes finds alternatives to the AIC in the literature, such as the lesser-known Kullback Information Criterion (KIC). This version is used because the discrepancy $D(\mathcal{M}_*\|\mathcal{M})$ is not symmetric in the PDF's f_*, f. A symmetrized version can be better for distinguishing between data models [288]. The unbiased estimator for the symmetrized version is given by

$$\text{KIC} = \chi^2 + 3\,k\,, \tag{12.7}$$

which disfavors overfitting more strongly than does the AIC.

Another alternative to the AIC, which is even better known, is the Bayes Information Criterion (BIC). The name is somewhat of a misnomer, however, because it is not based on information theory. The BIC emerges from an asymptotic $(n \to \infty)$ approximation to the outcome of a conventional Bayesian inference procedure for deciding between models [289]. This criterion is defined by

$$\text{BIC} = \chi^2 + (\ln n)\,k\,, \tag{12.8}$$

and suppresses overfitting very strongly when n is large. Several strong arguments have been made in favor of using the BIC for cosmological model selection and it has become quite popular over the past decade [290, 291].

It is not clear, however, that any one of these three information criteria should always be preferred over the others. Note, e.g., that the authors who popularized the use of Equation (12.5) [286], argue that AIC should be strongly preferred to BIC as a tool for model selection. They suggest that AIC may even be interpreted in Bayesian terms, as being the consequence of imposing a nonuniform but reasonable choice of prior distribution on the set of candidate models. On the flip side, others [292] have argued that they are both valuable tools.

The use of information criteria is clearly not a precise science, but either one of them is preferred over merely carrying out χ^2 minimization when the models being compared have different numbers of parameters. Nonetheless, to be safe, one ought to use all three whenever possible, and base the model selection on the 'preponderance' of evidence.

12.1.2 Two-point Diagnostics

The statistical method we have just described typically involves the use of a parametric model fit. A very different approach for evaluating the viability of a chosen cosmology avoids such formulations altogether, and instead relies on the use of 'two-point' diagnostics, which differ from the parametric fitting in several distinct ways.

A two-point diagnostic allows one to analyze n measurements of a particular variable in a pairwise fashion, via the construction of $n(n-1)/2$ comparisons between pairs of data. For example, in a sample of 30 measurements of some variable $V(z)$, one produces 435 comparisons and tests how well each pair of points fits the model. They also allow an assessment of how closely the stated error bars actually fit a normal distribution.

If, as frequently happens, the data are not perfectly Gaussian, a second non-Gaussian approach may be invoked, based on 'median statistics' with no reliance on error propagation [293]. With this method, one uses the fact that any single measurement within a truly random distribution has a 50% chance of being above the true median of that set. No assumption needs to be made concerning whether the distribution is normal or not. Thus, if n measurements are taken and rank ordered, the probability that the true median lies between measurements i and $i + 1$ is the binomial distribution:

$$P_i = \frac{2^{-n} n!}{i!(n-i)!} \, . \tag{12.9}$$

The essence of this test is therefore to measure the deviation of the median away from its expected value, typically zero in the case of two-point diagnostics in cosmology. And to do this, it is necessary to compute the size of the 68% confidence region of possible median values. Unlike individual measurements, however, this approach is not strictly valid for two-point diagnostics, so additional Monte-Carlo simulations need to be made with mock data in order to determine the precise number of steps defining this '1σ' region.

If the two-point diagnostic is designed to have an expectation value of zero for the 'true' cosmology, other models will have non-zero medians. The inferred 68% confidence region allows one to estimate the likelihood that these alternative values are nonetheless still consistent with zero, providing a ranking of the competing models based on their proximity to the 'true' cosmology.

Suppose we have n measurements of the variable $V(z)$ at redshifts z_i, with $i = 1 \ldots n$. For each pair of values $V(z_i)$ and $V(z_j)$, we define $\Delta z_{ij} \equiv z_i - z_j$, choosing z_i to always be greater than z_j, and $\Delta V_{ij} \equiv V(z_i) - V(z_j)$. When using weighted mean statistics (as opposed to median statistics), the variance $\sigma^2_{\Delta V_{ij}}$ of ΔV_{ij} is calculated via standard error propagation from the actual measurements $V(z_i)$.

Following the conventional procedure in weighted mean statistics, we would next define the weighted mean function

$$\Delta V_{\text{w.m.}} = \sigma^2_{\Delta V_{\text{w.m.}}} \sum_{i=1}^{n-1} \sum_{j=i+1}^{n} \frac{\Delta V_{ij}}{\sigma^2_{\Delta V_{ij}}}, \tag{12.10}$$

with corresponding variance

$$\sigma^2_{\Delta V_{\text{w.m.}}} = \left(\sum_{i=1}^{n-1} \sum_{j=i+1}^{n} \frac{1}{\sigma^2_{\Delta V_{ij}}} \right)^{-1}. \tag{12.11}$$

In the case of two-point diagnostics, however, this determination of the error significantly underestimates the true uncertainty in the mean. One can understand this quite easily based on the following argument [294].

The problem with this expression for the covariance is that it does not adequately take into account the fact that the mean of a two-point diagnostic has a heavier contribution from the highest and lowest redshifts, but far less from the middle ones in the sample, because the latter are sometimes added and sometimes subtracted when comparing pairs of data. The correct way to calculate the variance of a two-point diagnostic must therefore include a careful application of standard error propagation to the weighted mean function.

In this situation, the weighted mean in Equation (12.10) must include the multiplicative terms that affect each of the original n measurements. It needs to be rearranged as follows:

$$\Delta V_{\text{w.m.}} = \left(\sum_{i=1}^{n} \alpha_i V(z_i) \right) \left(\sum_{i=1}^{n-1} \sum_{j=i+1}^{n} \frac{1}{\sigma^2_{\Delta V_{ij}}} \right)^{-1}, \tag{12.12}$$

where

$$\alpha_i \equiv \sum_{j=1}^{i-1} \frac{1}{\sigma^2_{\Delta V_{ij}}} - \sum_{j=1+i}^{N} \frac{1}{\sigma^2_{\Delta V_{ij}}}. \tag{12.13}$$

Each α_i is the sum of every term in the numerator of the weighted mean that multiplies $V(z_i)$. The correct variance of the weighted mean is therefore

$$\sigma^2_{\Delta V_{\text{w.m.}}} = \left(\sum_{i=1}^{n} \alpha_i^2 \sigma^2(z_i) \right) \left(\sum_{i=1}^{n-1} \sum_{j=i+1}^{n} \frac{1}{\sigma^2_{\Delta V_{ij}}} \right)^{-2}, \tag{12.14}$$

which is always greater than the variance naively expected from Equation (12.11). Equations (12.12) and (12.14) will be featured in several kinds of data analysis later in this chapter.

Whereas the weighted mean assumes that all errors are Gaussian, median statistics is independent of the measurement distribution. As noted earlier, the probability that a given observation falls above the 'true' median (i.e., the value one gets with a very large sample) is simply given by the binomial distribution in Equation (12.9). In spite of making no assumptions concerning the error distribution, however, this method does assume that all the measurements are completely uncorrelated. Of course, this cannot be true in the case of a two-point diagnostic, since each of the n data points affects the rest of the $n-1$ two-point values. The net impact of this effect is that the 68% confidence region is significantly greater than one calculated for uncorrelated measurements.

The cleanest, though perhaps not most elegant, way to handle this is via Monte-Carlo simulations customized to one's choice of two-point function. Using the actual measurements $V(z_i)$ and their $\sigma_{V(z_i)}$ errors, one samples the distribution of V to create mock catalogs of the two-point diagnostic for each model. With a sufficiently large number of such renderings, one may then compute the 68% confidence region of the median by brute force. Then, once the real data are used to determine the actual median for each model, its confidence range is estimated based on the Monte-Carlo simulations. We shall see how all of these steps work in practice with several of the data sets considered below.

12.2   REDSHIFT-DEPENDENT EXPANSION RATE

Since the Hubble rate depends on the expansion parameter $a(t)$ according to $H(z) = \dot{a}/a$ (Equation 4.28), and $(1+z) = a(t_0)/a(t)$ (Equation 8.9), where $a(t_0)$ is the expansion factor today, $H(z)$ may be measured directly from the time-redshift derivative dt/dz using

$$H(z) = -\frac{1}{1+z}\frac{dz}{dt} \, . \tag{12.15}$$

Thus, if one can determine the age difference, Δt, of sources separated by an incremental redshift difference Δz, one can measure the redshift-dependent Hubble parameter $H(z)$ using the approximation

$$H(z) \approx -\frac{1}{1+z}\frac{\Delta z}{\Delta t} \, . \tag{12.16}$$

The uniqueness of this cosmic probe is that it circumvents the limitations associated with the use of integrated histories, such as the luminosity and angular-diameter distances, given that it relies entirely on the local rate of expansion at the sampled redshift.

Some galaxies evolve passively on a time scale much longer than the age difference between them and their nearby neighbors. These are among the best studied *cosmic chronometers* one can find today [295]. Empirical evidence suggests that less than 1% of the stellar mass in the most massive among them formed at $z < 1$ [296, 297, 298], and that star formation ceased by redshift $z \sim 4$ in galaxy clusters [299], while all galaxies with stellar mass over $\sim 5 \times 10^{11} \, M_\odot$ ended their star formation activity by $z \sim 2$ [300]. The data therefore imply that the stellar population in these galaxies has been aging steadily from $z \sim 3$ to today.

Thus, one may safely assume that galaxies in the highest density regions of clusters have been aging passively since $z \sim 3$, tracing the so-called 'red envelope,' constituting the oldest stars in the Universe at every redshift. It is not surprising, therefore, to see these structures used as reliable cosmic chronometers in numerous studies [301, 302], culminating with the most recent measurements at $z \sim 2$ [303].

Figure 12.1 shows a catalog of cosmic chronometer measurements compiled from various sources over the redshift range $0 \lesssim z \lesssim 2$. One of the most useful features of these data is that they are model independent. Since the expression used to obtain

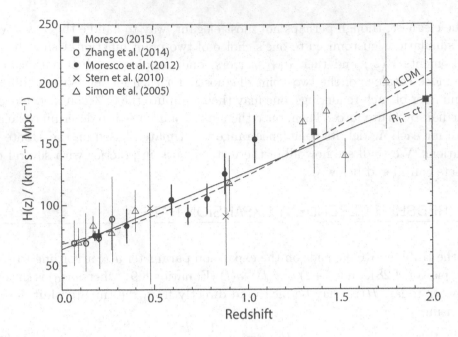

Figure 12.1 Measurement of the Hubble expansion rate, $H(z)$, based on the use of cosmic chronometers up to $z \sim 2$. These data are used to test the predictions of an optimized wCDM cosmology without the $R_h = ct$ constraint (dashed), and an analogous FLRW model with the inclusion of $R_h = ct$ (solid). The best-fit parameters for wCDM are $H_0 = 68.0 \pm 7.9$ km s^{-1} Mpc^{-1}, $\Omega_m = 0.31 \pm 0.16$ (see definition in Equations 6.3 and 6.4), and $w_{de} = -0.91^{-0.94}_{+0.42}$, all fully consistent with their *Planck* values (Planck Collaboration 2018). When the $R_h = ct$ condition is imposed, there is only one free parameter: $H_0 = 63.3 \pm 7.7$ km s^{-1} Mpc^{-1}. Based solely on their reduced χ^2_{dof} values, 0.56 (24 dof) for $R_h = ct$ and 0.58 (22 dof) for wCDM, the fits are hardly distinguishable. The Bayes Information Criterion, however, which also takes the number of free parameters into account, favors the $R_h = ct$ constraint over wCDM with a likelihood of $\sim 94.3\%$ versus $\sim 5.7\%$. (Adapted from ref. [304])

them (Equation 12.15) was not integrated over the expansion history of the Universe, there was no need to assume a model in order to extract the values of $H(z)$. This is an important property of these measurements that is not shared by many other kinds of observation. For example, another method used to 'measure' $H(z)$ involves the identification of baryon acoustic oscillations (BAO) and the Alcock-Paczyński distortion from galaxy clustering (see Section 12.6 below). But this approach depends on how 'standard rulers' evolve with redshift, rather than how cosmic time changes with z, so one needs to presume a background cosmology in order to calculate the distortions [305]. As such, the values of $H(z)$ obtained in this fashion are model-dependent and cannot be used to 'test' alternative cosmologies without an appropriate recalibration. Unfortunately, the measurements of $H(z)$ obtained with these varied methods—some model-independent, others not—are often combined to produce an overall $H(z)$ versus z diagram. But its diagnostic power is, of course, limited. Insofar as the redshift-dependent expansion rate is concerned, only the cosmic chronometer measurements (Figure 12.1) are truly model-independent and therefore suitable for comparing and testing different models.

When one imposes the $R_h = ct$ constraint on the FLRW cosmology, Equations (4.28) and (8.9) indicate that the Hubble variable scales with redshift according to the very simple relation

$$H^{R_h=ct}(z) = H_0(1 + z) \, . \tag{12.17}$$

With H_0 the only free parameter available now, one may adjust the corresponding theoretical curve in Figure 12.1 to fit the data by sliding it in the vertical direction, but its gradient is fixed. The comparison between the theoretical prediction and the measurements of $H(z)$ is therefore quite probative because these data extend to redshift ~ 2, which includes the region where the ΛCDM model requires a transition from decelerated to accelerated expansion somewhere near $z \lesssim 1$ (see Section 12.3.4 below).

In contrast, ΛCDM is characterized by a much larger number of free parameters, in part due to our current ignorance regarding the nature of dark energy. At the very least, one ought to include H_0, the fractional energy density Ω_m for matter (see Equations 6.3 and 6.4), the dark energy equation-of-state parameter $w_{de} = p_{de}/\rho_{de}$, and the spatial curvature constant k (see Section 9.1). Strictly speaking, dark energy is a cosmological constant in the basic version of the standard model, for which $w_{de} = -1$. To keep this discussion as streamlined as possible, one therefore optimizes just three parameters for ΛCDM: H_0, Ω_m and k. But sometimes it is also useful to relax the constraint on w_{de}, in which case the model is labeled wCDM (to distinguish it from the version with a true cosmological constant) and one assumes spatial flatness in this case, optimizing the three parameters H_0, Ω_m and w_{de}. The most general form of the Hubble parameter in this application may be found from Equations (4.28) and (9.1), and the definitions of ρ and p in Section 9.1, and is written

$$H^{\Lambda CDM}(z) = H_0 \left[\Omega_m(1+z)^3 + \Omega_r(1+z)^4 + \Omega_{de}(1+z)^{3(1+w_{de})} \right]^{1/2} \tag{12.18}$$

though, given the relative unimportance of radiation for redshifts $z \lesssim 2$, the Ω_r term is typically ignored in this cosmological test.

Following the procedure described in Section 12.1.1, one infers the following optimized parameter values based solely on the cosmic-chronometer data shown in Figure 12.1: $H_0 = 63.3 \pm 7.7$ km s^{-1} Mpc^{-1} when imposing the $R_h = ct$ constraint; $H_0 = 68.0 \pm 7.9$ km s^{-1} Mpc^{-1}, $\Omega_m = 0.31 \pm 0.16$, and $w_{de} = -0.91^{-0.94}_{+0.42}$ for wCDM; and $H_0 = 73.3 \pm 3.5$ km s^{-1} Mpc^{-1}, $\Omega_m = 0.28 \pm 0.04$ and $\Omega_{de} = 0.60 \pm 0.13$ for ΛCDM. The corresponding best fits for $R_h = ct$ and wCDM are shown in Figure 12.1, and the information criteria are summarized in Table 12.1. All the results are reported with 1σ standard errors calculated from the χ^2-distribution in each case.

The quality of the fits assessed via their χ^2_{dof} values suggests that the redshift-dependent Hubble parameter predicted by an FLRW cosmology with the $R_h = ct$ constraint accounts for the measured expansion rate of the Universe at least as well

TABLE 12.1 Optimized Fits of $H(z)$

Cosmology	$\chi^2_{\rm dof}$	AIC	KIC	BIC
$R_h = ct$	0.566	82.9%	93.0%	94.3%
wCDM	0.579	17.1%	7.0%	5.7%
ΛCDM	0.580	17.1%	7.0%	5.7%

as wCDM and ΛCDM, particularly since it has only one free parameter. The comparison is much more definitive than this, however, when model selection is based on a comparison of the AIC, KIC, and BIC. On statistical grounds, the FLRW expansion guided by $R_{\rm h} = ct$ is far more likely to be correct than that of the other two cases considered here.

The fact that all three statistics show a consistent result is quite important because, whereas the BIC is based on Bayesian statistics, the AIC and KIC are not. The number n (= 25) of data points here is large enough that any chosen Bayesian 'priors' over the parameters of the individual models drop out [292]. The prior distribution over the choice of models, however, does not. The model distribution is often assumed quite reasonably to be 'flat,' giving each of them an equal a priori likelihood of being selected. But a case can sometimes be made that one of them should be assigned a higher 'a priori preference' over the others.

When this occurs, the relative likelihood of two cosmologies being compared is multiplied by a Bayes factor drawn from the prior distribution of the models. Thus, if only the BIC were being used for this model selection, one would need to assess beforehand whether either wCDM, ΛCDM, or an FLRW cosmology constrained by $R_{\rm h} = ct$, is to be preferred. As is well known, this is the problem with attempting to prioritize models before testing them against the data, because different investigations may be based on different subjective assumptions. But given that all three information criteria may be considered together, one can avoid such biases based on subjective points of view by simply letting the data speak for themselves and assume that the prior distribution over the choice of models is flat.

Then, the fact that the AIC and KIC have their own foundations and are not derived from Bayesian statistics strengthens the overall assessment of the outcome of this analysis summarized in Table 12.1. Based on the data available today, the measured redshift-dependent expansion rate favors the FLRW cosmology constrained by $R_{\rm h} = ct$ over the basic ΛCDM model and its variant wCDM.

This inference may be tested further using a two-point diagnostic, following the complementary procedure described in Section 12.1.2 above which, in addition, provides a more robust assessment of the measurement errors. A diagnostic appropriate

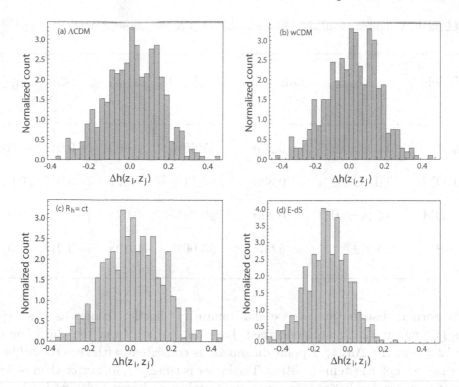

Figure 12.2 Histogram of the $\Delta h(z_i, z_j)$ 2-point diagnostic values (Equation 2.19) calculated for the cosmic-chronometer measurements of the Hubble expansion rate $H(z)$ in Figure 12.1. Panel (a) the ΛCDM cosmology; panel (b) the standard model with an unconstrained dark-energy equation-of-state parameter w (called wCDM); panel (c) the analogous FLRW model with the $R_{\rm h} = ct$ constraint; and panel (d) the Einstein-de Sitter universe. (Adapted from ref. [294])

for any cosmology may be defined as

$$\Delta h(z_i, z_j) \equiv \left(\frac{H_i}{H(z_i)} - \frac{H_j}{H(z_j)} \right) , \qquad (12.19)$$

where H_i is the Hubble constant measured at redshift z_i, and $H(z_i)$ is its value predicted by the assumed model (see Equations 12.17 and 12.18).

The unweighted distribution of $\Delta h(z_i, z_j)$ values is shown in Figure 12.2, which also includes the result for the Einstein-de Sitter universe to provide a broader perspective (see Section 5.3). Since the latter cosmology contains only matter (i.e., $\Omega_{\rm m} = 1$), whose density scales as $(1 + z)^3$, it is trivial to see from Equation (4.28) that, in this case,

$$H^{\rm E-dS}(z) = H_0(1 + z)^{3/2} . \qquad (12.20)$$

The Hubble constant H_0 cancels out in the expression for $\Delta h(z_i, z_j)$, so its actual value is inconsequential here. Where relevant, the rest of the parameters are taken from the optimizations summarized in Table 12.1 above.

A careful inspection of the four panels in Figure 12.2 reveals some departures from a pure Gaussian shape, e.g., a greater central peaking than one would expect

TABLE 12.2   Weighted Mean and Median Statistics (Cosmic Chronometers)

Model	$\Delta h_{\text{w.m.}}$	Offset from zero (w.m.)	$|N_\sigma| < 1$	$\Delta h_{\text{m.s.}}$	68% Range (m.s.)
$R_{\text{h}} = ct$	0.008 ± 0.028	0.27σ	87.36%	0.009	$(-0.012, +0.022)$
ΛCDM	0.012 ± 0.031	0.39σ	90.11%	0.021	$(-0.015, +0.022)$
wCDM	0.014 ± 0.031	0.44σ	90.34%	0.025	$(-0.018, +0.021)$
E-dS	-0.109 ± 0.021	5.06σ	85.06%	-0.108	$(-0.018, +0.012)$

from a normal distribution, as we shall quantify shortly. Insofar as the weighted means (Equation 12.12) are concerned, however, we see from the third column in Table 12.2 that the $\Delta h_{\text{w.m.}}$ 2-point diagnostic is consistent with zero to within 1σ in every case, except Einstein-de Sitter. The latter is ruled out at better than $\sim 5\sigma$, and its negative value implies that measurements taken at larger redshifts in this model would require smaller estimates of H_0 than the corresponding observations at low redshifts. Overall, the outcome based on $\Delta h_{\text{w.m.}}$ slightly favors an FLRW expansion constrained by $R_{\text{h}} = ct$ over ΛCDM and wCDM, though the distinction is not as conclusive as that drawn from the information criteria shown in Table 12.1.

As we probe the four panels shown in Figure 12.2 further, however, it becomes quite apparent that the measurement uncertainties deviate noticeably from true Gaussian errors (see column 4 in Table 12.2). In every case, the percentage of pairs N_σ with a $\Delta h(z_i, z_j)$ within 1σ of the weighted mean exceeds the expected 68% if the errors associated with the measured H_i values are truly Gaussian. This centralized 'peaking' implies that the published errors must be larger than their true value, or that there exists some systematic effect that shifts most measurements by a comparable amount. Of course, this also implies that the actual error (Equation 12.14) in the weighted mean $\Delta h_{\text{w.m.}}$ is probably smaller than that shown in column 2 of Table 12.2, so the offsets in column 3 are likely larger than the fractional values indicated here.

In such situations, the use of median statistics, as described in Section 12.1.2, is indispensable, chiefly because one does not need to know whether the distribution is Gaussian. The outcome of this analysis is summarized in columns 5 and 6 of Table 12.2. The true median for Einstein-de Sitter is entirely inconsistent with zero, confirming rather definitively that this cosmology is ruled out by the measured cosmic expansion rate.

Equally significant is the fact that an FLRW expansion constrained by $R_{\text{h}} = ct$ is the only model for which a zero median lies within the 68% confidence interval. The medians in both ΛCDM and wCDM are inconsistent with zero to some degree. Since median statistics is independent of whether or not the data are randomly distributed, these results are somewhat stronger than those based on the weighted mean alone. Nevertheless, all of the results discussed here concerning the measured expansion rate of the Universe yield a set of consistent conclusions, including (1) that Einstein-de Sitter is firmly rejected, and (2) that $R_h = ct$ is favored over the standard model, both the basic flat version with a cosmological constant and its variant with an unrestricted dark-energy equation-of-state.

12.3 THE HUBBLE DIAGRAM

An entirely different measure of the cosmic expansion is based on the integrated luminosity distance $d_L(z)$ as a function of redshift. The Hubble diagram, derived from $d_L(z)$, provides us with the cumulative expansion as a function of the 'look-back' time, and is drawn from the observation of 'standard candles'—sources whose luminosity is known in their respective rest frames.

Consider an object with known rest-frame luminosity L emitting photons at time t_e at a comoving distance (see Equations 8.4 and 8.5)

$$r_e \equiv c \int_{t_e}^{t_0} \frac{dt}{a(t)} \tag{12.21}$$

where, as usual, $a(t)$ is the expansion factor and t_0 is the cosmic time today. To keep our discussion as straightforward as possible, we have assumed spatial flatness (i.e., $k = 0$), consistent with the preponderance of observational evidence today (see, e.g., Chapters 6 and 9). If our detector has a proper cross-sectional area A, then the fraction of all isotropically emitted photons captured by our instrument is the ratio of the solid angle subtended by A at proper distance $a(t_0)r_e$ to 4π:

$$\text{frac} = \frac{1}{4\pi} \frac{A}{a^2(t_0)\, r_e^2} . \tag{12.22}$$

According to Equations (8.8) and (8.9), however, each photon's energy $h\nu_e$ is redshifted during its transit to Earth by an amount

$$h\nu_0 = h\nu_e \frac{a(t_e)}{a(t_0)} . \tag{12.23}$$

In addition, the time interval δt_e during which these photons are emitted at the source is dilated at our detector, so the radiation is captured over a corresponding time interval

$$\delta t_0 = \delta t_e \frac{a(t_0)}{a(t_e)} . \tag{12.24}$$

As such, the total power P measured by our instrument is the absolute rest-frame luminosity L, times the fraction 'frac' in Equation (12.22), times the redshift factor $a^2(t_e)/a^2(t_0)$:

$$P = L \frac{a^2(t_e)}{a^2(t_0)} \frac{A}{4\pi\, a^2(t_0)\, r_e^2} . \tag{12.25}$$

We detect an apparent flux $F = P/A$, i.e.,

$$F = \frac{L\,a^2(t_e)}{4\pi\,a^4(t_0)\,r_e^2}, \tag{12.26}$$

and, generalizing from the conventional definition of 'luminosity distance' in Euclidean space,

$$d_L^{\text{Eucl}} \equiv \left(\frac{L^{\text{Eucl}}}{4\pi\,F^{\text{Eucl}}}\right), \tag{12.27}$$

we infer that the corresponding expression in FLRW cosmology must be

$$d_L(t_e) \equiv a^2(t_0)\frac{r_e}{a(t_e)}. \tag{12.28}$$

Choosing the normalization $a(t_0) = 1$, it is not difficult for us to see from Equation (12.21) that

$$r_e = c\int_a^1 \frac{da}{a^2\,H(a)}. \tag{12.29}$$

Thus, changing the integration variable using Equation (8.9), with

$$dz = -\frac{da}{a^2}, \tag{12.30}$$

one finds that the comoving distance may also be written

$$r_e = c\int_0^z \frac{dz'}{H(z')}, \tag{12.31}$$

and

$$d_L(z) \equiv (1+z)r_e(z). \tag{12.32}$$

In spatially flat ΛCDM, the scaled energy density is $\Omega = \Omega_{\text{m}} + \Omega_{\text{r}} + \Omega_\Lambda = 1$, leading to the expression for $H^{\Lambda\text{CDM}}$ in Equation (12.18). In contrast, the Hubble parameter in the FLRW cosmology constrained by $R_{\text{h}} = ct$ is much simpler, and is given by Equation (12.17). Together, Equations (12.31), (12.17) and (12.18) yield the luminosity distance we shall need to construct the Hubble diagram from standard candles:

$$d_L^{\Lambda\text{CDM}}(z) = \frac{c}{H_0}(1+z)\int_0^z \frac{dz'}{E^{\Lambda\text{CDM}}(z')}, \tag{12.33}$$

where

$$E^{\Lambda\text{CDM}}(z) \equiv H^{\Lambda\text{CDM}}(z)/H_0, \tag{12.34}$$

and

$$d_L^{R_{\text{h}}=ct}(z) = \frac{c}{H_0}(1+z)\ln(1+z). \tag{12.35}$$

The distance modulus, $\mu \equiv m - M$, used to construct the Hubble diagram, is formally defined in terms of the apparent magnitude m of the source and the absolute magnitude M of the class to which it belongs. It is related to the luminosity distance (in parsecs) by the expression

$$\mu(z) = 5 \log_{10}\left[\frac{d_L(z)}{10\ \mathrm{pc}}\right].$$ (12.36)

Below, we shall construct the Hubble diagram, $\mu(z)$ versus z, using two different kinds of source: active galactic nuclei (AGN); and Type Ia SNe, which were instrumental in the discovery of dark energy [39, 40, 41].

12.3.1 Active Galactic Nuclei

The discovery of high-redshift quasars at $z > 5 - 6$ has been an enduring mystery in astronomy, given that the emergence of $10^{9-10}\ M_\odot$ supermassive black holes only ~ 900 Myr after the Big Bang, and only ~ 500 Myr after their apparent seeding via Population II and III stellar explosions, is inconsistent with the timeline in ΛCDM (see Section 9.3). As discussed in Chapter 9, the only way to rectify these inconsistencies, while preserving the standard model's timeline, is to invoke an anomalously high accretion rate or the generation of exotically massive seeds, neither of which is fully consistent with our current astrophysical understanding of how black holes form and grow.

In spite of this theoretical deficiency, however, bright quasars may be seen at redshifts well beyond the current reach of other sources, including Type Ia SNe (Section 12.3.4), which are restricted [306] to redshifts $\lesssim 2$. Thus, they probe the geometry of the Universe at luminosity distances unreachable by other means—a highly desirable feature that has motivated their use in the construction of a Hubble diagram. This promise has yet to be fully realized, however, due to the uneven quality of the available samples. But the situation has changed quite dramatically in recent years [307], with the careful selection of 1598 sources distributed in redshift up to $z \sim 6$.

The most accurate method for sampling the redshift-distance relationship using quasars derives from a correlation seen between their UV and X-ray monochromatic luminosities, first discovered over three decades ago [308, 309]. This relation appears to depend primarily on the geometry of the central engine, and is independent of evolution. The attention it has received recently is due to the availability of the aforementioned high-quality sample of suitable sources, which seem to avoid producing an excessive dispersion in the correlation.

The generally accepted origin of the UV spectrum is photon emission by an accretion disk, while the X-ray component appears to be due to Compton upscattering in an overlying, hot corona [204]. Most of the quasars in the parent sample have been identified from the cross-correlation of the XMM-*Newton* Serendipitous Source Catalogue Data Release 7 [310] with the Sloan Digital Sky Survey (SDSS) quasar catalogs from Data Release 7 [311] and 12 [312]. The high-quality catalog of 1598 quasars is a subset of this sample, whose reliable measurements of the intrinsic X-ray and UV emissions avoid possible contaminants and unknown systematics.

The empirical relation between the quasar's UV (disk) and X-ray (coronal) spectral components is parametrized using the ansatz

$$\log_{10} L_X = \gamma \log_{10} L_{UV} + \beta , \qquad (12.37)$$

where L_X and L_{UV} are the rest-frame monochromatic luminosities at 2 keV and 2,500 Å, respectively, and the slope γ seems to lie in the range $\sim 0.5 - 0.7$ [308, 313, 314]. The data contain the fluxes

$$F \equiv \frac{L}{4\pi \, d_L} , \qquad (12.38)$$

however, rather than the absolute, model-dependent luminosities, so a more practical version of Equation (12.37) is the modified form

$$\log_{10} F_X = \tilde{\beta} + \gamma \log_{10} F_{UV} + 2(\gamma - 1) \log_{10} d_L , \qquad (12.39)$$

where the constant $\tilde{\beta}$ subsumes the slope γ and the intercept β:

$$\tilde{\beta} \equiv \beta + (\gamma - 1) \log_{10} 4\pi . \qquad (12.40)$$

To test a given model using the quasar sample and its presumed correlation in Equation (12.39), one must first optimize the parameters γ and $\tilde{\beta}$ using the measured values z_i, $F_{UV,i}$ and $F_{X,i}$ (for $i = 1$–1598), and then compare its prediction with the Hubble diagram constructed from these data. As noted earlier (Section 12.1.1), the very large number of data points ($n = 1598 \gg 1$) suggests that the most appropriate information criterion to use in this case is the BIC. Writing

$$\exp(-\text{BIC}/2) \equiv n^{-k/2} L^* , \qquad (12.41)$$

one optimizes the fit to the data by maximizing $L^* = e^{\Pi}$, which is calculated in terms of the likelihood function

$$\Pi \equiv \sum_{i=1}^{n} \left\{ \frac{[\log_{10} F_{X,i} - \Phi(F_{UV,i}, d_L [z_i])]^2}{\tilde{\sigma}_i^2} + \ln\left(\tilde{\sigma}_i^2\right) \right\} . \qquad (12.42)$$

In this expression, which is clearly related to the χ^2 function in Equation (12.2), the variance $\tilde{\sigma}_i^2 \equiv \delta^2 + \sigma_i^2$ is given in terms of an unknown global intrinsic dispersion, δ, and the measurement error σ_i in $F_{X,i}$. The error in $F_{UV,i}$ is insignificant compared to σ_i and is thus typically ignored. The appearance of δ in the variance is a reflection of the fact that, in spite of the quasar sample being carefully selected, it is nonetheless subject to an irreducible intrinsic randomness, whose magnitude must be found via the optimization procedure along with the values of the other unknowns. The model-dependent function Φ is derived from Equation (12.39) and is defined as

$$\Phi(F_{UV,i}, d_L [z_i]) \equiv \tilde{\beta} + \gamma \log_{10} F_{UV,i} + 2(\gamma - 1) \log_{10} d_L(z_i) . \qquad (12.43)$$

The luminosities L_X and L_{UV} for the 1598 quasars, and the optimized correlation function (Equation 12.37) derived from them, are shown in Figure 12.3 for the

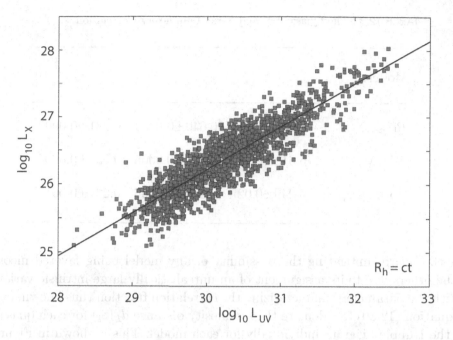

Figure 12.3 Rest-frame correlation of the monochromatic UV and X-ray luminosity densities of the highly selected 1598 Risaliti and Lusso (2019) quasar sample. The data and fit shown here have been optimized for the FLRW cosmology constrained by $R_h = ct$, which has no free parameters. A fiducial Hubble constant $H_0 = 70$ km s^{-1} Mpc^{-1} has been chosen for the purpose of display only. Its actual value does not affect the optimization of the key parameters in the correlation function, $\gamma = 0.64$, $\tilde{\beta} = -13.621$, and $\delta = 0.231$. (Adapted from refs. [307, 315])

FLRW cosmology constrained by $R_h = ct$. The Hubble constant does not affect this procedure, but has been assigned the 'fiducial' value 70 km s^{-1} Mpc^{-1} for display purposes throughout this section. The corresponding correlation plot for ΛCDM is so similar to this that there is no need to show it separately.

In order to provide a baseline comparison with a fit that is not derived from either $R_h = ct$ or ΛCDM, this analysis has also been carried out using a model-independent 'cosmographic' third-order polynomial function representing the luminosity distance, written as

$$d_L^{cos} = \ln(10)\frac{c}{H_0}\left(u + a_2 u^2 + a_3 u^3\right), \qquad (12.44)$$

where $u \equiv \log_{10}(1+z)$, and the two constants a_2 and a_3 must be optimized along with the other free parameters (see Table 12.4) [307]. Throughout this section, we shall therefore assess the likelihood of all three models based on how well they account for the high-z quasar Hubble diagram.

A summary of the best-fit parameters inferred from the three optimizations appears in Table 12.3, which shows a remarkable consistency in the correlation function from one application to the next. This lends support to the idea that the ansatz in Equation (12.37) does in fact represent a reliable standard candle for model testing. In addition, the intrinsic dispersion δ is inferred to have a uniform value of 0.231

TABLE 12.3   The Quasar UV and X-ray Luminosity Correlation

Model	$\tilde{\beta}$	γ	δ
$R_h = ct$	6.618±0.011	0.640±0.0004	0.231±0.0003
ΛCDM	6.618±0.012	0.639±0.0005	0.231±0.0004
Cosmographic	6.249±0.02	0.626±0.0006	0.231±0.001

across the board, mitigating the possibility of any model being favored incorrectly over the others due to its assignment of an unrealistically large intrinsic variance.

With the parameters characterizing the correlation function thus known, one may use Equation (12.39) to calculate the luminosity distance $d_L(z_i)$ for each quasar, and build the Hubble diagram individually for each model. This is shown in Figure 12.4 for the FLRW cosmology constrained by $R_h = ct$, together with the theoretical prediction, indicated by the solid black curve. The corresponding figures for ΛCDM and the cosmographic model are almost identical to this, save for small changes in the recalibrated data.

Finally, one may determine the quality of the fit in the Hubble diagram, implied by the reduced χ^2 values in Table 12.4, and calculate the corresponding BIC likelihoods. Clearly, the χ^2_{dof} hardly changes from one cosmology to the next, reaffirming the view that the simple use of χ^2 is often inadequate for model selection in cosmology. Indeed, the cosmographic polynomial function appears to fit the data slightly better than the other two, but this merely reflects its greater flexibility adjusting to the noise due to its larger number of free parameters.

The main conclusion one may draw from Table 12.4 is that the FLRW cosmology constrained by $R_h = ct$ is favored by the high-z quasar Hubble diagram over both the 1-parameter, flat ΛCDM cosmology and the empirical cosmographic polynomial. Studying these results more closely, it is also quite apparent that the differences in the BIC, and their associated likelihoods, are almost entirely due to the different number of adjustable free parameters: $R_h = ct$ has none, ΛCDM has one, and the cosmographic model has two.

Figure 12.5 shows all three theoretical Hubble diagrams together, allowing us to see—for the first time—how the luminosity distance predicted by ΛCDM compares with that of an FLRW cosmology constrained by $R_h = ct$ over such a large range in redshifts ($0 \lesssim z \lesssim 6$). The outcome is quite revealing: the curves are hardly distinguishable all the way to $z \sim 6$, the extent of these observations. One can see a slight deviation at $z \sim 2$, but nowhere else. Given that the $R_h = ct$ cosmology has no free parameters for this analysis, one concludes that the FLRW cosmology constrained by $R_h = ct$ is an 'attractor' for the optimization of ΛCDM [316]. Together with the results in Section 12.2, these results paint a picture in which the largely empirical

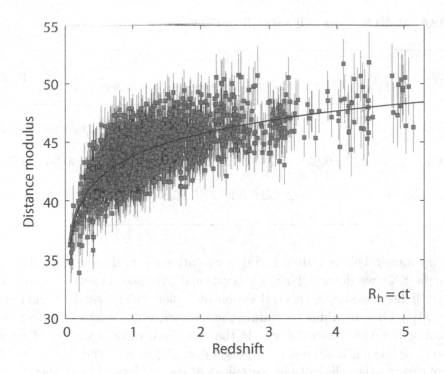

Figure 12.4 Hubble diagram constructed from the 1598 highly selected quasars in the Risaliti and Lusso (2019) sample. The optimization is based on the correlation shown in Figure 12.3, corresponding to the FLRW cosmology constrained by $R_h = ct$. The $\chi^2_{\rm dof}$ for this fit is 1.005, with a relative BIC likelihood $\sim 88.7\%$ of this being the correct model. (Adapted from refs. [307, 315])

parametrization of ΛCDM accounts very well for many of the observations because it has sufficient flexibility to mimic the much more tightly constrained observational signatures in $R_h = ct$.

At redshift $z \sim 6$, the Universe was roughly 900 Myr old in ΛCDM, and about 1.9 Gyrs in the FLRW cosmology constrained by $R_h = ct$. The analysis of high-z quasars therefore covers a considerable fraction of the visible Universe. Over much of this evolutionary history, the luminosity distance predicted by the latter cosmology, without any free parameters, is favored by the data when the BIC is used for model selection.

12.3.2 Type Ia Supernovae

Type Ia supernovae (Type Ia SNe) occupy an exalted position in modern cosmology, for it was the Hubble diagram constructed using their observed spectral properties that led to the discovery of dark energy [39, 40, 41]. Our understanding of their progenitor history has undergone a significant evolution over the past several decades, however, and our simple view of how these explosive events occur and what they tell us has been somewhat revised.

TABLE 12.4   Model Comparison Based on High-z Quasars

Model	Ω_m	a_2	a_3	χ^2_{dof}	BIC	Prob.
$R_h = ct$	—	—	—	1.005	1,625.86	88.70%
ΛCDM	0.31±0.05	—	—	1.004	1,630.26	9.79%
Cosmographic	—	2.93±0.33	2.65±0.80	1.003	1,633.96	1.51%

It is no longer believed that a single evolutionary path is responsible for all the Type Ia SNe we detect. Their observational diversity includes a variation in luminosity, lightcurve shape, spectral evolution, color and host-galaxy environment [317, 318, 319]. They are almost certainly thermonuclear explosions of carbon-oxygen white dwarfs, but the consensus now is that they represent a variety of explosion mechanisms, following a diversity of progenitor histories. Some have even raised doubts concerning the reliability of cosmological results derived from them.

Demonstrating that Type Ia SNe may be used as standard candles has been crucial for their application to cosmological measurements. They appear to explode with a luminosity ranging over a factor of only two, probably because of the narrow range of triggering masses and the amount of nuclear material burned during the event. The most significant step in modeling their luminosity to turn them into effective standard candles was the introduction of the lightcurve shape versus luminosity correlation, commonly referred to as the 'Phillips relationship' [320]. As of today, with a wide range of corrections applied to this relation, Type Ia SNe can provide individual distances with an error as small as $\sim 6\%$ [194, 321].

But moving beyond this limit is challenging, mainly due to our ignorance of the detailed underlying physics governing Type Ia supernova explosions. That the progenitor star is compact has been confirmed observationally by its non-detection in deep pre-supernova imaging of SN 2011fe, one of the closest, unobscured events ever seen [322]. The Type Ia supernova progenitor is almost certainly a member of a binary system, in which the white dwarf may accrete from its companion, or somehow be disturbed, and thereby eventually explode. The precise configuration of this system, however, remains unknown.

Nevertheless, the majority of Type Ia SNe do display similar characteristics, suggesting some uniformity in most of the explosions. Evidence for this includes the fact that the Phillips relation has been quite successful in producing a Hubble diagram with a relatively small scatter. The problem is that several sub-classes of Type Ia SNe have been identified that behave differently [323], some with a modified Phillips relation, and others that appear to be superluminous [319]. If our theoretical understanding of all these variations eventually improves, a cleaner sample may be produced that could reduce the measurement errors below the current $\sim 6\%$.

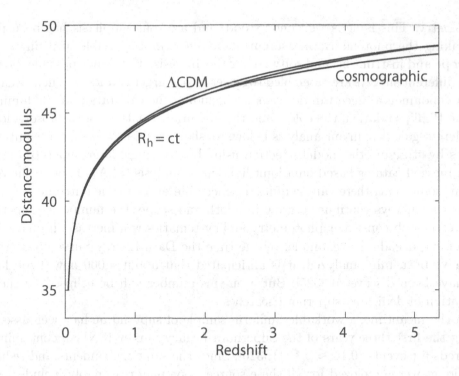

Figure 12.5 Best-fit distance moduli for (i) the FLRW cosmology constrained by $R_h = ct$, (ii) ΛCDM and (iii) the cosmographic model, all optimized using the 1598 highly-selected quasars in the Risaliti and Lusso (2019) sample. The ΛCDM curve lies slightly above the other two at $z \sim 2$, while the cosmographic curve is the lowest at $z > 3$. The $R_h = ct$ curve is the same as that shown in Figure 12.4. Though the curves deviate slightly from each other at various redshifts, the data are also re-calibrated individually for each model, with the effect that all three χ^2_{dof}'s are virtually identical (see Table 12.3). The principal difference between these three cases is the number of free parameters used for the optimization, which is incorporated into the BIC. (Adapted from ref. [315])

With the methodology we have available today, mitigating the negative impact of the remaining sources of uncertainty is quite challenging when one attempts to use Type Ia SNe for the optimization of model parameters in ΛCDM. It is even more daunting to use them for model selection, especially when the models being tested are not nested (Section 12.1.1). The major difficulty arises from the need to enlargen the catalog as much as possible in order to improve the statistics, which is typically done by merging different sub-samples. Yet each of these has its own unique systematics and photometric peculiarities, the combination of which affects the model fitting differently in each case.

And while the statistical errors have now been reduced to less than half of the total uncertainty, with promise to decrease even further as the samples continue to grow, the systematic problems may be too difficult to overcome [324]. Not surprisingly, the chief obstacle encountered when merging different samples is the accuracy of the photometry. This issue is well understood, and a strong emphasis has been placed on making the photometric measurements as uniform as possible, utilizing a single telescope and instrument, and software for the analysis. This is not possible, however, when different surveys are assembled together into a larger catalog. In such situations, major fundamental uncertainties remain, including the calibration of the luminosity via the Phillips relation, the colors and the potential variations of the absorption law in different galaxies. In our analysis below, we shall examine some of these mitigating factors by studying the model selection using both a single large supernova sample, and a merged catalog based on a joint lightcurve analysis (JLA). Though the results overlap somewhat, there can be quite distinct differences in the outcome.

As the surveys continue to grow in depth and scope, the number of supernovae discovered with consistent photometry and systematics will increase dramatically in the coming decade. For example, by the time the Dark Energy Survey observations will have been fully analyzed, it is anticipated that about 4,000 new Type Ia SNe will have been discovered [325]. But even this number will be eclipsed by the next generation of dedicated supernova searches.

In the meantime, a workable, uniform sample of supernovae has been assembled during the first three years of the Supernova Legacy Survey (SNLS), containing 252 high-redshift events $(0.15 \lesssim z \lesssim 1)$ [326]. Since the same instruments and reduction techniques were employed for all these sources, one may reasonably include a single intrinsic dispersion to address the irreducible scatter that inevitably appears in the Hubble diagram constructed from this sample. An additional positive feature of this catalog is that it covers that crucial redshift range $(z \lesssim 1)$ within which the standard model suggests the Universe underwent a transition from decelerated to accelerated expansion.

Several lightcurve fitters may be used with data such as these, but for simplicity, we shall here summarize the results based on just one of the more commonly employed algorithms, known as SiFTO [327]. For this application, the distance modulus is defined for each supernova as the linear combination

$$\mu_B = m_B + \alpha \cdot (s - 1) - \beta \cdot \mathcal{C} - M_B, \tag{12.45}$$

where m_B is the peak rest-frame B-band magnitude, s is the stretch (a measure of lightcurve shape), $\mathcal{C}$ is the color (peak rest-frame $B - V$), and M_B is the absolute magnitude of a Type Ia supernova.[1]

When corrected for shape and color, the luminosities of Type Ia SNe in this sample have a dispersion of $\sim 15\%$. The unknowns α, β, M_B are often referred to as 'nuisance' parameters, as they cannot be inferred independently of an assumed cosmology. They must be optimized simultaneously with the cosmological parameters.

This dependence of the lightcurve fitter on the model itself has been a persistent, though unavoidable, problem with supernova work, because it is virtually impossible

---

[1]The SNLS supernova compilation of 252 SNe is currently available in the University of Toronto's Research Repository. It includes the following information for each supernova: m_B (with corresponding standard error σ_{m_B}), s (with σ_s), $\mathcal{C}$ (with $\sigma_{\mathcal{C}}$), and the covariances between $m_B, s,$ and $\mathcal{C}$. (https://tspace.library.utoronto.ca/handle/1807/24512)

TABLE 12.5   Model Fits Based on the SNLS Sample

	ΛCDM	$R_\mathrm{h} = ct$
α	1.275 ± 0.120	1.175 ± 0.115
β	2.637 ± 0.155	2.605 ± 0.149
M_B	-19.165 ± 0.081	-18.959 ± 0.011
σ_int	0.103 ± 0.010	0.106 ± 0.010
Ω_m	0.365 ± 0.137	—
Ω_Λ	0.846 ± 0.353	—
$-2 \ln L$	-238.40	-231.85
BIC	-205.67	-210.03
Likelihood	10.1%	89.9%

to use supernova measurements in unbiased, comparative studies without introducing some model dependence. Since three (sometimes four) nuisance parameters are recalibrated from one model to the next, the data become compliant to the underlying theory [328, 329, 330]. For example, it is not inconceivable that two different models may fit their respective data sets equally well, because the data themselves change when one cosmology is replaced by another.

When an assumed intrinsic dispersion is not known a priori and must be estimated along with the other parameters, the preferred method of optimization is based on maximizing the likelihood function, which treats all of the unknowns on an equal footing [331, 332]. The joint likelihood function for estimating the cosmological parameters, the model-specific optimized nuisance parameters α, β, M_B, and the sample-wide intrinsic dispersion σ_int, based on a flat Bayesian prior, is [333]

$$L = \prod_i \frac{1}{\sqrt{2\pi(\sigma_{\mathrm{lc},i}^2 + \sigma_\mathrm{int}^2)}} \exp\left\{-\frac{(\mu_{B,i} - \mu_\mathrm{th}[z_i])^2}{2(\sigma_{\mathrm{lc},i}^2 + \sigma_\mathrm{int}^2)}\right\}$$

$$\propto \exp\left(-\chi^2/2\right). \tag{12.46}$$

Each individual distance modulus $\mu_{B,i}$ depends on α, β, M_B, and the theoretical distance modulus $\mu_\mathrm{th}(z_i)$ depends on the cosmological parameters (Equation 12.36). In this expression,

$$\sigma_{\mathrm{lc},i}^2 = \sigma_{m_B,i}^2 + \alpha^2 \sigma_{s,i}^2 + \beta^2 \sigma_{\mathcal{C},i}^2 + C_{m_B s \mathcal{C},i} \tag{12.47}$$

is the supernova-specific variance, written in terms of the standard errors of the peak magnitude and lightcurve parameters of the supernova, $\sigma_{m_B,i}$ and $\sigma_{s,i}$. The quantity $\sigma_{\mathcal{C},i}$ comes from the covariances among $m_B, s, \mathcal{C}$, and likewise depends quadratically on α, β. Once the maximum of the likelihood function is known, the BIC is calculated from L according to Equation (12.8), i.e., BIC $= -2 \ln L + (\ln n)k$.

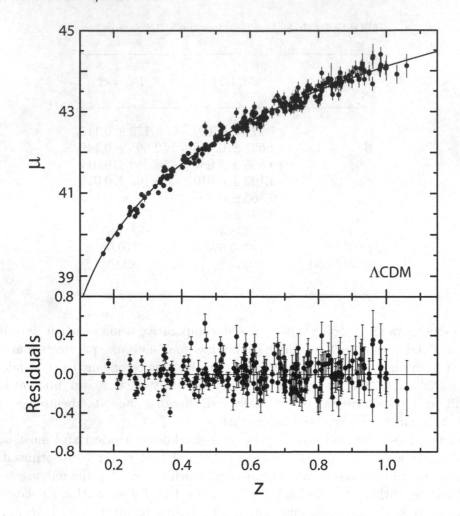

Figure 12.6 The Hubble diagram and its residuals constructed for ΛCDM, based on the SNLS sample of 234 Type Ia SNe. The solid curve in the upper panel represents the best-fit model with $\Omega_m = 0.365 \pm 0.137$, $\Omega_\Lambda = 0.846 \pm 0.353$, and dark energy in the form of a cosmological constant. (Adapted from ref. [332])

We learned in Sections 12.1.1 and 12.3 that it is also essential with model selection to be absolutely clear about the number of required parameters, which figure quite prominently in the estimated BIC used to prioritize the models. In ΛCDM there is some flexibility in the choice of free parameters, depending on how dark energy is characterized. One may adjust Ω_m, k (or Ω_{de}), w_{de} and H_0 where, as usual, a subscript 'de' denotes a quantity pertaining to dark energy when it is not necessarily a cosmological constant. Additional free parameters, such as the baryon density, are relevant when analyzing other kinds of data, such as the CMB fluctuations, but are generally not required to construct the supernova Hubble diagram.

As we have seen in Chapter 11, it is still not known whether the inflationary concept—as we invoke it today—will survive in the long run. Without it, however, ΛCDM has serious internal inconsistencies. One is therefore compelled to formulate

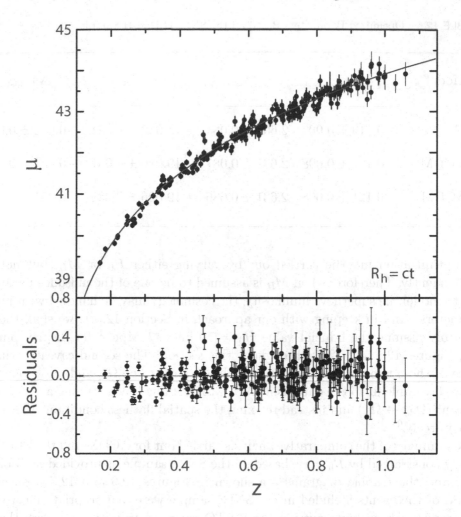

Figure 12.7 Same as Figure 12.6, except now for the FLRW cosmology constrained by $R_{\rm h} = ct$. There are no cosmological parameters to optimize for this fit. (Adapted from ref. [332])

the standard model assuming that inflation established at least some of its initial conditions, which means that spatial flatness ($k = 0$) is a reasonable prior. But even inflation says nothing about the equation-of-state for dark energy. Using empirically derived priors from other observations is not appropriate for supernova work, because their values change from source to source—and from instrument to instrument. As a concrete example, consider that the values of $w_{\rm de}$ and $\Omega_{\rm m}$ found by WMAP [106] are quite different from those inferred by *Planck* [29]. Thus, the most basic ΛCDM model one may use for supernova work has three essential parameters: $\Omega_{\rm m}$, $w_{\rm de}$ and H_0. It is the optimization of these three variables, or something like them, that seems to produce the most favorable outcome when fitting the data with the luminosity distance predicted by ΛCDM.

There are several caveats one must address with this framework, however. The first is that the actual value of H_0 is not independent of M_B (Equation 12.45).

TABLE 12.6 Optimized Parameters Based on the SNLS+SDSS-II Sample

Model	α	β	M_B	ΔM_{host}
$R_{\text{h}} = ct$	0.119 ± 0.007	2.600 ± 0.084	-18.932 ± 0.021	-0.052 ± 0.026
wCDM	0.121 ± 0.008	2.631 ± 0.087	-19.020 ± 0.037	-0.051 ± 0.028
ΛCDM	0.121 ± 0.008	2.631 ± 0.086	-19.026 ± 0.035	-0.051 ± 0.028

The optimization may be carried out by varying either H_0 or M_B, but not both independently. Therefore, when M_B is assumed to be one of the nuisance parameters used to model the supernova luminosity, the cosmology may be handled with just two parameters. And in keeping with our approach in Section 12.3.1, we shall therefore uniformly assume the fiducial value $H_0 = 70$ km s^{-1} Mpc^{-1} for display purposes (e.g., Figures 12.6 and 12.7) throughout this section. The second caveat is that the choice of these two parameters does not always have to be Ω_{m} and w_{de}. Sometimes, a more favorable fit to the data using ΛCDM is obtained by assuming a cosmological constant ($\Omega_{\text{de}} \rightarrow \Omega_\Lambda$) and instead relaxing the spatial flatness condition, allowing Ω_Λ to vary freely.[2]

A summary of the comparative analysis carried out for ΛCDM and the FLRW cosmology constrained by $R_{\text{h}} = ct$, based on the SNLS sample, is provided in Table 12.5 [332], and the Hubble diagrams are shown in Figures 12.6 and 12.7, respectively. Several of the events included in the SNLS sample were inappropriate for a variety of reasons, such as lying outside the SiFTO range of validity, and were therefore excluded from these diagrams [326].

A close inspection of Figures 12.6 and 12.7 reveals that the distance moduli are somewhat different when the nuisance parameters are optimized using different models, yet both ΛCDM and the FLRW cosmology constrained by $R_{\text{h}} = ct$ fit the data very well, chiefly because the reduced data themselves change from one cosmology to the next, echoing the result from the previous subsection. The last entry in Table 12.5 shows, however, that $R_{\text{h}} = ct$ is favored over basic ΛCDM when a single, uniform sample of Type Ia SNe is used in the analysis. In light of the fact that there are no adjustable model parameters to optimize the fit in Figure 12.7 (once H_0 is subsumed into M_B), this comparison again suggests that $R_{\text{h}} = ct$ acts as an 'attractor' for the parametrization in ΛCDM.

Even so, this result is far from compelling, given that the sample size is—by today's standards—relatively small. Merging subsamples to construct a larger supernova catalog is fraught with seemingly insurmountable issues, as noted earlier, but

---

[2] As noted earlier, this situation would give rise to several inconsistencies with the inflationary paradigm, but is nonetheless a useful approach to take when our interest is primarily in understanding how well Type Ia SNe constrain the basic cosmological model on their own.

TABLE 12.7   Model Selection Based on the SNLS+SDSS-II Sample

Model	Ω_{m}	Ω_{de}	w_{de}	χ^2_{dof}	BIC
$R_{\mathrm{h}} = ct$	—	—	—	1.04 (609 dof)	−514.57
wCDM	$0.203^{+0.137}_{-0.196}$	$0.445^{+0.265}_{-0.205}$	$-0.956^{+0.276}_{-0.384}$	1.01 (606 dof)	−511.43
ΛCDM	0.250 ± 0.152	0.480 ± 0.202	-1 (fixed)	1.01 (607 dof)	−517.86

recent progress has mitigated the negative impact of systematic effects by introducing a new cross-correlation of the SNLS (at $0.1 \lesssim z \lesssim 1$) and the Sloan Digital Sky Survey (SDSS; at $0.05 \lesssim z \lesssim 0.4$) [334]. Using the joint lightcurve analysis (JLA) [324], based on the use of tertiary standard stars observed by *both* experiments as a reference, a common point-spread function, and the same reduction technique, this combined effort has produced a photometric accuracy approaching ~ 5 mmag which, in principle, is quite precise compared to the system-wide dispersion, which is about 20 times bigger (see discussion below). This combined sample should therefore have a consistent calibration and systematics but, as we shall see shortly, there still appears to be a relic offset left over between SNLS and SDSS-II in the data.

Instead of SiFTO, the JLA collaboration opted to use another well-known lightcurve fitter known as SALT2 [335]. This seemed to be a more appropriate choice for this collection of supernovae, given that SALT2 is data-driven, and does not appear to introduce any significant bias between the low-redshift and high-redshift distances, an essential factor when dealing with a large sample spanning this redshift range. The ansatz for the distance modulus, however, remains the same as that given in Equation (12.45), except that the effect due to the shape of the lightcurve is written $\alpha \cdot X_1$, instead of $\alpha \cdot (s - 1)$, where X_1 is the 'shape' factor. An allowance is also made for the assumed host galaxy mass, introduced as an adjustment ΔM_{host} to M_B for supernovae in host galaxies with a mass $> 10^{10} \, M_\odot$.

The nuisance parameters for three models, optimized by maximizing the likelihood function with the combined SNLS+SDSS-II supernova sample, are shown in Table 12.6. As before, the basic ΛCDM model assumes a cosmological constant for dark energy though, in this case, a second version (wCDM) is also shown, in which w_{de} is allowed to vary freely [336]. The corresponding cosmological parameters are given in Table 12.7.

The weight of evidence from Tables 12.5 and 12.7 presents a somewhat confusing outcome, though perhaps not so surprisingly, given the complications we have considered in this section. At least with the surveys available at this time, the model selection based on a single, uniform supernova sample (e.g., SNLS) appears to be inconsistent with model comparisons provided by merged catalogs (e.g., SNLS+SDSS-II), even with the benefit afforded this process by the JLA.

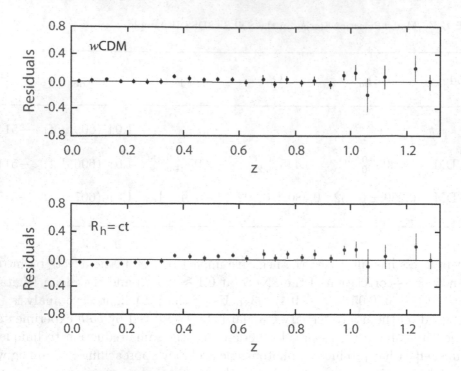

Figure 12.8 Residuals in the Hubble diagrams constructed for wCDM and the FLRW cosmology constrained by $R_h = ct$, using the merged SNLS+SDSS-II sample (see Tables 12.6 and 12.7). For added clarity, the residuals are shown as averages over redshift bins of 0.05. The Hubble diagrams themselves are very similar to those in Figures 12.6 and 12.7 and are not shown here. (Adapted from ref. [332])

According to the results summarized in Table 12.7, the FLRW cosmology constrained by $R_h = ct$ is mildly preferred over wCDM, with a relative likelihood of $\sim 83\%$ versus $\sim 17\%$. With one fewer parameter than wCDM, ΛCDM is, however, preferred over $R_h = ct$ with similar likelihoods, $\sim 84\%$ versus $\sim 16\%$. Yet the χ^2_{dof} values for the best-fit models, whether based on SNLS on its own, or in combination with SDSS-II, are hardly distinguishable. At least in the case of Type Ia SNe, model selection is heavily influenced by the number of free parameters.

Perhaps the difference between the analysis of SNLS on its own versus the SNLS+SDSS-II sample is due to the effect that produces an artifact appearing in the residuals (averaged over redshift bins of 0.05) of the Hubble diagrams constructed from the merged catalog (Figure 12.8). The residuals in both panels appear to be somewhat lower at $z \lesssim 0.35$ than those at $z > 0.35$, with an average difference of ~ 40 mmag for wCDM and about 80 mmag for $R_h = ct$. One also finds for the 4 highest redshift bins an average residual magnitude of 139 mmag for $R_h = ct$ and a somewhat worse 143 mmag for wCDM. An important clue appears to be that the residuals do not exhibit a monotonic trend, suggesting a systematic offset across $z \sim 0.35$.

This 'transition' redshift coincides with the division between the bulk of the SNLS sources and the majority of the SDSS-II events ($0.05 \lesssim z \lesssim 0.4$). It appears that the joint calibration of these two samples may not have been completely self-consistent after all. This artifact may thus be due to an unresolved measurement offset in their magnitude. Some quantitative evidence in support of this conclusion is the fact that the implied magnitude offset ($\sim$ 80–140 mmag) is comparable to the measured sample-wide 'intrinsic' dispersion $\sigma_{\mathrm{coh}} = 106 \pm 6$ mmag estimated by the JLA collaboration [324]. Also, a possible reason that the offset is slightly smaller for wCDM than $R_{\mathrm{h}} = ct$ may be the additional free parameter in the former that allows a greater flexibility in shaping the best-fit curve. Nevertheless, the key point here is that the offset is independent of redshift above and below the crossover at $z \sim 0.35$.

Given these mixed results, model selection based on the Hubble diagram constructed from Type Ia SNe is not yet as conclusive as one finds with high-z quasars. Even the JLA appears to have succeeded only partially in eliminating the sample-specific systematics. But such mitigating factors will almost certainly be addressed by several upcoming surveys. The principal advantage of quasars is that they stretch the redshift range to $\sim$ 6, well beyond our ability to observe Type Ia SNe ($\lesssim 2$). But with a large, homogeneous catalog, based on consistent photometry and calibration across the whole redshift range, Type Ia SNe will eventually provide a definitive answer to the question of whether or not the Universe is expanding according to FLRW subject to the $R_{\mathrm{h}} = ct$ constraint.

12.4   THE ANGULAR-DIAMETER DISTANCE

The luminosity distance (Equation 12.32), $d_L(z)$, is commonly used in cosmology for measurements involving sources identified as 'standard candles.' We have seen in the previous sections how its application to high-z quasars and Type Ia SNe allows us to construct the Hubble diagram for comparison with model predictions. By comparison, the angular-diameter distance, $d_A(z)$, applicable to objects or aggregates of objects (such as galaxy clusters) known to be 'standard rulers,' is used less frequently due to the paucity of such sources and complications arising from the evolution of their size with redshift. But there is growing interest in considering this new measure of distance and we shall next assess its value for cosmological measurements going forward.

Suppose we observe light rays traveling radially towards us from the edge of a distant source with proper diameter D. The light was emitted at time t_e, from a comoving distance r_e given in Equation (12.21). If the source subtends an angle $\delta\theta$ at the origin, then we must have

$$D = \delta\theta\, a(t_e)\, r_e \,, \tag{12.48}$$

under the assumption that $D \ll a(t_e)\, r_e$. The angular diameter of the source may therefore be written

$$\delta\theta = \frac{D}{a(t_e)\, r_e} \,. \tag{12.49}$$

In Euclidean geometry, Equation (12.49) would define the distance, d, to the source as

$$\delta\theta = \frac{D}{d} \,. \tag{12.50}$$

By analogy, we define the angular diameter distance d_A in FLRW by the expression

$$d_A = \frac{D}{\delta\theta} , \tag{12.51}$$

so that

$$d_A = a(t_e)\, r_e . \tag{12.52}$$

Notice that the angular diameter distance is therefore the proper distance to the source at the time the light was emitted.

Comparing Equations (12.28) and (12.52), one sees that the angular diameter and luminosity distances are related via the simple expression

$$d_A = \left[\frac{a(t_e)}{a(t_0)}\right]^2 d_L \tag{12.53}$$

so, using Equation (8.9), one may write

$$d_A(z) = \frac{1}{(1+z)^2}\, d_L(z) . \tag{12.54}$$

Finally, combining this expression with Equations (12.33) and (12.35), one finds that the angular diameter distance in ΛCDM and $R_{\rm h} = ct$ may be formulated as

$$d_A^{\Lambda\text{CDM}}(z) = \frac{c}{H_0} \frac{1}{(1+z)} \int_0^z \frac{dz'}{E^{\Lambda\text{CDM}}(z')} , \tag{12.55}$$

and

$$d_A^{R_{\rm h}=ct}(z) = \frac{c}{H_0} \frac{1}{(1+z)} \ln(1+z) , \tag{12.56}$$

respectively, where $E^{\Lambda\text{CDM}}$ is defined in Equation (12.34).

Given the simple relation in Equation (12.54), one may wonder why we even need the second formulation of distance expressed as d_A. The straightforward reason is that d_L and d_A are used for two entirely different types of cosmological measurement. Whereas the luminosity distance is used with measurements of flux from a standard candle, the angular diameter distance comes into play when we measure the angular size of a standard ruler in the sky. Very famously, while the luminosity distance increases indefinitely with redshift, the angular diameter distance typically increases at first, reaches a maximum, and then decreases to zero as our observations approach the Big Bang. As we shall see shortly, however, this behavior is not true in all cosmologies and may be used as a diagnostic to distinguish models that do predict such a turnover from those that do not.

The physics underlying Equation (12.54) is actually more elaborate than this synopsis would imply. Known as the 'cosmic distance duality' (CDD) relation, it is based on the reciprocity theorem derived by Etherington almost a century ago [337]. It holds true as long as (i) the cosmic spacetime is based on Riemannian geometry, (ii) photons propagate along null geodesics, and (iii) the photon number is conserved.

There are many reasons why the CDD could be violated in nature, however. For example, it could break down if the spacetime is not described by a metric theory of gravity, though this now appears to be very unlikely [338], or because photons arriving from presumed standard candles are altered along the line of sight by absorption or scattering effects [339].

Many have attempted to validate the CDD using, e.g., angular diameter distances extracted from galaxy clusters and luminosity distances from Type Ia SNe [340, 341]. But the cosmology dependence of these measurements is not trivial, and the presumption of a specific cosmological model can bias the inferred distances, particularly if the model is incomplete (or wrong). As such, some (or all) violations of the CDD claimed by many previous studies may simply be due to problems with the models themselves [342]. More recent tests of the CDD, relying on model-independent measurements, have largely confirmed its validity [343].

In this section, we shall describe a recent application of the angular diameter distance to the observed compact structures in radio quasars which, under certain restrictions, appear to function as a true standard ruler. Previous cases where some progress has been made with $d_A(z)$ include (i) baryon acoustic oscillations (BAO) seen in large-scale structure (see Section 11.1); (ii) the Sachs-Wolfe induced 1° fluctuations seen in the CMB [106, 29, 256]; and (iii) strong lensing systems, with and without time delays (see Section 12.5). As we shall describe below, several recent advances in our understanding of radio galaxy cores present us with what appears to be an even more reliable measuring rod, with a negligible evolution in the redshift range $0 \lesssim z \lesssim 3$. In principle, these sources allow us to make precise cosmological measurements over a larger fraction of the visible Universe than any of the other methods using $d_A(z)$ thus far.

These developments have been facilitated by a deeper understanding of synchrotron self-absorption near supermassive black holes [344, 204] and the discovery of crucial constraints on their spectral index and luminosity that allow us to select a sample of quasars with a radio emitting region of fixed length [345, 346, 347].

The idea of using standard rulers to measure the geometry of the Universe was proposed half a century ago by Fred Hoyle (1915–2001), though finding suitable sources to implement his concept took several decades more [349]. Indeed, the earliest cosmological tests using the angular size of kpc-scale radio sources and galaxies were unsuccessful due to the lack of a reliable standard ruler [350]. The effort to improve this technique continued, however, motivated by a subsequent study of double-lobed quasars showing that their apparent angular size remains constant with increasing angular diameter distance over the redshift range $1 \lesssim z \lesssim 2.7$. This is what one would expect in an FLRW cosmology without significant evolution (see Figure 12.9 below for a related example) [351].

We now know that ultracompact radio sources are more likely to be standard rods than the large-scale jets, principally because their emission is dominated by self-absorbed synchrotron processes, that form opaque features with angular diameters in the milliarcsecond range, corresponding to a linear size ~ 10 parsecs [352]. They are much smaller than the kpc-scale structures and have a reasonably stable environment from source to source [353]. The compact structures evolve primarily under

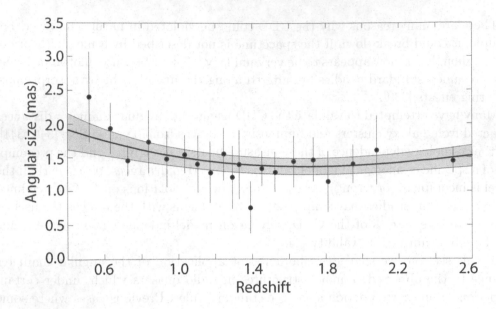

Figure 12.9 Angular size of 140 compact quasar cores divided into bins of seven, as a function of redshift. Each datum corresponds to the median value in each bin. The reconstructed angular-size function using Gaussian processes is represented by the thick, solid curve, and its 1σ variation is indicated by the shaded region. (Adapted from ref. [348])

the influence of the black hole itself, characterized by just a few physical parameters, such as its mass and spin. Moreover, their dynamical timescale is tens of years, much shorter than a Hubble time, and are therefore free of long-term evolutionary effects.

Studies of compact radio jets tend to be based on the large imaging survey completed with Very Large Baseline Interferometry (VLBI) in the 1990's. But a persistent problem with this sample is the broad range of hosts, from radio galaxies to BL Lacs and other quasar sub-categories. Their systematic differences are therefore difficult to disentangle from true cosmological effects. This has changed quite dramatically with the recent mitigation of this confusion via the discovery of a well-defined set of strategies that one may use to refine the survey sample [346]. As we shall demonstrate shortly, this effort has significantly reduced the scatter, allowing us to study the geometry of the Universe in unprecedented ways.

It is now understood that the dispersion in the linear size of these compact cores is greatly reduced by retaining only those sources with spectral indices $-0.38 < \alpha < 0.18$ [345, 347]. Moreover, the core size, ℓ_{core}, apparently depends quite strongly on the luminosity at both the low and high ends. Adopting the parametrization $\ell_{core} = \ell_0 L^\beta (1+z)^n$, where ℓ_0 is a scaling constant, one finds that only a subsample of *intermediate*-luminosity radio quasars with $10^{27}\,\mathrm{W/Hz} < L < 10^{28}\,\mathrm{W/Hz}$ have a core size only negligibly dependent on L and z. For this category, $\beta \approx 10^{-4}$ and $|n| \approx 10^{-3}$, yielding a reduced sample of compact radio structures with a reliable, fixed linear size.

When these constraints are applied to the original VLBI catalog of 613 sources, the final reduced sample contains 140 quasars. These are shown in Figure 12.9, binned into groups of seven, with the median value representing the angular size in each bin [354]. Their 1σ errors have been estimated assuming a Gaussian distributed variation in each bin.

These data display a rather unique dependence on redshift unlike any of the other integrated quantities, such as $d_L(z)$, in that a clear flattening and reversal are seen at $z \sim 1.4 - 1.8$. We shall understand shortly why this behavior alone completely rules out the Milne universe (see Sections 5.1 and 5.42), one of the best studied cosmological models. It is quite straightforward to compare the prediction of each cosmology with the angular-size data shown in this figure, but fitting procedures typically require the pre-assumption of a parametric form tailored to each specific model. At least in this case, however, one can learn a great deal about the geometry of the Universe by first using a Gaussian Processes (GP) approach, avoiding the use of parametric functions that may not be reasonable representations of the actual redshift dependence of the measurements.[3]

In modeling a function $f(x)$ rigorously without the use of a prior parametric form, the GP procedure assumes that the n observations of a data set $y = \{y_1, y_2, ..., y_n\}$ have been sampled from a multivariate Gaussian distribution. The mean of the GP partnered to the data is taken to be zero. The drawback with this process, however, is that one must face certain ambiguities (discussed below). Nevertheless, reasonable steps may be taken to ensure that the outcome of the reconstruction is not heavily dependent on the choice of GP components.

The first potential difficulty arises because the values of $f(x)$ evaluated at different points x_1 and x_2 are not independent of each other. The procedure calls for the introduction of a covariance function $k(x_1, x_2)$ to handle this linkage, but $k(x_1, x_2)$ is not unique or well known. In fact, there may be a broad range of such covariances. Most applications adopt a squared exponential,

$$k(x_1, x_2) = \sigma_f^2 \exp\left\{-\frac{(x_1 - x_2)^2}{2l^2}\right\} , \qquad (12.57)$$

which is infinitely differentiable. It may be used to reconstruct not only the function representing the data, but also its derivative. The so-called hyperparameters σ_f and l do not specify the form of the function itself but, rather, its 'bumpiness.' The length l is a measure of the distance in x over which the reconstructed function varies significantly, while the dependence in the y direction is scaled by the signal variance σ_f.

The GP reconstructed function shown in Figure 12.9 was optimized using the kernel in Equation (12.57). One of the most important goals with these data is to measure the turning point, z_{max}, at which the angular size stops decreasing and then increases again with redshift. So to ensure that this inferred quantity is not being unduly affected by the choice of covariance function, one may also carry out a parallel

---

[3]A detailed account of this implementation may be found in ref. [355] and other sources cited therein. Additional helpful references include ref. [356, 357].

TABLE 12.8 z_{max} for Four Cosmological Models

Model	z_{max}	$\|z_{max} - z_{max}^{obs}\|/\sigma$	Probability (%)
$R_h = ct$	1.718	0.09	92.8
Planck ΛCDM	1.594	0.53	59.6
Einstein-de Sitter	0.682	5.09	~ 0
Milne universe	∞	∞	0

simulation based on a very different kind of kernel, known as a Matérn covariance function (specifically Matérn92), whose explicit form is

$$k(x_1, x_2) = \sigma_f^2 \exp\left\{-\frac{3|x_1 - x_2|}{l}\right\}\left(1 + \frac{3|x_1 - x_2|}{l} + \frac{27|x_1 - x_2|^2}{7l^2} + \frac{18|x_1 - x_2|^3}{7l^3} + \frac{27|x_1 - x_2|^4}{35l^4}\right\}. \tag{12.58}$$

A direct comparison between these two choices has shown that a given kernel may change the probabilities (see Table 12.8) by a few points, but also that the model selection is qualitatively unaltered [348]. The rank ordering of models we shall discuss below is completely unaffected by the choice of $k(x_1, x_2)$.

A second potential ambiguity may arise with the hyperparameters themselves. A generally agreed upon procedure is to train them by maximizing the likelihood that the reconstructed function $f(x)$ reproduces the measured values at the data points x_i. Of course, a purely Bayesian analysis would marginalize the hyperparameters instead of optimizing them. As it turns out, however, the marginal likelihood in applications such as this is sharply peaked, so optimization is an excellent approximation to marginalization. For the data shown in Figure 12.9, there is therefore no freedom to choose l and σ_f separately once one carries through with the optimization procedure described above.

The measured value of z_{max} from the GP reconstructed curve in Figure 12.9 is 1.70 ± 0.20. The physics behind this phenomenon is actually easy to understand [155]. The angular-diameter distance is based on the measurement of a lateral proper size of an object we see in projection as it was when it emitted the light approaching us now. But all sources were closer to us at $t_e < t_0$, so the *apparent* angular size θ_{core} of compact quasar cores increases as $z \to \infty$. And since $d_A(z) \sim \theta_{core}^{-1}$, the angular-diameter distance thus gets smaller with increasing redshift.

As long as ℓ_{core} is a true standard ruler (at least on average), one does not need to know its actual value to identify z_{max} because this measurement is based on the ratio of scales at different redshifts. One also does not need to know H_0, which does

TABLE 12.9   Model Selection Using Compact Quasar Cores

Model	Ω_{m}	η	χ^2_{dof}	BIC	Probability
ΛCDM	$0.24^{+0.1}_{-0.09}$	0.58 ± 0.05	0.31	11.6	19.8%
$R_h = ct$	—	$0.5^{+0.03}_{-0.02}$	0.31	8.8	80.2%

not impact the location of the turning point in $d_A(z)$. The GP reconstruction in Figure 12.9 is thus completely free of any cosmological model and assumptions, and the turning point z_{max} therefore provides an unambiguous test of the theoretical predictions.

Table 12.8 lists the values of z_{max} predicted by four cosmological models relevant to this analysis. All of them are found by identifying the turning point in the angular diameter distance: Equation (12.56) for the FLRW cosmology constrained by $R_h = ct$, and Equation (12.55) for ΛCDM and the Einstein-de Sitter universe (Section 5.3). The assumed parameters are: the *Planck* optimization for the standard model [29], i.e., $\Omega_{\mathrm{m}} = 0.315 \pm 0.007$, $\Omega_{\mathrm{r}} = (5.48 \pm 0.001) \times 10^{-5}$ and $\Omega_{\Lambda} = 1.0 - \Omega_{\mathrm{m}} - \Omega_{\mathrm{r}}$; and $\Omega_{\mathrm{m}} = 1$ with $\Omega_{\Lambda} = \Omega_{\mathrm{r}} = 0$ for the latter (see Equations 6.1 and 6.3 for the definition of Ω_i in terms of the critical density ρ_{c}). The corresponding expression for the Milne universe is also easy to derive, starting from the Friedmann Equation (4.28) with $\rho = 0$ and $k = -1$, then finding the comoving distance r_e in Equation (12.21), and finally expressing $d_L(z)$ and $d_A(z)$ in terms of the redshift (Equations 12.28 and 12.54):

$$d_A^{\mathrm{Milne}}(z) = \frac{c}{H_0} \frac{1}{1+z} \sinh\left[\ln(1+z)\right] . \tag{12.59}$$

Unlike the angular diameter distance in the other cosmologies tabulated here, this expression has no turning point so $z_{\mathrm{max}} = \infty$ for the Milne universe.

In Table 12.8 we see each model's prediction compared with the measured value, $z_{\mathrm{max}} = 1.70 \pm 0.20$, together with their difference as a fraction of the measurement error, $\sigma_{z_{\mathrm{max}}}$. From this ratio, one infers the probability that the predicted turning point is consistent with its measured value, assuming a Gaussian distribution. The percentages quoted here are absolute, in the sense that they are not based on a comparison of relative probabilities. Each model is being compared directly with the data, independently of the other cosmologies.

It may be surprising to find how strongly the various models are differentiated on the basis of z_{max} alone, even without considering the quality of the fit at all available redshifts, which we shall do shortly. A principal concern with tests of the expansion dynamics using Type Ia SNe is the hypothesized transition from deceleration to acceleration at $z \sim 0.7$ (see Section 12.3.2). The compact structure in quasar cores plays a comparably important role in revealing the geometry of the Universe at a second, critical transition redshift, z_{max}, where $d_A(z)$ turns over. This redshift is so different

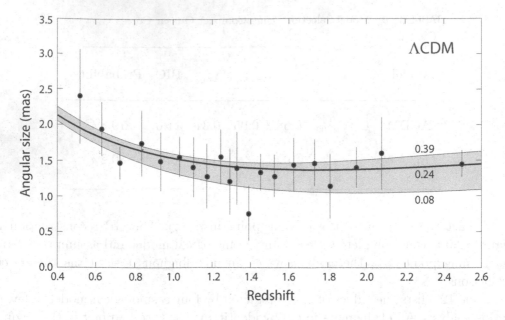

Figure 12.10 Same as Figure 12.9, except now showing the optimized flat ΛCDM fit (solid curve), with angular size constant $\eta = 0.58 \pm 0.05$ and $\Omega_m = 0.24^{+0.1}_{-0.09}$. The shaded region is bounded by the theoretical curves corresponding to $\Omega_m = 0.08$ and 0.39, defining the 1σ region when η is held constant. (Adapted from ref. [354])

between competing cosmologies that its measured value already strongly favors only two of the models listed in Table 12.8: the FLRW expansion constrained by $R_h = ct$, followed by *Planck* ΛCDM. The Milne model, in particular, is completely ruled out by the flattening observed in the angular size versus redshift diagram (Figure 12.9).

Although *Planck* ΛCDM is somewhat disfavored compared to $R_h = ct$, the expression for $d_A^{\Lambda\text{CDM}}(z)$ has enough flexibility that one should consider an alternative parameter optimization to see if its status relative to $R_h = ct$ can be raised in Table 12.8. One can easily show that flat ΛCDM with $\Omega_m = 0.23$ predicts a turning point at $z_{\max} = 1.70$, fully consistent with the measured value, and a probability exceeding 92.8%. The downside of this option, however, is that this scaled matter density would be in tension at more than 6.5σ with the *Planck* value. One could argue that dark energy is not a cosmological constant ($w_{\text{de}} \neq -1$), and optimize w_{de} along with the other free parameters. But with each modification to the standard model, one is receding farther and farther from the concordance cosmology, leading to a concern that finding consistency with $z_{\max}$ may not be worth damaging the optimization of ΛCDM's fits to other data, particularly those pertaining to the CMB.

Before discussing the individual model optimizations using the data in Figure 12.9, we remark on how different the outcomes are for the Milne universe and $R_h = ct$. Both predict a linear expansion rate, $a(t) \propto t$, and are sometimes confused with each other in the primary literature. But whereas Milne is an empty universe driven purely by spatial curvature ($k = -1$), the FLRW cosmology constrained by $R_h = ct$ is not empty and has a very unique equation-of-state—the zero active mass

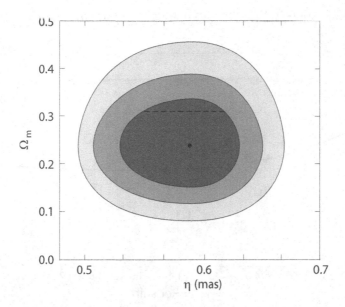

Figure 12.11 The 1σ, 2σ and 3σ confidence regions (dark to light) associated with the optimized parameters η and $\Omega_{\rm m}$ in flat ΛCDM (see Table 12.9). The dashed line indicates the *Planck* value $\Omega_{\rm m} = 0.31$ (Planck Collaboration 2018) consistent with the 1σ region. (Adapted from ref. [354])

condition, $\rho + 3p = 0$. It is this difference that directly alters the angular diameter distance from Equation (12.59) to (12.56). The diametrically opposite results shown in Table 12.8 result from these very diverse functional forms of $d_A(z)$.

For ΛCDM, we now put

$$\theta_{\rm core}(z) = \eta \, \frac{1+z}{F(z)} \,, \tag{12.60}$$

where $\eta \equiv \ell_{\rm core} H_0/c$ and

$$F(z) \equiv \int_0^z \frac{du}{\left[\Omega_{\rm m}(1+u)^3 + 1 - \Omega_{\rm m}\right]^{1/2}} \,. \tag{12.61}$$

Using standard χ^2 minimization, one may optimize the values of η and $\Omega_{\rm m}$ to produce the best fit, shown as a solid, black curve in Figure 12.10. The resulting optimized parameters are given in Table 12.9, and the corresponding 1σ, 2σ and 3σ confidence regions are plotted in the $\eta - \Omega_{\rm m}$ phase plane shown in Figure 12.11. The quality of the fit in Figure 12.10 is highlighted by the 1σ confidence (shaded) region constructed by keeping the best-fit value of η fixed, while allowing $\Omega_{\rm m}$ to vary. Clearly, ΛCDM accounts for the redshift-dependent compact quasar core size very well, given that its best-fit parameter values are fully consistent with the *Planck* optimization at better than 1σ.

Turning to the FLRW cosmology constrained by $R_{\rm h} = ct$, one finds from Equation (12.56) that the corresponding expression for $\theta_{\rm core}(z)$ in Equation (12.60) contains the integral

$$F(z) = \ln(1+z) \,. \tag{12.62}$$

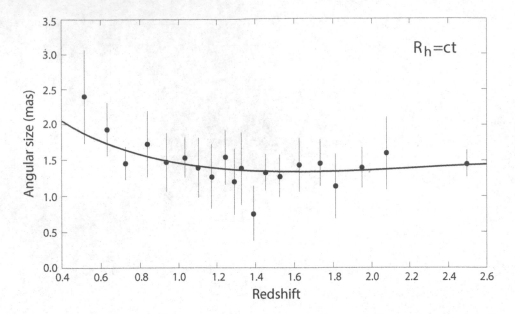

Figure 12.12 Same as Figure 12.9, except now showing the prediction of the FLRW cosmology constrained by $R_h = ct$. When the Hubble constant H_0 and the compact structure size are marginalized, there are no parameters with which to optimize the fit. Even so, this curve matches the data better than those of other models, including ΛCDM. (Adapted from ref. [354])

The optimization of the fit now involves just the parameter η, and is shown along with the data in Figure 12.12. The reduced $\chi^2_{\rm dof}$'s for these two fits are identical (see Table 12.9), but this was achieved without any model parameters in the case of $R_h = ct$, an influential factor in model selection based on information criteria (Section 12.1.1). The relative BIC's shown in this table affirm the prioritization produced previously on the basis of $z_{\rm max}$ alone (Table 12.8). Once again, the observations are telling us that an FLRW cosmology constrained by $R_h = ct$ accounts for the data better than the native formulation in ΛCDM with its unconstrained constituents in the cosmic fluid.

The identification of compact quasar cores as standard rulers has opened up an exciting new chapter in observational cosmology. These may be used to map the geometry of the Universe well beyond the reach of Type Ia SNe, even sampling the epoch during which the apparent size of sources increases with redshift. This unique effect is unseen with any other type of measurement, promising to make the use of $d_A(z)$ more relevant and common as the quality of the observations continues to improve.

12.5 STRONG GRAVITATIONAL LENSING

In fact, the angular diameter distance (Equation 12.51) is critical to yet another type of observation, specifically the bending of light produced by distant sources through the gravitational lens of intervening objects, such as an elliptical galaxy. Modern surveys can now easily measure this effect with sources at redshifts of ~ 3 or more, producing catalogs with hundreds of lens-source combinations.

The principle behind this technique comes from the well understood behavior of null geodesics through a strong gravitational field. The deflection of light produced by this effect is a classic test of general relativity. We shall here consider the simplest situation in which the metric of the lens is the Schwarzschild solution (Equation 3.19), and further simplify its coefficients by considering the weak-field limit, writing

$$ds^2 = \left(1 + \frac{2\Phi}{c^2}\right) c^2\, dT^2 - \left(1 - \frac{2\Phi}{c^2}\right) dr^2 - r^2\, d\Omega^2 \,, \tag{12.63}$$

where

$$\Phi \equiv -\frac{GM}{r} \tag{12.64}$$

is the Newtonian gravitational potential for mass M, such that $|\Phi| \ll c^2/2$. We continue to follow the convention established in Chapter 3, in which we use a capital letter to represent time in the accelerated frame.

It may seem somewhat confusing to discuss 'strong' gravitational lensing using the 'weak-field' limit, but these two designations correspond to different aspects of the problem. The latter means we are considering the gravitational influence of a mass well beyond its Schwarzschild radius, while the former distinguishes this lensing of a single background source by a single foreground lens from weak lensing due to the passage of light rays through an inhomogeneous medium.

The trajectory of a photon may be calculated from the null geodesic equation, equivalent to Equation (2.41) for a massive particle, though with τ replaced by an affine parameter we shall call λ, to characterize its advancement along the path:

$$\frac{d^2 X^\alpha}{d\lambda^2} + \Gamma^\alpha{}_{\beta\gamma} \frac{dX^\beta}{d\lambda} \frac{dX^\gamma}{d\lambda} = 0 \,. \tag{12.65}$$

The Christoffel symbols for this metric are given in Equation (3.10), with

$$B = \left(1 + \frac{2\Phi}{c^2}\right) \,, \tag{12.66}$$

and

$$A = \left(1 - \frac{2\Phi}{c^2}\right) \,. \tag{12.67}$$

In addition, since we are now dealing with a null geodesic, rather than a time-like one, we also have the constraint

$$0 = \frac{dX^\mu}{d\lambda} \frac{dX^\nu}{d\lambda} g_{\mu\nu}(x) \,, \tag{12.68}$$

where $g_{\mu\nu}(x)$ are the metric coefficients in Equation (12.63).

The expression in Equation (12.65) yields four simultaneous equations for the coordinates (cT, r, θ, ϕ). Without any loss of generality, one may integrate the θ equation immediately by setting $\theta = \pi/2$, i.e., using the fact that the photon's trajectory lies within a two-dimensional plane. And since this spacetime is isotropic, with a non-spinning central source, the coordinate frame may be rotated as needed to ensure that the $\hat{z}$-axis is always perpendicular to this plane.

The T and ϕ equations are also very simple, yielding two first integrals of the motion:

$$\left(1 - \frac{2\Phi}{c^2}\right) \frac{dT}{d\lambda} = \tilde{E}/c^2 , \qquad (12.69)$$

and

$$r^2 \frac{d\phi}{d\lambda} = \tilde{L} , \qquad (12.70)$$

where the two constants $\tilde{E}$ and $\tilde{L}$ are, respectively, the conserved 'energy-at-infinity' and conserved angular momentum. Combining these two expressions with the null condition in Equation (12.68), one then finds that

$$\tilde{E}^2 \left(1 - \frac{2\Phi}{c^2}\right)^{-1} - \left(\frac{dr}{d\lambda}\right)^2 \left(1 - \frac{2\Phi}{c^2}\right)^{-1} - \frac{c^2 \tilde{L}^2}{r^2} = 0 , \qquad (12.71)$$

from which one then extracts the photon's equations of motion:

$$\left(\frac{dr}{d\lambda}\right)^2 = \tilde{E}^2 - V_{\text{eff}}(r) , \qquad (12.72)$$

and

$$\frac{d\phi}{d\lambda} = \frac{\tilde{L}}{r^2} . \qquad (12.73)$$

The effective potential for the photon is written

$$V_{\text{eff}} \equiv \frac{c^2 \tilde{L}^2}{r^2} - \frac{2GM\tilde{L}}{r^3} . \qquad (12.74)$$

Though we do not have any reason to consider the analogous equations of motions for massive particles in this book, it is nevertheless useful to compare this effective potential for light with that arising in a timelike setting [69]. The centrifugal barrier $\tilde{L}^2/r^2$ is common to both potentials, but the 'Newtonian' attraction ($\sim -1/r$) is missing for light. This is consistent with what one would expect in the Newtonian limit, where photons do not 'feel' gravity since they are massless. But the short range attractive effect ($\sim -1/r^3$) due to general relativity applies to both photons and massive particles. This is the term responsible for the bending of light around sources of gravity.

To fully understand this phenomenon, one combines Equations (12.72) and (12.73) to arrive at an orbit equation written solely in terms of r and ϕ, using

$$\frac{dr}{d\lambda} = \frac{dr}{d\phi} \frac{d\phi}{d\lambda} = \frac{dr}{d\phi} \frac{\tilde{L}^2}{r^2} . \qquad (12.75)$$

Defining the 'impact parameter'

$$b \equiv \frac{c^2 \tilde{L}}{\tilde{E}} , \qquad (12.76)$$

one finds that

$$\frac{dr}{d\phi} = \pm r^2 \left[\frac{1}{b^2} - \frac{1}{r^2} \left(1 - \frac{2GM}{c^2 r} \right) \right]^{1/2} , \qquad (12.77)$$

which may be further simplified to the most commonly used form of the orbit equation (with $u \equiv 1/r$),

$$\frac{d\phi}{du} = \pm \left[\frac{1}{b^2} - u^2 + \frac{2GM}{c^2} u^3 \right]^{-1/2} . \qquad (12.78)$$

It is instructive to first integrate this equation ignoring the relativistic term $2GMu^3$, which gives

$$\phi + \phi_0 = \arcsin(bu) , \qquad (12.79)$$

or

$$b = r \sin(\phi + \phi_0) . \qquad (12.80)$$

This describes a straight line in the $\theta = \pi/2$ plane, with a distance of closest approach b from the mass M, as one would expect for the trajectory of a light ray in Newtonian mechanics.

The relativistic correction $2GMu^3/c^2$ alters the path, causing the ray to deflect around the mass. Squaring Equation (12.78),

$$\left(\frac{du}{d\phi} \right)^2 - \frac{1}{b^2} - u^2 + \frac{2GM}{c^2} u^3 = 0 , \qquad (12.81)$$

and differentiating this expression once more with respect to ϕ, one finds that

$$2 \frac{du}{d\phi} \left(\frac{d^2u}{d\phi^2} + u - \frac{3GM}{c^2} u^2 \right) = 0 , \qquad (12.82)$$

which is solved either by $du/d\phi = 0$, or by

$$\frac{d^2u}{d\phi^2} + u = \frac{3GM}{c^2} u^2 . \qquad (12.83)$$

We may ignore the first possibility, which corresponds to circular orbits. The second equation may be solved perturbatively in the weak-field limit, by putting

$$u(\phi) = u_0(\phi) + \Delta u(\phi) , \qquad (12.84)$$

implying that $\Delta u(\phi)$ is a small correction to the unperturbed $u_0(\phi)$. Thus, perturbing Equation (12.83), one has

$$\Delta \frac{d^2u}{d\phi^2} + \Delta u = \frac{3GM}{c^2} u_0^2 , \qquad (12.85)$$

where u_0 is the unperturbed solution given in Equation (12.80). That is,

$$\Delta \frac{d^2u}{d\phi^2} + \Delta u = \frac{3GM}{c^2 b^2} \left[1 - \cos^2 (\phi + \phi_0) \right] , \qquad (12.86)$$

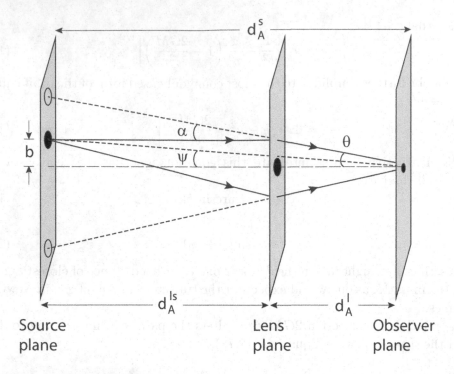

Figure 12.13 Schematic diagram showing the source-lens system, in terms of the angular diameter distances d_A^s and d_A^l from the observer, and the relative angular diameter distance d_A^{ls} between the lens and source. The angles θ and ψ, and the deflection angle α, are all related to the impact parameter b, i.e., the shortest distance between the light ray and the center of the lens.

which has the following solution:

$$\Delta u(\phi) = \frac{3GM}{2c^2b^2}\left\{1 + \frac{1}{3}\cos\left[2\left(\phi + \phi_0\right)\right]\right\}. \tag{12.87}$$

Thus, orienting the coordinate system so that $\phi_0 = 0$, one finds from Equation (12.84) that the complete (perturbed) light path is given by the expression

$$u_{\text{pert}}(\phi) = \frac{1}{b}\left\{\sin(\phi) + \frac{3GM}{2c^2b}\left[1 + \frac{1}{3}\cos(2\phi)\right]\right\}. \tag{12.88}$$

According to Equation (12.80), $\phi \to \pi$ when the photon approaches from $r \to -\infty$, and $\phi \to 0$ as it recedes to $r \to +\infty$. The deflection angle may be found by examining ϕ in Equation (12.88) under the same conditions. Since the deflection is small, one expects $\sin(\phi_\infty) \approx \phi_\infty$ and $\cos(2\phi_\infty) \approx 1$. Therefore, on either side of the mass M, the deviation away from 0 or π would be

$$\phi_\infty = \frac{2GM}{c^2b}, \tag{12.89}$$

and the total deflection angle relative to the incoming ray is therefore

$$\alpha \equiv 2\phi_\infty = \frac{4GM}{c^2b}. \tag{12.90}$$

The angle α is shown schematically in Figure 12.13, which illustrates the source-lens geometry that we shall now discuss.

From this diagram, one may clearly see that $b = \theta\, d_A^l$, so

$$\alpha = \frac{4GM}{c^2 \theta\, d_A^l} \,. \tag{12.91}$$

On the source plane, one also has

$$\theta\, d_A^s = \psi\, d_A^s + \alpha\, d_A^{ls} \,, \tag{12.92}$$

which is known as the 'lens equation.' When the source lies directly behind the lens, one may simply put $\psi = 0$, and therefore θ takes on a special value known as the *Einstein radius*,

$$\theta_{\mathrm{E}} \equiv \left(\frac{4GM}{c^2} \frac{d_A^{ls}}{d_A^l d_A^s} \right)^{1/2} . \tag{12.93}$$

A beautiful example of such a source-lens system is shown in Figure 12.14, produced by the Atacama Large Millimeter/sub-millimeter Array (ALMA) observatory [358]. At millimeter wavelengths, we do not see very much emission from the intervening lens galaxy, which lies very close to the center of this image, but we see the almost perfect ring of lensed dust emission produced by the distant source.

Of course, the mass of the lens is not concentrated at a point. The distribution of mass in an isolated galaxy may be approximated more accurately as a 'singular isothermal (gas) sphere' (SIS) [359] with a small core radius, outside of which the density approaches a power law [360] of the form $\rho \propto 1/r^2$. The solution to Poisson's equation (2.143) in such a system yields a radially-dependent mass

$$M(r) = \frac{2\sigma^2 r}{G} \,, \tag{12.94}$$

with density

$$\rho(r) = \frac{\sigma^2}{2\pi G r^2} \,, \tag{12.95}$$

in terms of the velocity dispersion (rms deviation from the mean) σ, here assumed to be isotropic. To find the deflection angle, one must first project the density along the line-of-sight, thereby obtaining the surface mass density $\Sigma(\xi)$ as a function of projected radius ξ in the lens plane:

$$\Sigma(\xi) = \int_{-\frac{\pi}{2}}^{\frac{\pi}{2}} \frac{\sigma^2}{2\pi G \xi}\, d\phi = \frac{\sigma^2}{2G\xi} \,. \tag{12.96}$$

The net deflection angle at a given impact parameter b arises from the contributions of all the mass elements up to b, where

$$M(b) = 2\pi \int_0^b \Sigma(\xi)\xi\, d\xi = \pi b\, \frac{\sigma^2}{G} \,. \tag{12.97}$$

Figure 12.14 The Einstein ring formed of the starburst galaxy SDP.81 (at $z = 3.042$) by an intervening lens galaxy (at $z = 0.299$) and imaged by the Atacama Large Millimeter/sub-millimeter Array (ALMA). The bright central region of the ring reveals the glowing dust in the lensed source, while the surrounding portions trace the millimeter-wavelength radiation emitted by CO_2 and H_2O molecules. The 'Einstein radius' of the ring is approximately 1.55 arcseconds. (Image courtesy of Ralph Bennett and the ALMA partnership, ESO/NAOJ/NRAO)

The deflection angle corresponding to this impact parameter is

$$\alpha(b) = \frac{4GM(b)}{c^2 b} = 4\pi \left(\frac{\sigma}{c}\right)^2 . \tag{12.98}$$

Thus, from Equations (12.92) and (12.98), with $\psi = 0$, one obtains the Einstein radius of an SIS given by the simple formula

$$\theta_E = 4\pi \left(\frac{\sigma_{SIS}}{c}\right)^2 \mathcal{D} , \tag{12.99}$$

in terms of the distance ratio defined as

$$\mathcal{D} \equiv \frac{d_A^{ls}}{d_A^s} . \tag{12.100}$$

With the catalog of such systems expanding rapidly, gravitational lensing is becoming a powerful diagnostic tool for studying the cosmic expansion. When

Einstein first proposed this effect, he believed it would be very difficult to observe [361]. That is certainly true, but in recent decades we have seen a rapid progression in the detection of strongly lensed sources, starting with the Lenses Structure and Dynamics (LSD) survey [362]. An additional 57 confirmed lenses were found by its successor, the Sloan Lens Advanced Camera for Surveys (SLACS) [363], followed by an additional 40 systems a decade later [364]. Several more surveys have since added another 50 or so lens-source systems, and the bigger consortia, such as the Dark Energy Survey (DES), have already started to compile hundreds of possible candidates awaiting followup observations and confirmation [365]. The sample available for cosmological testing today includes lenses from $z = 0.06$ to 1, and sources from $z = 0.2$ to 3.6. Nine of them lie beyond $z \sim 3$. Perhaps more than many of the other kinds of observation we have been discussing in this book, the rapid acquisition of new entries for this catalog is crucial for model selection because the strength of the statistical analysis relies on both the number of sources and their redshift range.

There is a complication in using Equation (12.99), however, because σ_{SIS} is not directly measurable. The actual velocity dispersion σ_{ap} is typically measured within a given aperture and converted to a velocity dispersion within a circular aperture of half the effective radius of the lens galaxy, yielding the 'observed' velocity dispersion

$$\sigma_0 = \sigma_{\mathrm{ap}} \left(\frac{\theta_{\mathrm{eff}}}{2\theta_{\mathrm{ap}}} \right)^{-0.04} , \qquad (12.101)$$

where θ_{eff} is the half-light radius of the lens and θ_{ap} is the instrument aperture size used for the measurement [366]. The dispersion σ_0 appears to be a reasonable representation of the system-wide approximation σ_{SIS} [367]. With this substitution, $\mathcal{D}$ constitutes an observable quantity when written

$$\mathcal{D}^{\mathrm{obs}} = \frac{c^2 \theta_{\mathrm{E}}}{4\pi \, \sigma_0^2} . \qquad (12.102)$$

For the model-dependent prediction in Equation (12.100), we shall need the relative angular diameter distance $d_A(z_1, z_2)$ between two redshifts z_1 and z_2, in addition to the more commonly used distance between us and the source (Equation 12.54). Its derivation is very similar to that of $d_A(z)$, except that the limits of integration are, of course, different. In flat wCDM, on finds

$$d_A^{w\mathrm{CDM}}(z_1, z_2) = \frac{c}{H_0} \frac{1}{1 + z_2} \int_{z_1}^{z_2} \frac{dz'}{E^{w\mathrm{CDM}}(z')} , \qquad (12.103)$$

where

$$E^{w\mathrm{CDM}}(z) \equiv \sqrt{\Omega_{\mathrm{m}}(1 + z)^3 + \Omega_{\mathrm{de}}(1 + z)^{3(1 + w_{\mathrm{de}})}} , \qquad (12.104)$$

and $w_{\mathrm{de}} \equiv p_{\mathrm{de}}/\rho_{\mathrm{de}}$ is the dark-energy equation-of-state parameter. The other symbols have their usual meaning. In principle, wCDM therefore has three free parameters one may use to optimize the fit to the lens-source data, but since the observed quantity is the ratio $\mathcal{D}$, rather than an individual angular diameter distance $d_A(z)$, the Hubble

TABLE 12.10 Best-fitting Parameters (for $\mathcal{D}^{\text{obs}} \leq 1$)

Model	Ω_{m}	w_{de}	X	χ^2_{dof}
wCDM	$0.33^{+0.13}_{-0.15}$	$-1.29^{+0.97}_{-6.09}$	12.2%	0.998
ΛCDM	$0.29^{+0.12}_{-0.08}$	-1 (fixed)	12.2%	0.999

constant H_0 cancels out and is irrelevant for this purpose. The optimization of wCDM using these data must therefore rely on only two free parameters, Ω_{m} and w_{de}.

The same expression (Equation 12.103) may be used for the Einstein-de Sitter universe, except that $\Omega_{\text{m}} = 1$ and $\Omega_{\text{de}} = 0$. For the Milne universe, the corresponding expression is

$$d_A^{\text{Milne}}(z_1, z_2) = \frac{c}{H_0}\frac{1}{1+z_2}\sinh\left[\ln\left(\frac{1+z_2}{1+z_1}\right)\right] , \qquad (12.105)$$

while for the FLRW cosmology constrained by $R_{\text{h}} = ct$, it is simply

$$d_A^{R_{\text{h}}=ct}(z_1, z_2) = \frac{c}{H_0}\frac{1}{1+z_2}\ln\left(\frac{1+z_2}{1+z_1}\right) . \qquad (12.106)$$

The theoretical predictions are compared to the data based on the expected ratio

$$\mathcal{D}^{\text{th}} = \frac{d_A(z_1, z_2)}{d_A(0, z_2)} , \qquad (12.107)$$

where z_1 and z_2 are, respectively, the lens and source redshifts. In $R_{\text{h}} = ct$, for example, one has

$$\mathcal{D}^{R_{\text{h}}=ct}(z_l, z_s) = 1 - \frac{\ln(1+z_l)}{\ln(1+z_s)} . \qquad (12.108)$$

Other than wCDM, none of these cosmologies has any free parameters for the model selection. In principle, gravitational lensing thus promises to provide a powerful test of various theoretical scenarios.

The results summarized here correspond to an overall sample size of 158 confirmed lens-source systems [367, 368], all with spectroscopically measured redshifts, and with accurately estimated Einstein radii θ_{E} with a relatively uniform error $\sigma_{\theta_{\text{E}}}$ of only 5% [369]. This catalog must be further refined in order to carefully exclude lens systems with $\mathcal{D}^{\text{obs}} > 1$, which are clearly unphysical (see Equation 12.100), and the analysis needs to include a dispersion σ_{SIS} to characterize possible random deviations from the simple isothermal sphere lens model. Standard error propagation then yields the uncertainty associated with each measurement, $\mathcal{D}^{\text{obs}}$, according to the expression

$$\sigma_{\mathcal{D}^{\text{obs}}} = \mathcal{D}^{\text{obs}}\sqrt{\left(\frac{\sigma_{\theta_E}}{\theta_E}\right)^2 + \left(\frac{2\sigma_{\sigma_0}}{\sigma_0}\right)^2 + X^2} , \qquad (12.109)$$

TABLE 12.11   Model Selection with 129 Lenses

Model	χ^2_{dof}	BIC	Relative Likelihood
$R_{\text{h}} = ct$	1.020	131.559	73.032%
ΛCDM	0.999	133.748	24.443%
wCDM	0.998	138.484	2.290%
Milne	1.109	143.063	0.232%
E-dS	1.194	151.314	0.004%

where X is a unitless composite error term representing the SIS scatter σ_{SIS}.

The model tests are based on maximization of the joint likelihood function, assuming a flat Bayesian prior, expressed in the form

$$\mathcal{L} = \prod_i \frac{1}{\sqrt{2\pi}\,\sigma_{\mathcal{D}_i^{\text{obs}}}} \exp\left(-\frac{\chi_i^2}{2}\right) \tag{12.110}$$

where, for each measurement i,

$$\chi_i^2 \equiv \frac{\left(\mathcal{D}_i^{\text{obs}} - \mathcal{D}_i^{\text{th}}\right)^2}{\sigma_{\mathcal{D}_i^{\text{obs}}}^2}. \tag{12.111}$$

In the first step, this procedure is applied solely to the wCDM and ΛCDM models to confirm that the outcome is consistent with the *Planck* measurements [29]. As one may see in Table 12.10, both Ω_{m} and w_{de} comply with the requirements of the concordance model, particularly ΛCDM.

With this confirmation in place, one may proceed to use the lens-source catalog for model selection, based on the Bayes information criterion, which is appropriate for a sample of this size: 129 systems after the unphysical ones have been removed (see Section 12.1.1). Table 12.11 summarizes the χ^2_{dof}, the BIC, and *relative* likelihood (calculated from Equation 12.5) of each cosmology being compared.

Without the inclusion of an intrinsic σ_{SIS}, the model fits to the data would be unsatisfactory. For example, in the case of ΛCDM, the reduced χ^2_{dof} would exceed 2.5, signaling a clear breakdown of the simplified lens model. The SIS may be sufficient at this level of accuracy, but no doubt as the sample size increases by orders of magnitude with future dedicated surveys, a better approach to modeling the lens distribution simultaneously with the cosmology will be required.

This caveat notwithstanding, the prioritization evident in Table 12.11 confirms the outcome we have seen throughout this chapter—that ΛCDM (or its variant wCDM) generally accounts for the data rather well. But a conclusion we cannot avoid is that the FLRW cosmology constrained by $R_{\rm h} = ct$ is preferred by the observations over all the other models we have considered thus far, including ΛCDM, which is also an FLRW cosmology, though without the zero active mass equation-of-state.

12.6   THE COSMIC EQUATION OF STATE

Our non-exhaustive survey of several key cosmological observations provides strong support for the inference drawn in Chapter 9 concerning the emergence of the $R_{\rm h} = ct$ constraint. Without the theoretical argument in Chapter 10, the adoption of FLRW to describe the spacetime on its own does not tell us much about the equation-of-state, $p = w\rho$, in terms of the total pressure p and energy density ρ. The best we can do today in the context of ΛCDM, from an empirical standpoint, is to assume that ρ must contain matter $\rho_{\rm m}$ and radiation $\rho_{\rm r}$, which one sees directly, and a mysterious 'dark' energy $\rho_{\rm de}$, whose existence is implied by a host of observations, notably Type Ia SNe. But the evidence presented in this chapter, and earlier in Chapter 9, tells us that the zero active mass condition, $\rho + 3p = 0$, needs to be incorporated into our theoretical modeling.

This is comforting in a way because the indication is not that ΛCDM is wrong, but that it merely lacks an important ingredient. After all, the standard model has succeeded remarkably well over the past three decades accounting for a broad range of cosmological measurements, so its empirical parametrization, based on the equation-of-state $p = \rho_{\rm r}/3 - \rho_\Lambda$, cannot be too far off the mark. Given that a one-to-one comparison with the even simpler equation-of-state $p = -\rho/3$ always seems to show the data favoring the latter, however, it is becoming increasingly clear that $R_{\rm h} = ct$ is an *attractor* for the optimization of the parameters in ΛCDM (or its more general version, wCDM).

A very simple analogy is the following. Suppose we were interested in fitting a straight line between two points on a Cartesian plane but—perhaps due to lack of information—the only available resource was a low-order polynomial. For example, if we were trying to fit the data using ΛCDM with three free parameters, say H_0, $\Omega_{\rm m}$ and Ω_Λ, this would be equivalent to using a cubic spline. The optimization would approximate the straight line, but not fit it perfectly because none of these parameters are allowed to be zero. Nevertheless, the straight line is the attractor because that is always going to be the ideal fit in this simple optimization exercise.

Let us therefore invert the process and assume that FLRW constrained by $R_{\rm h} = ct$ is the correct cosmology, and see whether ΛCDM (with its assumed parametrization) behaves as if it is attempting to mimic the hypothesized exact expansion profile in the former. The first question one may ask is "what is so special about the value $\Omega_{\rm m} \sim 0.27$ required for ΛCDM to adequately fit the data?" Consider the ratio $\mathcal{I} \equiv ct_0/R_{\rm h}(ct_0)$ which, according to Equation (9.3), may be written

$$\mathcal{I} \equiv \int_0^1 \frac{u\,du}{\sqrt{\Omega_{\rm r} + \Omega_{\rm m}u + \Omega_{\rm de}u^{1-3w_{\rm de}}}}. \tag{12.112}$$

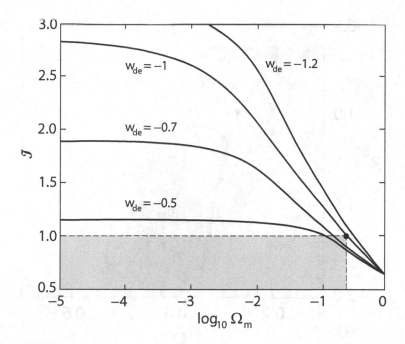

Figure 12.15 The ratio $\mathcal{I} \equiv ct_0/R_h(t_0)$ (the inverse of the quantity in Equation 9.2), as a function of Ω_m and w_{de} in wCDM. When $R_h(ct_0) = ct_0$ and $w_{de} = -1$, Ω_m must have the unique value 0.27 (shown as a black dot). (Adapted from ref. [316])

This quantity is plotted in Figure 12.15 as a function of Ω_m for various choices of w_{de}, assuming spatial flatness. Clearly, $\mathcal{I}$ may have a broad range of values but, for any given w_{de}, only one specific Ω_m satisfies the condition $\mathcal{I} = 1$. In the special case of a cosmological constant, for which $w_{de} = -1$, that value is 0.27. Thus, in the context of $R_h = ct$, an optimized matter density of ~ 0.3 in the standard model is not random at all, but is in fact required to mimic the zero active mass equation-of-state in the cosmic fluid.

This conclusion finds additional support in an apparent correlation between the optimized parameters, such as Ω_m and w_{de}. None of the individual observations thus far has produced fits that are so precise as to suggest unique values for these variables. As we understand it, the reason for this is that, other than the Sachs-Wolfe effect, which is responsible for the largest CMB fluctuations (see Section 11.1), none of the remaining mechanisms producing structure of one kind or another depends on the expansion history of the Universe. This leads to an inevitable degeneracy among the possible choices of parameter values characterizing the CMB [370].

There is certainly a well-defined region in the w_{de}–Ω_m plane within which the optimized parameter values are most likely to be found (see Figure 12.16). There is also an unmistakable correlation between the possible variation of w_{de} and changes in Ω_m. The solid black curve in this figure shows the loci of (Ω_m, w_{de}) points permitted by the constraint $\mathcal{I} = 1$. The fact that the WMAP result (black dot) falls almost exactly on this curve echoes what we already learned from Figure 12.15, and is already very suggestive. But the fact that the second datum (shown as a star), produced more

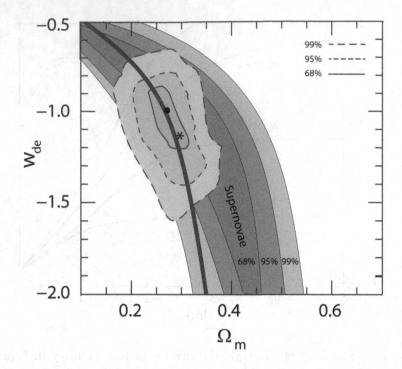

Figure 12.16 The value w_{de} as a function of Ω_m (solid black curve) when the condition $R_h(ct_0) = ct_0$ is imposed on wCDM, together with the observed constraints on the dark-energy equation-of-state for a spatially flat Universe. The bands represent the Type Ia supernova confidence levels (Suzuki et al. 2012), while the closed contours include results from CMB anisotropies, measurements of the Hubble constant, and large-scale structure (Melchiorri et al. 2003). The black dot is the WMAP measurement (Bennett et al. 2013), while the star corresponds to the *Planck* values, $\Omega_m = 0.31$ and $w_{de} = -1.13$. The value $\Omega_m = 0.27$ is realized only when $w_{de} = -1$. (Adapted from ref. [316])

recently by *Planck*, also falls on this curve is even more striking. Whereas $\Omega_m = 0.27$ is linked to a cosmological constant, the *Planck* measurement of $\Omega_m = 0.31$ is associated with $w_{de} = -1.13^{-0.13}_{+0.10}$. The implication of these two measurements is that, while the inferred value of Ω_m may vary from one instrument to the next, the co-optimization of w_{de} apparently changes in such a way as to preserve the condition $\mathcal{I} = 1$.

On its own, the standard model cannot explain why the most preferred region in this space is confined by the limits $-1.38 < w_{de} < -0.82$ and $0.22 < \Omega_m < 0.35$, and why this oblong region is slanted to couple the higher values of w_{de} with the smaller values of Ω_m. The $R_h = ct$ curve offers an explanation, however, because this is the region consistent with the $\mathcal{I} = 1$ requirement. The theoretical curve passes directly through the middle of the observationally permitted confidence region, even tracking its *orientation*. Thus, while the data lack sufficient precision to suggest a unique value for Ω_m, its optimization invariably produces a co-optimization of w_{de} consistent with the condition $\mathcal{I} = 1$.

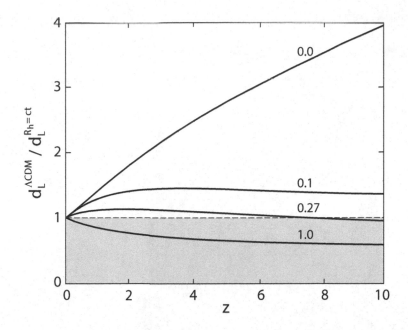

Figure 12.17 Ratio of luminosity distance in ΛCDM to that in the FLRW cosmology constrained by $R_{\mathrm{h}} = ct$, as a function of redshift for different values of Ω_{m}. All of these curves assume a cosmological constant (with $w_{\mathrm{de}} = -1$). The value $\Omega_{\mathrm{m}} = 0.27$, which allows ΛCDM to satisfy the condition $R_{\mathrm{h}}(t_0) = ct_0$, best approximates the condition $d_L^{\Lambda\mathrm{CDM}} = d_L^{R_{\mathrm{h}}=ct}$ over a very large range in redshift. (Adapted from ref. [316])

This perspective offers us a glimpse into the fundamental reasons why ΛCDM often fits the data very well, while apparently also mimicking the predictions of $R_{\mathrm{h}} = ct$. We begin to understand that, while the empirically motivated choice of density $\rho = \rho_{\mathrm{r}} + \rho_{\mathrm{m}} + \rho_{\mathrm{de}}$ is not entirely consistent with the equation-of-state $p = -\rho/3$ from one moment to the next, it nonetheless produces the same expansion as $R_{\mathrm{h}} = ct$ over a Hubble time—but only if $\Omega_{\mathrm{m}} \sim 0.27$.

Figure 12.17 provides a final clue in favor of this interpretation. It shows the ratio of luminosity distances $d_L^{\Lambda\mathrm{CDM}}/d_L^{R_{\mathrm{h}}=ct}$ versus redshift for various choices of Ω_{m}, assuming a cosmological constant. It is quite clear that the value of Ω_{m} satisfying the condition $\mathcal{I} = 1$ also corresponds to the version of ΛCDM with a luminosity distance that most closely tracks its counterpart in $R_{\mathrm{h}} = ct$. For many kinds of observation in the local Universe, such as Type Ia SNe and gamma ray bursts, the optimized ΛCDM therefore mimics $R_{\mathrm{h}} = ct$ at individual redshifts as well as globally over a Hubble time.

CHAPTER 13

Structure Formation

T HE stress-energy tensor $T^{\mu\nu}$ in the standard model [76] is largely *empirical*, without much input from theory. This lack of a conceptual basis is reflected in its name, which is 'descriptive'—Λ plus cold dark matter—rather than designating a specific physical process or underlying physical law. Nevertheless, this model derives considerable observational support from its successful accounting of the measured matter power spectrum (see Figure 13.3 below), one of the key indicators used to assess the formation of large-scale structure.

Our current view holds that structure formation began with primordial quantum fluctuations, seeded in an inflationary field shortly after the Big Bang (Section 6.7) [125, 371, 372, 373, 374, 126, 133]. According to this paradigm, self gravity grew these quantum fluctuations into the inhomogeneous Universe we see today after they classicalized in ways yet to be understood, though it is believed that this occurred once they crossed the Hubble horizon during the inflated expansion. As we shall see shortly, the pivotal feature of this process that argues strongly in favor of inflationary ΛCDM is that the small-scale (i.e., large k) modes would have re-entered first, during the radiation-dominated expansion, and hence were suppressed, while the large-scale (small k) modes re-entered later, during the matter-dominated era, and continued to grow. This distinction uniquely shapes the matter power spectrum in ways that are otherwise very difficult to replicate.

These inhomogeneities are studied observationally in the CMB (Section 6.4) and in the distribution of galaxy clusters [375] which, together, yield the matter power spectrum. The existence of dark matter in these fluctuations is strongly supported by an additional critical observation of galaxy rotation curves [34, 376, 377] and weak lenses [378, 379, 380, 381, 382]. Any attempt at modeling the formation of structure must therefore include both dark and baryonic matter, and radiation. But since dark matter is presumably collisionless and non-interacting, its fluctuations must have grown solely under the influence of gravity, without any subjugation from the radiation pressure. Standard astrophysical principles suggest that, once the dark-matter perturbations grew beyond a critical limit, they would have become bound [383]. Baryons, on the other hand, were tightly coupled to the radiation and could not form similarly bound objects. But as they gradually dissociated themselves from the photon field, they would have started to accrete into the gravitational potential wells created earlier by the condensing dark matter. In the standard model, the large-scale structure we see today would have been created by this infalling baryonic matter. Detailed simulations have shown that without the participation of dark matter, structure formation would have been delayed due to the baryon-radiation coupling, creating an inconsistency with the observations (see Section 1.2 and Figure 1.4) [33, 32].

We must now confront this compelling evidence with what at first appears to be the diametrically opposite indication we have been exploring in this book—that the cosmic spacetime must be consistent with the zero active mass condition, $\rho + 3p = 0$. An FLRW expansion constrained in this way could break the detailed large-scale structure scenario in many different ways. Chief among them, is the removal of a horizon crossing by quantum modes prior to decoupling, viewed as the most important factor shaping the matter power spectrum. But we shall learn in this chapter that an almost scale-free primordial power spectrum can be produced without inflation, and that the growth rate of large- and small-scale modes can be different even without the radiation- and matter-dominated dichotomy. We shall see that the observed matter power spectrum can be reproduced just as reliably when the expansion is constrained by $R_h = ct$ and, moreover, that several unresolved problems with the standard picture, such as the trans-Planckian inconsistency (Section 6.7), are removed along the way.

13.1 QUANTUM FLUCTUATIONS AT THE PLANCK SCALE

Three of the most discerning measurements we have today pertaining to the origin of quantum fluctuations are (1) the scalar spectral index, $n_s = 0.9649 \pm 0.0042$, in the primordial power spectrum $\mathcal{P}_\Theta^{\text{obs}}(k) = A_s(k/k_0)^{n_s-1}$ (Equation 6.153), (2) the amplitude $A_s = (2.1 \pm 0.04) \times 10^{-9}$ (Equation 6.154) and, more recently, (3) the hard cutoff $k_{\min} = 4.34 \pm 0.50/r_{\text{cmb}}$, in terms of the comoving distance r_{cmb} to the last scattering surface (Equation 11.13). None of these data argues *against* the influence of a scalar field in the early Universe, or anisotropies arising from quantum fluctuations seeded within it, but we can no longer take it for granted that the scalar field's potential must necessarily be inflationary.

Indeed, there have already been several attempts at producing quantum fluctuations without inflation, though they differ from the procedure we shall follow here, which is guided primarily by the $R_h = ct$ (i.e., zero active mass) constraint. For example, Bengochea et al. (2015) [384] adopted the Hollands-Wald concept, focusing primarily on the question of how classicalization might have occurred in such a scenario. The Hollands-Wald idea [385] for how quantum fluctuations are born is based on the assumption that semi-classical physics applies to phenomena on spatial scales larger than some fundamental length, so that modes effectively emerged into the semi-classical Universe only when their proper wavelength equaled this scale. As we have noted previously, the issue of how a homogeneous and isotropic quantum fluctuation is converted into actual inhomogeneities and anisotropies at the classical scale is common to all models invoking a quantum origin for the perturbations, and is yet to be resolved [386].

A non-inflationary mechanism for generating the power spectrum was also considered in ekpyrotic [387] and cyclic [388] models. But like the situation in inflation, these models have the common feature that quantum fluctuations exited and re-entered the Hubble horizon during their evolution. The cyclic model repeats its periods of expansion and contraction, the latter of which is identical to the ekpyrotic case. The singular distinction of the mechanism we shall be exploring here, based on the zero active mass equation-of-state, is that this is the only situation in which all proper distances and the Hubble radius expand at the same rate, so that quantum modes never cross the horizon. The mechanism for producing a near-scale free power spectrum is thus unique (and simpler), with the added benefit that the observed fluctuation scale in the CMB may be traced back directly to the Planck wavelength at the Planck time.

We shall be comparing the roles played by a non-inflaton scalar field and a conventional inflaton field, so we shall informally call the former a *numen* field, ϕ_n, in order to distinguish it from the latter. The field ϕ_n represents the earliest manifestation of substance in the Universe, characterized by an equation-of-state $\rho_{\phi_n} + 3p_{\phi_n} = 0$, in order to comply with the zero active mass condition. We shall strive to understand whether a non-inflationary numen field can account for the measured values of n_s, A_s and k_{min} as well, perhaps even better, than its inflationary counterpart. We begin by first deriving the perturbation growth equation for zero active mass, and study how a numen quantum fluctuation evolves as the Universe expands.

The essential steps in perturbing the FLRW metric have appeared frequently in both the primary and secondary literature, so we need not dwell on all the details here. We refer the reader to these excellent resources for a more complete derivation of all the necessary equations [69, 389, 390, 391, 392, 393, 394, 395, 396, 397, 398]. The perturbed FLRW spacetime metric for the linearized scalar and tensor fluctuations is given in Equation (6.120). We shall continue to use natural units with $c = \hbar = 1$ throughout this section.

Writing

$$\phi_n(t, \vec{x}) = \phi_{n0}(t) + \delta\phi_n(t, \vec{x}) \tag{13.1}$$

for small perturbations about the homogeneous numen field $\phi_{n0}(t)$, we identify the same gauge-invariant curvature perturbation Θ (Equation 6.121) as for the inflaton field, yielding the same equation of motion (Equation 6.123) for mode k, and definition for z (Equation 6.124). In addition, the background numen field has the energy density ρ_{ϕ_n} and pressure p_{ϕ_n} given in Equations (6.98) and (6.99), respectively. Up to this point, there is no distinction between the inflaton and numen fields. They separate from each other at the next step, however, when we introduce the zero active mass equation-of-state, $\rho_{\phi_n} + 3p_{\phi_n} = 0$.

From Equations (6.98) and (6.99), we find that

$$V(\phi_n) = \dot{\phi}_n^2 \,, \tag{13.2}$$

with the explicit solution

$$V(\phi_{\mathrm{n}}) = V_0 \exp\left\{-\frac{2\sqrt{4\pi}}{m_{\mathrm{P}}}\,\phi_{\mathrm{n}}\right\}. \tag{13.3}$$

The numen field is evidently a special member of a category of minimally coupled fields studied in the 1980's, designed to induce power-law inflation [399, 400, 401, 402]. Of course, the numen field's zero active mass makes it the only member of this class that actually does not inflate, since $a(t) = t/t_0$, in terms of the current age t_0 of the Universe.

With this expansion factor, the conformal time is thus

$$\tau(t) = t_0 \ln a(t)\,, \tag{13.4}$$

when the zero of τ is chosen to coincide with time t_0. Therefore,

$$z = m_{\mathrm{P}}\frac{a(t)}{\sqrt{4\pi}} \tag{13.5}$$

in Equation (6.124), written in terms of the Planck mass $m_{\mathrm{P}} \equiv 1/\sqrt{G}$. Thus,

$$\frac{z'}{z} = \frac{1}{t_0} \tag{13.6}$$

and

$$\frac{z''}{z} = \frac{1}{t_0^2}\,. \tag{13.7}$$

We next follow the usual procedure of writing Equation (6.123) in terms of the Mukhanov-Sasaki variable u_k in Equation (6.126), with which the curvature perturbation equation then becomes

$$u_k'' + \alpha_k^2 u_k = 0\,, \tag{13.8}$$

where

$$\alpha_k \equiv \frac{1}{t_0}\sqrt{\left(\frac{2\pi R_{\mathrm{h}}}{\lambda_k}\right)^2 - 1}\,, \tag{13.9}$$

and $\lambda_k \equiv 2\pi a(t)/k$ is the proper wavelength of mode k. The distinction between quantum fluctuations produced in an inflaton field and those seeded in a numen field may be seen through a direct comparison of Equation (6.127), for the former, with Equation (13.8), for the latter. In the expression for α_k, we have used the apparent (or gravitational) radius $R_{\mathrm{h}} \equiv c/H = ct$ (Chapter 7) which, as one can see, coincides with the Hubble horizon in a spatially flat Universe. The frequency α_k is critical to understanding how and why quantum fluctuations in the numen field are far better suited to the above measurements than those in an inflaton field.

In contrast to the pessimistic conclusion we drew for slow-roll inflationary potentials in Section 11.2 , the advantage provided by the frequency in Equation (13.9) may be realized immediately by considering the explanation it provides for the origin of k_{min}. Its most relevant departure from the corresponding frequency in inflation (which appears in Equation 6.127) is that both R_h and λ_k scale exactly the same way with time. The ratio R_h/λ_k or, equivalently, $kR_h/a(t)$, therefore remains constant for each mode k as the Universe expands. Thus, numen quantum fluctuations do not cross back and forth across the Hubble horizon. Once the wavelength of a mode is established upon exiting into the semi-classical Universe, it remains a fixed fraction of R_h as both grow proportionately in time.

Because α_k in Equation (13.8) is constant in time, this expression may be solved analytically, yielding

$$u_k(\tau) = \begin{cases} B(k)\, e^{\pm i\alpha_k \tau} & (2\pi R_h > \lambda_k) \\ B(k)\, e^{\pm |\alpha_k| \tau} & (2\pi R_h < \lambda_k) \end{cases}, \qquad (13.10)$$

showing that all modes u_k with a wavelength smaller than $2\pi R_h$ oscillate, while the super-horizon ones do not, echoing some of the physical characteristics we have found for a traditional inflaton field (Section 11.2). But note that in the case of a numen field, the mode with the longest wavelength relevant to the formation of structure is always the one for which $\lambda_k(t) = 2\pi R_h(t)$, which we identify as $k_{min}^n = 1/t_0$ for the numen field. Do not be fooled by its appearance, however, even though this equality may seem to constitute a coincidence. It has this form because $a(t)$ itself is normalized to 1 at t_0.

It is not difficult to understand why the observed k_{min} in Equation (11.13) must clearly represent the emergence of the first quantum mode out of the Planck regime, when we identify it with k_{min}^n. The Planck scale is the length λ_P for which the Compton wavelength

$$\lambda_C \equiv \frac{2\pi}{m_C} \qquad (13.11)$$

for mass m_C equals its Schwarzschild radius $R_h \equiv 2Gm_C$. That is,

$$\lambda_P \equiv \sqrt{4\pi G}. \qquad (13.12)$$

The trans-Planckian problem [138] with conventional inflation arises because λ_C grows as R_h shrinks, so quantum mechanics cannot provide a clear insight into how one should handle modes with wavelengths $\lambda_k < 2\pi\lambda_P$. (The factor 2π arises from the definition of λ_P in terms of R_h.) This is a serious problem for the standard model because the CMB fluctuation amplitude A_s requires quantum modes to have been seeded well before the Planck time

$$t_P = \lambda_P, \qquad (13.13)$$

when $\lambda_C \gg \lambda_k$.

The numen field avoids this fundamental inconsistency completely when we accept that quantum mechanics does not permit a semi-classical description of mode k when its wavelength λ_k is less than $2\pi\lambda_P$, and instead consider it to have exited into the semi-classical Universe at the time when $\lambda_k = 2\pi\lambda_P$. After this event, mode k

would have evolved according to the oscillatory solution in Equation (13.10). We shall shortly demonstrate that a near-scale-free power spectrum $\mathcal{P}_\Theta(k)$ with $n_s \sim 1$ was produced due to the fact that each succeeding k emerged at later times. We recognize this notion as a variation on the Hollands-Wald concept alluded to earlier though, in their case, the fundamental scale was not related to λ_P.

Indeed, there is a simple, compelling reason why the fundamental scale has to be λ_P for the numen field. According to the expression

$$k = 2\pi \frac{a(t)}{\lambda_k(t)} \, , \tag{13.14}$$

the observed value of $k_{\min}$ defines the time $t_{\min}$ at which the very first mode emerged into the semi-classical Universe (Section 11.2). Thus,

$$t_{\min} = \frac{4.34 \, t_P}{\ln(1 + z_{\text{cmb}})} \tag{13.15}$$

where, as usual, z_{cmb} is the redshift at the last scattering surface. We know that in ΛCDM this redshift would have been ~ 1081 (Section 6.4), so if we were to use this value in Equation (13.15), we would estimate that $t_{\min} \sim 0.63 \, t_P$. Of course, when we modify the expansion history according to the $R_h = ct$ constraint, the location of the last scattering surface will probably be different. Nevertheless, the value of $t_{\min}$ is only weakly dependent on z_{cmb} anyway. For example, even if we were to choose the extremely different value $z_{\text{cmb}} = 50$, the implied first emergence of a k-mode would then have occurred at $t_{\min} \sim 1.1 \, t_P$.

In an FLRW cosmology constrained by the zero active mass condition, the spatially largest numen mode measured in the CMB therefore had to emerge from the Planck regime at approximately the Planck time. The measured $k_{\min}$ belongs to that very first mode that physically exited from the Planck scale after the shortest measurable interval of time (i.e., $t_P \sim 10^{-43}$ sec) following the Big Bang. This situation is unique among cosmological scalar fields because the macroscopic anisotropies in the CMB are directly coupled to quantum fluctuations at the Planck scale. Moreover, the range of mode wavelengths and exit times are all fully consistent with our understanding of quantum mechanics today. At least conceptually, the numen mechanism for the creation of quantum fluctuations in the early Universe thus appears to be superior to the more conventional inflaton approach, because it appears to be more consistent with known physics.

Let us now follow the evolution of the quantum modes as they cross into the semi-classical Universe across the Planck scale and calculate their power spectrum. We can already see, at least intuitively, that this concept provides a natural mechanism for generating a power-law spectrum $\mathcal{P}_\Theta(k)$. The question is whether it is also effectively scale-free. Whatever the outcome, this approach does not rely on the use of a Bunch-Davies vacuum [137] below the Planck scale for the normalization $B(k)$ of the quantum fluctuations in Equation (13.10), where mode wavelengths shorter than λ_C would be difficult to interpret quantum mechanically. Nor does this process rely on the uncertain transition of quantum modes across the Hubble horizon, as one needs for an inflaton field.

The normalization $B(k)$ of the modes is typically calculated by minimizing the expectation value of the Hamiltonian. But this procedure is quite complicated in time-dependent spacetimes, such as inflationary ΛCDM, in which the curvature makes the frequencies time dependent. This is why one must seed the inflaton fluctuations in the remote conformal past, motivated by the argument that the modes we see today presumably started with a wavelength much smaller than the horizon. Another way of saying this is that in order to minimize the Hamiltonian unaffected by gravity, one must seek to do this in Minkowski space. Then one may simply write [134]

$$B(k) = \frac{1}{\sqrt{2k}} \tag{13.16}$$

(see also Equation 6.130). This is actually how one defines the Bunch-Davies vacuum, though as noted earlier, it is subject to a trans-Planckian inconsistency.

The numen field does not have this problem, even though the quantum modes emerged at the Planck scale, because the zero active mass condition makes the frame into which the modes emerged be geodesic. The Hubble frame was obviously expanding in this spacetime, but it always remained in free fall, with zero internal acceleration. One can see this quantitatively by direct inspection of the frequencies α_k in Equation (13.9), which shows that they are always time independent, since both R_h and λ_k are proportional to t. The numen quantum fluctuations therefore automatically have the normalization in Equation (13.16), *without* us having to invoke the Bunch-Davies vacuum.

After time t_{min}, which one may now identify with the Planck time t_P, the numen modes continued to exit the Planck domain sequentially, with a wavelength $\lambda_k = 2\pi\lambda_P$ and a comoving wavenumber k based on Equation (13.14). That is, $k = a(t)/\lambda_P$, so that mode k exited at time

$$t_k \equiv kt_0\lambda_P . \tag{13.17}$$

Thus, from Equation (13.10) and the definition of u_k in Equation (6.126), one finds that the Fourier mode Θ_k in the expansion (Equation 6.122) for the gauge-invariant curvature perturbation Θ (Equation 6.121) must satisfy the condition

$$|\Theta_k|^2 = \frac{2\pi}{m_P^2} \frac{1}{\alpha_k a^2} . \tag{13.18}$$

And with the definition of the power spectrum $\mathcal{P}_\Theta(k)$ in Equation (6.138), one finds that

$$\mathcal{P}_\Theta(k) = \frac{1}{(2\pi)^2} \left[\frac{a(t_k)}{a(t)}\right]^2 \left\{1 - \left(\frac{k_{min}^n}{k}\right)^2\right\}^{-1/2} , \tag{13.19}$$

for the numen quantum modes exiting into the semi-classical Universe across the Planck length scale.

These are homogeneous, isotropic quantum fluctuations that continued to oscillate and grow as the Universe expanded, but ultimately must have devolved into classical anisotropies. This process may have something to do with a phase transition, perhaps at the GUT scale (Sections 6.2 and 6.6), if the numen field is coupled to GUT particles.

How classicalization occurred is a problem with all models invoking a quantum origin for the perturbations, and has not yet been solved [403, 404, 386, 405, 406, 384]. A reasonable approach one may adopt is to assume that the dynamics of classicalization is associated with a particular length (i.e., energy) scale [407, 408] L_*, analogous to the Planck scale λ_P. In our application, this procedure will serve the principal purpose of establishing the amplitude of the numen quantum fluctuations.

After exiting the Planck domain, mode k would have reached the classicalization scale at

$$a(t_k^*) = \frac{L_* k}{2\pi} , \tag{13.20}$$

corresponding to time

$$t_k^* = t_0 L_* k . \tag{13.21}$$

Thus, one may re-write Equation (13.19) as

$$\mathcal{P}_\Theta(k) = \frac{1}{(2\pi)^2} \left[\frac{\lambda_P}{L_*}\right]^2 \left\{1 - \left(\frac{k_{min}^n}{k}\right)^2\right\}^{-1/2} , \tag{13.22}$$

the quantity one needs to compare directly with the observed power spectrum $\mathcal{P}_\Theta^{obs}(k) = A_s(k/k_0)^{n_s-1}$ (Equation 6.153). From the measured value of A_s, one finds that

$$L_* \sim 3.5 \times 10^3 \lambda_P , \tag{13.23}$$

and with the Planck scale set at

$$m_P \approx 1.22 \times 10^{19} \text{ GeV} , \tag{13.24}$$

one therefore infers a classicalization of the numen modes at roughly 3.5×10^{15} GeV, remarkably consistent with the energy scale in grand unified theories, as we speculated above. Of course, the physics describing this process would clearly lie beyond the standard model of particle physics, so this is merely conjectural at this point. Nevertheless, a working hypothesis suggests that numen quantum fluctuations exited out of the Planck domain and oscillated until $t \sim 3.5 \times 10^3 t_P$—the time corresponding to L_*—after which they devolved into GUT particles and the perturbation amplitude remained frozen thereafter.

This association with a possible GUT transition is already quite promising, but a more tangible benefit offered by the numen scenario is the power spectrum associated with these modes. Equation (13.22) suggests that the numen scalar curvature perturbations are almost scale-free. More quantitatively, consider the standard definition of the index,

$$n_s = 1 + \frac{d \ln \mathcal{P}_\Theta(k)}{d \ln k} = 1 - \frac{2}{2(k/k_{min}^n)^2 - 1} . \tag{13.25}$$

Clearly, n_s is slightly less than 1. Very interestingly from a fundamental physics point of view, the deviation from a pure scale-free spectrum appears to be due to the difference between k and α_k in Equation (13.9), which ultimately arises from the Hubble expansion term Θ_k' in the growth Equation (6.123).

All in all, our introduction to the exploration of structure formation in an FLRW cosmology constrained by the zero active mass condition appears to be quite auspicious. We cannot be certain that the primordial spectrum was actually produced in this fashion, though the signs suggest an overall improvement over the current inflationary paradigm. The observed cutoff $k_{\min}$ finds a natural explanation in this context, representing the first quantum fluctuation to have emerged into the semi-classical Universe after the shortest measurable time delay following the Big Bang. Most importantly, this process completely avoids the trans-Planckian inconsistency, thereby finding much support from quantum mechanics as we know it. Probing the physics prior to t_{P}, however, should nonetheless be a principal focus of future work addressing the nature of ϕ_{n} and its emergence into the semi-classical Universe.

But already the scenario we have described here passes a crucial observational test related to a well-known argument often used in support of inflation. The super-horizon freeze-out mechanism for quantum fluctuations in an inflaton field provides a viable explanation—actually, the only explanation known thus far—for the coherence observed in the CMB fluctuations [409]. If all of the Fourier modes of a given wavenumber did not have an identical phase, we would not see the pattern of anisotropies visible at the level of 1 part in 100,000. One would merely see 'white' noise. Inflation offers a meaningful explanation because all of the modes of a given k cross the Hubble horizon at the same time. But something very similar to this happens with the numen quantum fluctuations as well. All modes of a given k have the same phase because they emerge from the Planck domain at the same time t_k. Indeed, the numen mechanism is somewhat simpler than that of inflation because it requires fewer steps. At the same time, it avoids at least some of the challenging hurdles faced by slow-roll inflaton fields.

13.2 STRUCTURE FORMATION

In order to follow the evolution of the fluctuations once they classicalize, we return now to Equation (6.120) for the perturbed FLRW metric describing the linearized scalar and tensor components. The data available today are restricted solely to the evolution and growth of the scalar modes. In addition, vector perturbations die away as the Universe expands, so if we focus our attention just on the scalar perturbations, the linearized perturbed FLRW metric relevant for this analysis simplifies to the form

$$ds^2 = (1 + 2\Phi)\,dt^2 - a^2(t)(1 - 2\Psi)\delta_{ij}\,dx^i\,dx^j \qquad (13.26)$$

where, as usual, the indices i and j denote spatial coordinates.

Using this simplified metric, one may then compute the Ricci tensor (Equation 2.152):

$$R_{00} = \nabla^2 \Phi + 3\frac{d^2\Psi}{d\tau^2} + 3\mathcal{H}\left[\frac{d\Phi}{d\tau} + \frac{d\Psi}{d\tau}\right] - 3\frac{d\mathcal{H}}{d\tau}, \tag{13.27}$$

$$R_{0i} = 2\partial_i\frac{d\Psi}{d\tau} + 2\mathcal{H}\,\partial_i\Phi, \tag{13.28}$$

$$R_{ij} = \left[\frac{d\mathcal{H}}{d\tau} + 2\mathcal{H}^2 - \frac{d^2\Psi}{d\tau^2} + \nabla^2\Psi - 2\left(\frac{d\mathcal{H}}{d\tau} + 2\mathcal{H}^2\right) \times\right.$$
$$\left.(\Psi + \Phi) - \mathcal{H}\frac{d\Phi}{d\tau} - 5\mathcal{H}\frac{d\Psi}{d\tau}\right]\delta_{ij} + \partial_i\partial_j(\Psi + \Phi), \tag{13.29}$$

and the Ricci scalar (Equation 2.154), noting the slight change in notation from R to $\mathcal{R}$ to avoid confusion with the proper distance R in FLRW:

$$a^2\mathcal{R} = 6\left(\frac{d\mathcal{H}}{d\tau} + \mathcal{H}^2\right) - 2\nabla^2\Phi + 4\nabla^2\Psi -$$
$$12\left(\frac{\mathcal{H}}{d\tau} + \mathcal{H}^2\right)\Phi - 6\frac{d^2\Psi}{d\tau^2} - 6\mathcal{H}\left(\frac{d\Phi}{d\tau} + 3\frac{d^2\Psi}{d\tau^2}\right). \tag{13.30}$$

The Einstein tensor (Equation 2.155) follows immediately:

$$G_{00} = 3\mathcal{H}^2 + 2\nabla^2\Psi - 6\mathcal{H}\frac{d\Psi}{d\tau}, \tag{13.31}$$

$$G_{0i} = 2\partial_i\frac{d\Psi}{d\tau} + 2\mathcal{H}\,\partial_i\Phi, \tag{13.32}$$

$$G_{ij} = -\left(2\frac{d\mathcal{H}}{d\tau} + \mathcal{H}^2\right)\delta_{ij} + \left[\nabla^2(\Phi - \Psi) + 2\frac{d^2\Psi}{d\tau^2}\right.$$
$$+2\left(2\frac{d\mathcal{H}}{d\tau} + \mathcal{H}^2\right)(\Psi + \Phi) + 2\mathcal{H}\frac{d\Phi}{d\tau} + 4\mathcal{H}\frac{d\Psi}{d\tau}\bigg]\delta_{ij}$$
$$+\partial_i\partial_j(\Psi + \Phi). \tag{13.33}$$

In these expressions, $d\tau \equiv dt/a(t)$ is the usual conformal time variable and we have defined the Hubble parameter

$$\mathcal{H} \equiv \frac{1}{a}\frac{da}{d\tau}. \tag{13.34}$$

To complete the linearization of Einstein's Equations (2.167), we must also perturb the stress-energy tensor (Equation 4.27), which we shall write in the form

$$T^\alpha{}_\beta = (\rho + \delta\rho + p + \delta p)(u^\alpha + \delta u^\alpha)(u_\beta + \delta u_\beta) + (p + \delta p)g^\alpha{}_\beta, \tag{13.35}$$

where ρ and $\delta\rho$ are, respectively, the total and perturbed energy density, including all of the species; p and δp are similarly the total and perturbed pressure; and u_α and δu_α are the total and perturbed four-velocity. In addition, the four-velocity must satisfy the condition

$$g_{\alpha\beta}(u^\alpha + \delta u^\alpha)(u^\beta + \delta u^\beta) = 1, \tag{13.36}$$

from which one may obtain the individual components of δu^α using the metric coefficients in Equation (13.26).

Substituting all of these pieces into Einstein's equations, one may then compute the spatial components from the general expression

$$G_{ij} = -8\pi G \, g_{i\alpha} T^\alpha{}_j \, . \tag{13.37}$$

One finds

$$\left[\nabla^2(\Phi - \Psi) + 2\frac{d^2\Psi}{d\tau^2} + 2\left(2\frac{d\mathcal{H}}{d\tau} + \mathcal{H}^2\right)(\Psi + \Phi) + 2\mathcal{H}\frac{d\Phi}{d\tau}\right.$$

$$\left. + 4\mathcal{H}\frac{d\Psi}{d\tau}\right]\delta_{ij} + \partial_i\partial_j(\Psi + \Phi) = 8\pi G a^2(\delta p - 2\Psi p)\delta_{ij} \, . \tag{13.38}$$

Immediately, we see that $\Psi = -\Phi$ when $i \neq j$, and so we adopt this condition throughout the derivation. The rest of the components in Einstein's equations then yield

$$2\frac{d^2\Phi}{d\tau^2} + 4\left(2\frac{d\mathcal{H}}{d\tau} + \mathcal{H}^2\right)\Phi + 6\mathcal{H}\frac{d\Phi}{d\tau} = 8\pi G a^2(\delta p - 2p\,\Phi) \, , \tag{13.39}$$

$$2\nabla^2\Phi - 6\mathcal{H}\frac{d\Phi}{d\tau} = 8\pi G a^2(\delta\rho + 2\rho\,\Phi) \, , \tag{13.40}$$

and

$$2\partial_i\frac{d\Phi}{d\tau} + 2\mathcal{H}\partial_i\Phi = 8\pi G a^2(\rho + p)v_i \, , \tag{13.41}$$

where v_i are the components of the three-velocity vector. Thus, using a Fourier decomposition as in Equation (6.122), one finds from these expressions the following dynamical equation for each mode k of the potential Φ:

$$k^2\Phi_k + 3\mathcal{H}\left(\frac{d\Phi_k}{d\tau} + \mathcal{H}\Phi_k\right) = 4\pi G a^2\delta\rho_k \, , \tag{13.42}$$

which describes the evolution of the perturbed gravitational potential due to the source term $\delta\rho$ driving its growth.

To solve this equation, one must also determine the expression governing the evolution of $\delta\rho$, but this is not straightforward when various influences can change either $\delta\rho$, δp, or both, with time. Certainly, if the cosmic fluid were dominated by only one species, with a conserved particle number and a fixed pressure in terms of its energy density, this exercise would be quite simple. For example, one could then merely use the perturbed form of Equation (4.31). Unfortunately, the fluctuation growth arched across several different epochs of species domination and/or particle non-conservation, regardless of whether or not the FLRW cosmology was constrained by the zero active mass condition. Even in basic ΛCDM, the expansion would have been dominated by a coupled baryon-radiation fluid at the beginning, transitioning into a matter-dominated phase later on. One must therefore derive the evolution equations for individual species by perturbing the Boltzmann equation for their respective distribution functions [410]. The combination of fluid elements depends on whether or not the zero active mass condition is adopted in the FLRW cosmology, however, so one must handle the perturbation of the Boltzmann equations differently in the two cases.

13.2.1 Perturbed Boltzmann Equation, ΛCDM

As alluded to earlier, the baryons in ΛCDM would have been coupled strongly to the radiation for at least the first 380,000 years, and could therefore not contribute to the growth of structure during this epoch. In ΛCDM, the initial growth of structure would therefore have been dominated by dark matter, which we now consider.

The Boltzmann equation describes the evolution of a distribution function in the 8-dimensional phase space composed of four spacetime coordinates, x^α, and the components of the 4-momentum, p^α. The number of independent degrees of freedom is reduced by 1, however, by the invariance of the contraction $g_{\alpha\beta}\,p^\alpha p^\beta = m^2$, where m is the rest mass of the particular particle we are considering. One typically chooses the independent variables to be x^α, the magnitude of the vector momentum, $p \equiv |\mathbf{p}|$, and its direction cosines $\hat{p}^i$, for $i = 1, 2, 3$.

Liouville's theorem then says that the distribution function $f_s(x^\mu, p, \hat{p}^i)$ for species 's' evolves according to the following evolution equation:

$$\frac{df_s}{d\lambda} = \frac{\partial f_s}{\partial x^0}\frac{\partial x^0}{\partial \lambda} + \frac{\partial f_s}{\partial x^i}\frac{\partial x^i}{\partial \lambda} + \frac{\partial f_s}{\partial p}\frac{\partial p}{\partial \lambda} + \frac{\partial f_s}{\partial \hat{p}^i}\frac{\partial \hat{p}^i}{\partial \lambda} = C[f_s]\,, \tag{13.43}$$

where λ is the affine parameter, and $C[f_s]$ is the source and/or collision term for this species, representing whatever interaction and number non-conservation this species experiences during the cosmic expansion. We use the affine parameter here, rather than the proper time, to keep this equation generally applicable to both massive and massless particle species. It is therefore necessary to define the momentum as $p^\alpha = dx^\alpha/d\lambda$, with $p^0 = dx^0/d\lambda$. The fourth term in this expression is of second order and may be ignored. Dividing both sides by p^0, one then finds that, to first order,

$$\frac{df_s}{d\tau} = \frac{\partial f_s}{\partial \tau} + \frac{\partial f_s}{\partial x^i}\frac{p^i}{p^0} + \frac{\partial f_s}{\partial p}\frac{\partial p}{\partial \tau} = \frac{C[f_s]}{p^0}\,, \tag{13.44}$$

where τ is the conformal time,

But $p^i/p^0 = (p\hat{p}^i)/E$, where E is the particle's energy. In addition, one may use the metric coefficients in Equation (13.26), with the definition of the Christoffel symbols in Equation (2.45), to derive the geodesic equation

$$\frac{dp}{d\tau} = -\mathcal{H}p + E\hat{p}^i\partial_i\Phi - p\frac{d\Phi\delta_{jl}}{d\tau}\hat{p}^j\hat{p}^l\,. \tag{13.45}$$

Then, substituting Equation (13.45) into (13.44), one gets

$$\frac{df_s}{d\tau} + p\frac{\hat{p}^i}{E}\frac{\partial f_s}{\partial x^i} + p\left(-\mathcal{H} + \frac{E}{p}\hat{p}^j\partial_j\Phi - \frac{d\Phi\delta_{ln}}{d\tau}\hat{p}^l\hat{p}^n\right)\frac{\partial f_s}{\partial p} = \frac{a}{E}(1-\Phi)C[f_s]\,. \tag{13.46}$$

Note that the term p^0 has been replaced with $(E/a)(1 + \Phi)$, using the perturbed gravitational potential, Φ.

Next, one must separate the distribution function into its unperturbed, f_{0s}, and perturbed, $\tilde{f}_s$, components:

$$f_s(\tau, x^i, p, \hat{p}^i) = f_{0s}(\tau, x^i, p, \hat{p}^i) + \tilde{f}_s(\tau, x^i, p, \hat{p}^i)\,. \tag{13.47}$$

Multiplying Equation (13.46) by $E(p)$ and using this substitution for f_s, one may then integrate the resulting expression over momentum space. Collecting the zero-order terms, one then finds

$$\int \frac{d^3p}{(2\pi)^3} E(p) \frac{df_{0s}}{d\tau} - \int \frac{d^3p}{(2\pi)^3} \mathcal{H}pE(p) \frac{\partial f_{0s}}{\partial p} = \int \frac{d^3p}{(2\pi)^3} a\, C[f_s]\,. \tag{13.48}$$

In the context of ΛCDM, the particle number is conserved during this early growth phase and the dark matter does not interact with anything else (other than through gravity). Thus, $C[f_s] = 0$ here, which allows one to easily solve Equation (13.48). Integrating the second term on the left-hand side by parts and neglecting the boundary term, one arrives at the expression

$$\int \frac{d^3p}{(2\pi)^3} E(p) \frac{df_{0s}}{d\tau} + 3\mathcal{H} \int \frac{d^3p}{(2\pi)^3} \left(E + \frac{p^2}{3E} \right) f_{0s} = 0\,. \tag{13.49}$$

Of course, these integrals represent the density and pressure of species 's,' according to the definitions

$$\rho_s \equiv \int \frac{d^3p}{(2\pi)^3} E(p) f_{0s}\,, \tag{13.50}$$

and

$$\mathcal{P}_s = \int \frac{d^3p}{(2\pi)^3} \frac{p^2}{3E} f_{0s}\,. \tag{13.51}$$

To avoid confusion with the momentum, p, we are changing the notation for pressure of species 's' in this section to $\mathcal{P}_s$. Then, applying Equations (13.49)–(13.51) to dark matter, one concludes that

$$\frac{d\rho_{\rm dm}}{d\tau} + 3\mathcal{H}(\rho_{\rm dm} + \mathcal{P}_{\rm dm}) = 0\,. \tag{13.52}$$

As one would expect, this is precisely Equation (4.31), which is valid when particle number is conserved and the species in question is non-interacting.

Next, if one multiplies Equation (13.46) and integrates over momentum space using the definitions of ρ_s and $\mathcal{P}_s$, the first-order terms now yield an expression for $\delta\rho_{\rm dm}$:

$$\frac{d(\delta\rho_{\rm dm})}{d\tau} + (\rho_{\rm dm} + \mathcal{P}_{\rm dm})\, \partial_i v_{\rm dm}^i + 3\mathcal{H}(\delta\rho_{\rm dm} + \delta\mathcal{P}_{\rm dm}) + 3(\rho_{\rm dm} + \mathcal{P}_{\rm dm})\frac{d\Phi}{d\tau} = 0\,. \tag{13.53}$$

The dark-matter density perturbation is defined as

$$\delta_{\rm dm} \equiv \frac{\delta\rho_{\rm dm}}{\rho_{\rm dm}}\,. \tag{13.54}$$

In addition, both $\mathcal{P}_{\rm dm}$ and $\delta\mathcal{P}_{\rm dm}$ are zero here. Consequently,

$$\frac{d\delta_{\rm dm}}{d\tau} = \frac{1}{\rho_{\rm dm}} \frac{d(\delta\rho_{\rm dm})}{d\tau} - \frac{\delta_{\rm dm}}{\rho_{\rm dm}} \frac{d\rho_{\rm dm}}{d\tau}\,, \tag{13.55}$$

and substituting this expression into Equation (13.53), one gets for each mode k

$$\frac{d\delta_{\mathrm{dm},k}}{d\tau} = -ku_{\mathrm{dm},k} - 3\frac{d\Phi_k}{d\tau} , \tag{13.56}$$

where $\partial_i v_k^i = ku_{\mathrm{dm},k}$, in terms of the velocity perturbation $u_{\mathrm{dm},k}$ of dark matter.

If one instead takes the second moment of Equation (13.46), by multiplying it with $p\hat{p}^i$ and contracting it with $i\hat{k}_i$ (these being the corresponding direction cosines of the wavevector), and integrating over momentum space, the result is

$$\frac{d(\rho_{\mathrm{dm}}\, u_{\mathrm{dm},k})}{d\tau} + 4\mathcal{H}\rho_{\mathrm{dm}}\, u_{\mathrm{dm},k} + k\Phi\rho_{\mathrm{dm}} = 0 . \tag{13.57}$$

And substituting for $d\rho_{\mathrm{dm}}/d\tau$ from Equation (13.55), one thus gets

$$\frac{du_{\mathrm{dm},k}}{d\tau} = -\frac{1}{a}\frac{da}{d\tau} u_{\mathrm{dm},k} - k\Phi_k . \tag{13.58}$$

Equations (13.56) and (13.58), together with the expression for gravitational-potential growth (13.42) applied to dark matter,

$$k^2\Phi_k + 3\mathcal{H}\left(\frac{d\Phi_k}{d\tau} + \mathcal{H}\Phi_k\right) = 4\pi\, Ga^2\rho_{\mathrm{dm}}\delta_{\mathrm{dm},k} , \tag{13.59}$$

constitute the complete set of relations one must solve to follow the growth of structure in the dark-matter dominated ΛCDM Universe.

For convenience, this last expression is usually written in the form

$$\frac{d\Phi_k}{d\tau} = -\left(1 + \frac{k^2}{3\mathcal{H}^2}\right)\mathcal{H}\Phi_k + \frac{4\pi Ga^2\rho_{\mathrm{dm}}}{3\mathcal{H}}\delta_{\mathrm{dm},k} . \tag{13.60}$$

The first term on the right-hand side of this equation is the so-called Hubble friction which, by virtue of the cosmic expansion, tends to suppress growth. The second term represents self-gravity, which makes the fluctuation $\delta_{\mathrm{dm},k}$ grow in time. Equations (13.56), (13.58) and (13.60) must be solved simultaneously, e.g., by following the procedure established in the 'Cosmological Initial Conditions and Microwave Anisotropy Codes' (COSMICS) constructed for this purpose [412]. The simulations begin with $\delta_{\mathrm{dm},,k} = 3\Phi_k/2$ and $u_k = k\tau\Phi_k/2$ and follow the re-entry of modes across the Hubble horizon at times consistent with k. In COSMICS, the conformal time at which this classical growth phase initiates is chosen to be $\tau_{\mathrm{init}} \equiv \min[10^{-3}k^{-1}, 10^{-1}h^{-1}\,\mathrm{Mpc}]$.

The solution to these equations is shown for three representative values of k in Figures 13.1 and 13.2. The first of these plots illustrates the evolution of the gravitational potential Φ_k, subject to the constraints that (1) all modes were frozen outside the horizon, (2) the small-scale modes (i.e., large k) re-entered during the radiation-dominated expansion and decayed at first, then grew only when matter started to dominate, and (3) the large-scale modes (i.e., small k) re-entered after matter dominated and so grew continuously thereafter. The matter power spectrum is calculated from the trajectories shown in Figure 13.2, and will be compared with the data in Section 13.2.3 below.

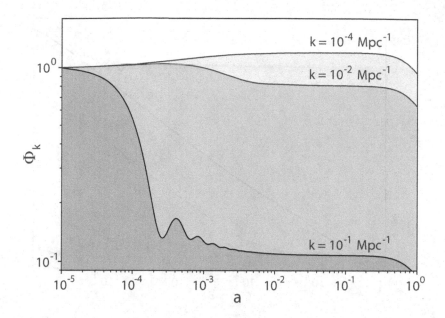

Figure 13.1 A numerical solution of Equation (13.60) for the perturbed gravitational potential Φ_k in ΛCDM, for modes $k = 10^{-4}$, 10^{-2}, and 10^{-1} Mpc^{-1}. (Adapted from ref. [411])

13.2.2   Perturbed Boltzmann Equation, Zero Active Mass

While the source and/or collision term $C[f_s]$ could be set to zero in ΛCDM, this is not the case when the FLRW cosmology is constrained by the zero active mass condition, $\rho + 3\mathcal{P} = 0$, as we shall now demonstrate. Putting

$$\rho = \rho_r + \rho_m + \rho_{de} \, , \tag{13.61}$$

and

$$\mathcal{P} = -\rho/3 = w_{de}\rho_{de} + \rho_r/3 \, , \tag{13.62}$$

under the usual assumption that $\mathcal{P}_{dm} = 0$ and $\mathcal{P}_r = \rho_r/3$, one immediately sees that

$$\rho_r = -3w_{de}\rho_{de} - \rho \, . \tag{13.63}$$

Throughout the cosmic evolution, the total energy density in this expression evolves according to

$$\rho(t) = \rho_c \, a(t)^{-2} \, , \tag{13.64}$$

as one may see from Equation (6.7), where ρ_c is the usual critical density defined in Equation (6.1).

Equation (13.63) expresses the radiation energy density in terms of dark energy and ρ at all times t but, at low redshifts, the CMB temperature ($T_0 \approx 2.72548$ K; see Section 1.1) implies a normalized radiation energy density $\Omega_r \approx 5 \times 10^{-5}$, which

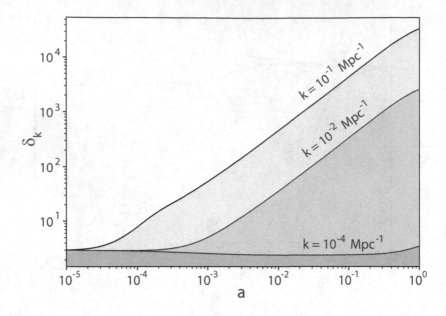

Figure 13.2 A numerical solution of Equation (13.56) for the dark matter fluctuation $\delta_{\mathrm{dm},k}$ in ΛCDM, for modes $k = 10^{-4}, 10^{-2}$, and 10^{-1} Mpc^{-1}. (Adapted from ref. [411])

is negligible compared to Ω_{m} and Ω_{de}. Therefore, w_{de} must be $\sim -1/2$ in order to produce a partitioning of the elements in the cosmic fluid in line with what we see in the local Universe. Adopting this value, one then has

$$\Omega_{\mathrm{de}} = -\frac{1}{3w_{\mathrm{de}}} = \frac{2}{3} \,, \tag{13.65}$$

and

$$\Omega_{\mathrm{m}} = \frac{1 + 3w_{\mathrm{de}}}{3w_{\mathrm{de}}} = \frac{1}{3} \,, \tag{13.66}$$

though locally the quantity $\Omega_{\mathrm{m}} = \Omega_{\mathrm{b}} + \Omega_{\mathrm{dm}}$ contains both dark and baryonic matter [413].

The radiation energy density increases with redshift, however, so it must have dominated over matter at $z \gg 1$. But in order to comply with Equation (13.62), $\rho = \rho_{\mathrm{r}} + \rho_{\mathrm{de}}$ must have contained both radiation and dark energy in the early Universe. At $z \gg 1$, one would thus have

$$\rho_{\mathrm{de}} = \frac{2}{1 - 3w_{\mathrm{de}}} \rho_{\mathrm{c}} (1 + z)^2 \,, \tag{13.67}$$

and

$$\rho_{\mathrm{r}} = \frac{3w_{\mathrm{de}} + 1}{3w_{\mathrm{de}} - 1} \rho_{\mathrm{c}} (1 + z)^2 \,, \tag{13.68}$$

implying a relative partitioning of $\rho_{\mathrm{de}} = 0.8\rho$ and $\rho_{\mathrm{r}} = 0.2\rho$, under the assumption that w_{de} has always had the value $-1/2$. These two limiting behaviors, at $z = 0$

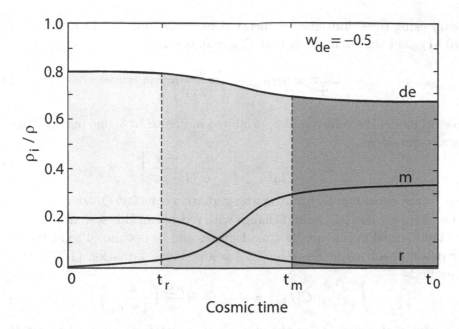

Figure 13.3 Schematic diagram showing a possible evolution of the various constituents ρ_i—dark energy (de), radiation (r) and matter (m)—as a function of time in an FLRW cosmology constrained by the zero active mass condition, $\rho+3\mathcal{P}=0$. The observations today suggest that $w_{\text{de}} = -0.5$, which then fixes $\rho_{\text{m}}/\rho = 1/3$ and $\rho_{\text{de}}/\rho = 2/3$ for $z \sim 0$. At $z \gg 1$, however, the dominance of radiation over matter requires a different partitioning, with $\rho_{\text{r}}/\rho = 0.2$ and $\rho_{\text{de}}/\rho = 0.8$. Radiation dominated over matter at $t < t_{\text{r}}$, while matter was dominant at $t > t_{\text{m}}$.

and $z \gg 1$, suggest that the various constituents in the cosmic fluid must have gone through a gradual transition starting with the early Universe, where $\rho_{\text{de}}/\rho = 0.8$, to the present, where $\rho_{\text{de}}/\rho = 2/3$. In addition, the radiation energy density that dominated at $z \gg 1$, with $\rho_{\text{r}}/\rho = 0.2$, would eventually have yielded to matter with $\rho_{\text{m}}/\rho = 1/3$ at late times. Figure 13.3 illustrates this evolution schematically (though not to scale), with the two limiting epochs shown at $t < t_{\text{r}}$ and $t > t_{\text{m}}$, and a transition period in between. The behavior shown here would be consistent with partial decay of the dark-energy field, suggesting a possible component in physics beyond the standard model.

An implied coupling between dark energy and dark matter is thus unavoidable if the FLRW cosmology is constrained by $R_{\text{h}} = ct$, and the Boltzmann equations must therefore incorporate the effect of a non-zero $C[f_s]$. Returning to Equation (13.48), the integration over momentum now yields the equation

$$\int \frac{d^3p}{(2\pi)^3} E(p) \frac{df_{0s}}{d\eta} + 3\mathcal{H} \int \frac{d^3p}{(2\pi)^3} \left(E + \frac{p^2}{3E} \right) f_{0s} = \int \frac{d^3p}{(2\pi)^3} aC[f_s] \,, \qquad (13.69)$$

and again using the definitions for energy density and pressure in Equations (13.50) and (13.51), and the assumption that $\mathcal{P}_{dm} = 0$, one gets

$$\frac{d\rho_{dm}}{d\tau} + 3\mathcal{H}\rho_{dm} = \int \frac{d^3p}{(2\pi)^3} aC[f_{dm}] \,. \tag{13.70}$$

Now, to model the behavior of ρ_i exhibited in Figure 13.3, one may use the simple empirical expression

$$\rho_{dm} = \frac{\rho_c}{3a^2} \exp\left[-\frac{a_*}{a}\frac{(1-a)}{(1-a_*)}\right], \tag{13.71}$$

where a_* is the expansion factor at matter-radiation equality. Of course, one does not yet know precisely how ρ_{dm} would change with redshift in this scenario, but this 'toy' model may be used to capture its gross features and be optimized to fit the observed matter power spectrum in Section 13.2.3 below. Using Equation (13.71) in (13.70), one then finds

$$\int \frac{d^3p}{(2\pi)^3} aC[f_{dm}] = \mathcal{H}\rho_{dm} + \mathcal{H}\frac{\rho_{dm}}{a}\left(\frac{a_*}{1-a_*}\right), \tag{13.72}$$

and it is straightforward to see that the above equation is satisfied to zeroth order only if

$$C[f_{0dm}] = \frac{\mathcal{H}E}{a}f_{0dm} + \frac{\mathcal{H}E}{a^2}f_{0dm}\left(\frac{a_*}{1-a_*}\right). \tag{13.73}$$

This equation describes a partial conversion of dark energy into dark matter, possibly via decay. The corresponding growth equation for dark energy must therefore contain a source/collision term that is the negative of this quantity:

$$\frac{d\rho_{de}}{d\tau} + 3\mathcal{H}(\rho_{de} + \mathcal{P}_{de}) = -\int \frac{d^3p}{(2\pi)^3}\left[\mathcal{H}Ef_{0dm} + \frac{\mathcal{H}E}{a}f_{0dm}\left(\frac{a_*}{1-a_*}\right)\right], \tag{13.74}$$

where

$$C[f_{0de}] = -\frac{\mathcal{H}E}{a}f_{0dm} - \frac{\mathcal{H}E}{a^2}f_{0dm}\left(\frac{a_*}{1-a_*}\right). \tag{13.75}$$

It is straightforward to see that Equation (13.53) must then be replaced with

$$\frac{d(\delta\rho_{dm})}{d\tau} + (\rho_{dm} + \mathcal{P}_{dm})\partial_i v^i_{dm} + 3\mathcal{H}(\delta\rho_{dm} + \delta\mathcal{P}_{dm}) + 3(\rho_{dm}$$
$$+\mathcal{P}_{dm})\frac{d\Phi}{d\tau} = \left[\mathcal{H} + \frac{\mathcal{H}}{a}\left(\frac{a_*}{1-a_*}\right)\right](\delta\rho_{dm} - \rho_{dm}\Phi), \tag{13.76}$$

from which one may then derive the expression corresponding to Equation (13.56), remembering that $\mathcal{P}_{dm} = \delta\mathcal{P}_{dm} = 0$:

$$\frac{d\delta_{dm,k}}{d\tau} = -ku_{dm,k} - 3\frac{d\Phi_k}{d\tau} - \mathcal{H}\left[1 + \frac{a_*}{a(1-a_*)}\right]\Phi_k \,. \tag{13.77}$$

Correspondingly, the second-moment Equation (13.57) becomes

$$\frac{d(\rho_{dm}u_{dm,k})}{d\tau} + 4\mathcal{H}\rho_{dm}u_{dm,k} + k\Phi_k\rho_{dm} = \mathcal{H}\left[1 + \frac{a_*}{a(1-a_*)}\right]\rho_{dm}u_{dm,k} \,. \tag{13.78}$$

And upon substitution of $d\delta_{\rm dm}/d\tau$ from Equation (13.55), one gets

$$\frac{du_{{\rm dm},k}}{d\tau} = -\frac{1}{a}\frac{da}{d\tau}u_{{\rm dm},k} - k\Phi_k \, . \tag{13.79}$$

A similar procedure may be followed to derive the analogous equations for dark energy, with the notable difference that $\mathcal{P}_{\rm de}$ and $\delta\mathcal{P}_{\rm de}$ are not zero. The dark-energy, first-moment equation corresponding to (13.76) may be written

$$\frac{d\delta\rho_{\rm de}}{d\tau} + \rho_{\rm de}(1+w_{\rm de})ku_{{\rm de},k} + 3\mathcal{H}\delta\rho_{\rm de}\left(1+\frac{\delta\mathcal{P}_{\rm de}}{\delta\rho_{\rm de}}\right)$$
$$+3\frac{d\Phi}{d\tau}\rho_{\rm de}(1+w_{\rm de}) = -\mathcal{H}\left[1+\frac{a_*}{a(1-a_*)}\right](\delta\rho_{\rm dm} - \rho_{\rm dm}\Phi) \tag{13.80}$$

where, as usual, $w_{\rm de} \equiv \mathcal{P}_{\rm de}/\rho_{\rm de}$ and $\delta\mathcal{P}_{\rm de}/\delta\rho_{\rm de}$ is a measure of the sound speed in the perturbation. In addition, from the definition $\delta_{\rm de} \equiv \delta\rho_{\rm de}/\rho_{\rm de}$ and the inferred equation-of-state $\mathcal{P}_{\rm de} = -\rho_{\rm de}/2$ (see above) for dark energy, one may write

$$\frac{d\delta_{\rm de}}{d\tau} = \frac{1}{\rho_{\rm de}}\frac{d(\delta\rho_{\rm de})}{d\tau} - \frac{\delta\rho_{\rm de}}{\rho_{\rm de}^2}\frac{d\rho_{\rm de}}{d\tau} \, , \tag{13.81}$$

whereupon Equation (13.80) then becomes

$$\frac{d\delta_{\rm de}}{d\tau} = -\frac{k}{2}\partial_i v_{\rm de}^i - 3\mathcal{H}\delta_{\rm de}\left(\frac{1}{2}+\frac{\delta\mathcal{P}_{\rm de}}{\delta\rho_{\rm de}}\right) - \frac{3}{2}\frac{d\Phi}{d\tau}$$
$$+\mathcal{H}\left[1+\frac{a_*}{a(1-a_*)}\right]\frac{\rho_{\rm dm}}{\rho_{\rm de}}(\Phi + \delta_{\rm de} - \delta_{\rm dm}) \, . \tag{13.82}$$

The sound speed in the coupled dark-energy/dark-matter fluid is not yet known but, as a practical approach, one may characterize it as follows:

$$c_s^2 \equiv \frac{\delta\mathcal{P}}{\delta\rho} = \frac{\delta\mathcal{P}_{\rm de}}{\delta\rho_{\rm dm} + \delta\rho_{\rm de}} = \frac{\delta\mathcal{P}_{\rm de}/\delta\rho_{\rm de}}{(1+\delta\rho_{\rm dm}/\delta\rho_{\rm de})} \, , \tag{13.83}$$

which gives

$$\frac{\delta\mathcal{P}_{\rm de}}{\delta\rho_{\rm de}} = c_s^2\left[1+\frac{2\rho_{\rm dm}}{\rho_{\rm de}}\right] \, . \tag{13.84}$$

Simulations have shown that the results of this analysis depend only weakly on its value [411], however, so one may simply assume that it is constant, limited to the range $0 < (c_s/c)^2 < 1$. Adopting the fraction $c_s^2 = c^2/2$ is a reasonable approximation. Thus, noting that $\delta_{\rm dm} \ll \delta_{\rm de}$ (see Figure 13.3), Equation (13.82) yields

$$\frac{d\delta_{{\rm de},k}}{d\tau} = -\frac{k}{2}u_{{\rm de},k} - \delta_{{\rm de},k}\left[\frac{3\mathcal{H}}{2} + 3\mathcal{H}c_s^2 + \frac{6\mathcal{H}c_s^2\rho_{\rm dm}}{\rho_{\rm de}}\right] - \frac{3}{2}\frac{d\Phi_k}{d\tau}$$
$$+\mathcal{H}\left[1+\frac{a_*}{a(1-a_*)}\right]\frac{\rho_{\rm dm}}{\rho_{\rm de}}(\Phi_k + \delta_{{\rm de},k} - \delta_{{\rm dm},k}) \, , \tag{13.85}$$

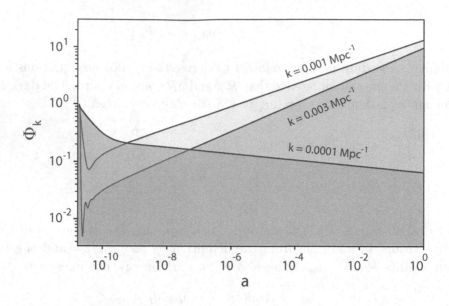

Figure 13.4 A numerical solution of Equation (13.87) for the perturbed gravitational potential Φ_k in the FLRW cosmology constrained by $R_\mathrm{h} = ct$, for modes $k = 10^{-4}$, 3×10^{-3}, and 10^{-3} Mpc^{-1}. (Adapted from ref. [411])

for each mode k. Finally, taking the second moment of Equation (13.46), and using Equation (13.81), one finds the expression (analogous to 13.58) for the dark-energy perturbed velocity:

$$\frac{du_{\mathrm{de},k}}{d\tau} = -\frac{5}{2}\mathcal{H}u_{\mathrm{de},k} - k\Phi + 2kc_s^2\left[1 + \frac{2\rho_{\mathrm{dm}}}{\rho_{\mathrm{de}}}\right]\delta_{\mathrm{de},k} -$$
$$\mathcal{H}\left[1 + \frac{a_*}{a(1-a_*)}\right]\frac{\rho_{\mathrm{dm}}}{\rho_{\mathrm{de}}}\left(u_{\mathrm{de},k} - 2u_{\mathrm{dm},k}\right). \quad (13.86)$$

The density $\delta\rho_k$ in Equation (13.42) includes contributions from both dark energy and dark matter, $\delta\rho_k = \delta\rho_{\mathrm{de},k} + \delta\rho_{\mathrm{dm},k}$, so the expression analogous to Equation (13.60) is

$$\frac{d\Phi_k}{d\tau} = -\left(1 + \frac{k^2}{3\mathcal{H}^2}\right)\mathcal{H}\Phi_k + \frac{4\pi Ga^2\rho_{\mathrm{dm}}}{3\mathcal{H}}\left(\delta_{\mathrm{dm},k} + \frac{\rho_{\mathrm{de}}}{\rho_{\mathrm{dm}}}\delta_{\mathrm{de},k}\right). \quad (13.87)$$

The coupled first-order Equations (13.85), (13.86) and (13.87) may be solved analogously to those in Section 13.2.1, and the results are shown in Figures 13.4 and 13.5.

The different behavior of δ_k in Figures 13.2 and 13.5 may be understood in terms of which forces are dominant at different times. Radiation had an impact on growth in ΛCDM due to the different times at which the modes re-entered the horizon. In the FLRW cosmology constrained by the zero active mass condition, none of the modes

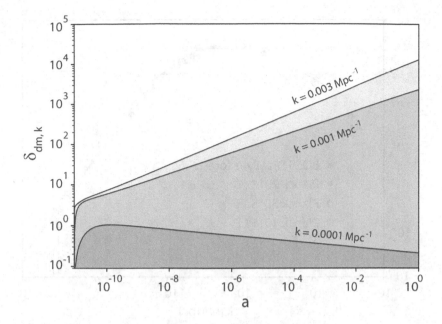

Figure 13.5 A numerical solution of the combined dark matter ($\delta_{\mathrm{dm},k}$ from Equation 13.77) plus dark energy ($\delta_{\mathrm{de},k}$ from Equation 13.82) fluctuation in the FLRW cosmology constrained by $R_{\mathrm{h}} = ct$, for modes $k = 10^{-4}$, 3×10^{-3}, and 10^{-3} Mpc^{-1}. (Adapted from ref. [411])

had to contend with the radiation-dominated suppression, but their growth rate was strongly influenced by the competition between the inward pull of gravity (second term on the right-hand side of Equation 13.87) and the Hubble friction due to the expansion of the Universe (first term to the right in this equation). As a result, the transition from suppression to growth would have occurred at different times due to the presence of k in the Hubble friction term, whose impact one can easily recognize in the trajectories shown in Figure 13.4.

Unlike the situation in ΛCDM, however, the starting point for the evolution illustrated in Figures 13.4 and 13.5 is not as easily fixed to a particular epoch in the Universe's early history, such as the time at which the modes re-entered the horizon, which does not happen when $\rho + 3p = 0$. Of course, the τ_{init} in this picture should correspond to the epoch at which the modes classicalized out of the quantum realm (Section 13.1). In the absence of a comprehensive physical explanation for this process, one should therefore be concerned about the possibility that results such as this, and those shown in Figures 13.7 and 13.8 below, could be overly dependent on the choice of initial conditions. But the simulations [411] have also revealed that the matter power spectrum is completely independent of τ_{init}, as long as $a(\tau_{\mathrm{init}}) \ll a_*$, where $a_* \sim 10^{-9}$ (Equation 13.71). One may view this value of a_* as an optimization based on 'fitting' the calculated power spectrum to the data in Figure 13.7.

The matter power spectrum calculated from the profiles shown in Figure 13.5 will be discussed and compared to the data in the next section.

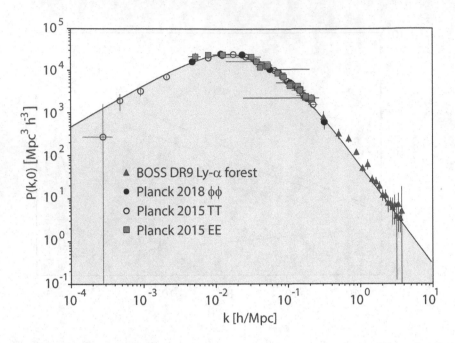

Figure 13.6 The observed matter power spectrum (defined in Equation 13.88) measured from the *Planck* temperature (TT: open circles), E-mode polarization (EE: grey squares) and lensing potential ($\phi\phi$: black solid circles) data, and from the BOSS Ly-α survey (grey triangles). The solid curve is the best fit ΛCDM model. The scaled Hubble constant $h \equiv H_0/(100 \text{ km s}^{-1} \text{ Mpc}^{-1})$ is assumed to have the *Planck* optimized value 0.6732. (Adapted from ref. [411])

13.2.3 The Observed Matter Power Spectrum

The primordial power spectrum $\mathcal{P}_\Theta(k)$ is not visible to us directly because it evolves as the Universe expands. This should be quite evident from the k-dependence of the growth equations (13.56, 13.58 and 13.60 for ΛCDM, and 13.77, 13.85 and 13.87 for $R_\mathrm{h} = ct$), which predict a shape that changes with redshift. The matter power spectrum we measure today is extracted principally from (a) the CMB, (b) clusters of galaxies and (c) the Ly-α forest. The inferred data are model dependent, however, because to evolve the k-dependent growth rate one must assume a specific background cosmology. The data must therefore be recalibrated individually for each model. For some measurements, such as (a) and (c), this procedure is straightforward. In (b), the situation is not as simple, particularly when the true cosmological effects need to be disentangled from redshift space distortions, which add yet another layer of model dependence. In this section, we shall therefore restrict our attention to the more directly measurable CMB anisotropies and Ly-α absorption-line spectra.

The *Planck* CMB data are shown in Figure 13.6 for the concordance model parameters in ΛCDM (see Section 9.1). These were extracted using temperature (TT), polarization (EE) and lensing ($\phi\phi$) observations, spanning mode numbers k from $\sim 10^{-4}\, h\, \mathrm{Mpc}^{-1}$ to $\sim 0.3\, h\, \mathrm{Mpc}^{-1}$. The measurements were converted into a power spectrum using the decomposition

$$P(k, z) \equiv T(k, z)^2 k \mathcal{P}_\Theta(k) , \tag{13.88}$$

in which $\mathcal{P}_\Theta(k)$ is the primordial spectrum and $T(k, z)$ is the matter transfer function incorporating the physics of growth towards the local Universe [414]. The transfer function is formally defined as $T(k, z) \equiv \delta_k(z)/\delta_k(z_{\text{init}})$, where z_{init} is the redshift corresponding to the conformal time (τ_{init}) at which the classical growth phase started (see Section 13.2.1). For simplicity, one typically assumes that $\mathcal{P}_\Theta(k) \propto k^\alpha$, with $\alpha = 0$, approximately consistent with the observations showing that $\alpha = n_s - 1$, with $n_s = 0.9649 \pm 0.0042$.

The angular power spectrum of the CMB depends on the primordial power spectrum $\mathcal{P}_\Theta(k)$ through the linear relation

$$C_\ell = \int_{-\infty}^{+\infty} W_\ell(k) k \mathcal{P}_\Theta(k) \, d\ln k , \tag{13.89}$$

where

$$\mathcal{C}_\ell \equiv \ell(\ell+1)\frac{C_\ell}{2\pi} , \tag{13.90}$$

in terms of the angular power, C_ℓ, of multipole ℓ (see Equation 6.66 and the discussion in Section 6.4). The transfer functions $W_\ell(k)$ depend on the cosmic matter budget and the reionization optical depth. With the decomposition in Equation (13.88), one may therefore write

$$C_\ell = \int_{-\infty}^{+\infty} \frac{W_\ell(k)}{T(k,0)^2} P(k, 0) \, d\ln k . \tag{13.91}$$

In broad terms, the matter power spectrum is extracted from an analytical comparison of Equations (13.90) and (13.91).

For each mode k, the CMB data plotted in Figure 13.6 show the median of the distribution $W_\ell(k)k$ and a horizontal error bar ranging from its 20th to 80th percentiles, representing the 1σ range. The power spectrum is

$$P_{\Lambda\text{CDM}}(k, 0) = T^2_{\Lambda\text{CDM}}(k, 0) k \mathcal{P}_\Theta(k) \tag{13.92}$$

where, in obvious notation, $T_{\Lambda\text{CDM}}(k, 0)$ is the matter transfer function in ΛCDM. To use these data in the FLRW cosmology constrained by $R_{\text{h}} = ct$, they must be recalibrated using the analogous matter transfer function for this model:

$$P_{R_{\text{h}}=ct}(k, 0) = P_{\Lambda\text{CDM}}(k, 0) \frac{T^2_{R_{\text{h}}=ct}(k, 0)}{T^2_{\Lambda\text{CDM}}(k, 0)} , \tag{13.93}$$

assuming a common primordial power spectrum. The recalibrated data for $R_{\text{h}} = ct$ are shown (using the same symbols) in Figure 13.7.

For wavenumbers greater than $\sim 10^{-1} \, h \, \text{Mpc}^{-1}$, the matter power spectrum is obtained from the Ly-α forest—a sequence of absorption lines seen in high redshift quasar spectra, produced by neutral hydrogen along the line-of-sight in a continuously

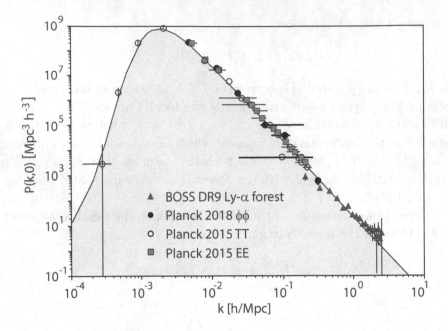

Figure 13.7 Same as Figure 13.6, except now for the FLRW cosmology constrained by $R_\mathrm{h} = ct$. (Adapted from ref. [411])

fluctuating photoionized intergalactic medium. Simulations have shown that the optical depth of the Ly-α absorption is related to the underlying mass density, allowing the matter power spectrum to be extracted from these absorption lines, assuming all of the relevant physics is included in the calculations [415, 416, 417, 418, 419]. For example, TreeSPH hydrodynamical simulations [420] have indicated that the optical depth through the Ly-α forest may be written

$$\tau(s) = A\rho_\mathrm{b}(s)^\beta \, , \tag{13.94}$$

where ρ_b is the baryon matter density, s is the line-of-sight distance towards the quasar and A is an amplitude that depends on the cosmology and physical state of the gas. It is not possible to determine A theoretically, however. Instead, its value is obtained empirically, by matching the simulated and observed Ly-α absorption profiles, and is therefore model dependent. In addition to this drawback, there are several other caveats with this approach to finding the matter power spectrum, including the fact that the hydrodynamic simulations using only dark matter do not include all of the relevant physics. It is also unclear how the uncertainties in the reionization history should be included in the analysis, nor how the ionizing background and its fluctuations propagate through the reconstruction of $P(k, z)$. Consequently, the data obtained in this fashion may not be as reliable as the others when one is comparing different cosmological models.

Nevertheless, a semi-robust method for incorporating the Ly-α measurements into the matter power spectrum may be described as follows [411]. The dependence of A on the cosmology arises from the need to match the observed and simulated absorption spectra. This creates an insurmountable hurdle when the necessary simulations

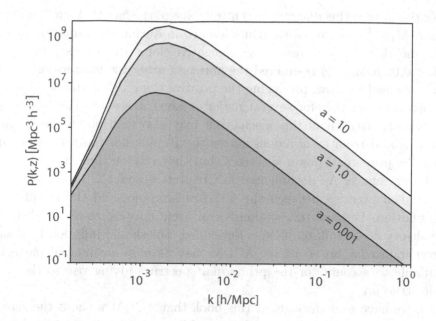

Figure 13.8 The simulated matter power spectrum $P(k, z)$ in the FLRW cosmology constrained by $R_h = ct$, at different values of the expansion factor: $a = 1$ (i.e., today), $a = 0.001$ in our past and $a = 10$ in our future. The shape remains qualitatively unchanged as the Universe evolves, though the peak of the distribution drifts slowly towards higher mode wavenumbers k as time advances. (Adapted from ref. [411])

have yet to be carried out in models other than ΛCDM. Instead, one may rely on a statistical approach to calibrate the matter power spectrum extracted from Ly-α, using the *Planck* CMB measurements. Assuming these two sets of data belong to the same sample, one may use the t-test to examine whether the former is consistent with that of the latter, and then use it to calibrate the Ly-α spectrum with its unknown amplitude against the spectrum measured absolutely using the CMB data.

This simple procedure requires a two-step process. One first uses χ^2 minimization to optimize the constants a_1, b_1, c_1 and D_1 in the polynomial function $f(k) = a_1 k^{-1} + b_1 k^{-2} + c_1 k^{-3} + D_1$ for the *Planck* data above the peak. An analogous optimization is carried out for Ly-α, using the polynomial $g(k) = a_2 k^{-1} + b_2 k^{-2} + c_2 k^{-3} + D_2$. Then, a relative calibration may be found between $f(k)$ and $g(k)$ by varying the 'normalization' constant D_2 until the p-value produced by the t-test lies above the 95% confidence level. The result of this exercise is a calibrated Ly-α matter power spectrum that does not rely on simulations, which may not be completely self-consistent anyway. The outcome shows that the p-values for ΛCDM and the FLRW cosmology constrained by $R_h = ct$ are, respectively, $\approx 99.8\%$ and $\approx 98.7\%$.

Figure 13.6 also shows the matter power spectrum $P(k, 0)$ calculated for ΛCDM using Equations (13.56), (13.58) and (13.60). The consistency between this fit and the data is quite remarkable, constituting very strong evidence in support of the standard model, as discussed earlier in this chapter. In particular, the existence and location of the peak in this diagram are entirely due to the fact that small-scale modes $(k > 0.02 \, h \, \mathrm{Mpc}^{-1})$ re-entered the Hubble horizon during the radiation-dominated expansion and then decayed, producing a negative slope. In contrast, the large-scale modes $(k < 0.02 \, h \, \mathrm{Mpc}^{-1})$ re-entered the horizon during the matter-dominated era, and thus continued to grow, producing the positive slope below the peak.

This picture is at risk for several major reasons, however. The first is simply that the evident strength of this mechanism may also turn out to be its Achilles' heel. The exit and re-entry of modes across the Hubble horizon are paramount for generating the primordial power spectrum, but they rely on the existence of a slow-roll inflationary epoch. As documented in Chapters 6 and 11, there now appear to be several inconsistencies between the CMB anisotropies and the requirements of slow-roll inflation. Perhaps this weakness can eventually be resolved when a more complete theory of the inflaton field is developed, but should inflation be disfavored by the ever-improving observations, ΛCDM may develop an internal inconsistency and be unable to account for the primordial spectrum giving rise to the observed large-scale structure.

Second, we have seen throughout this book that ΛCDM without the zero active mass constraint is becoming less and less tenable as observations probe deeper and farther into the Universe. The empirical evidence—not to mention a growing body of theoretical support—already suggests that a mitigation of the tension between the standard model and various types of observation is best addressed by restricting the equation-of-state to the well-motivated form, $\rho + 3p = 0$, from general relativity. One of the key questions posed in this chapter is whether the consistency shown by the measurements in Figure 13.6 and the growth of structure predicted by ΛCDM, is truly unique, or whether a comparable match can be obtained with an FLRW cosmology constrained by $R_\mathrm{h} = ct$. The answer appears to be that the latter can fit the data just as well as the standard model does (Figure 13.7), perhaps even better at small k's.

As noted earlier, the matter power spectrum in Figure 13.7 is shaped by the strong k-dependence of the Hubble friction versus gravitational influences in Equation (13.87), and does not rely on modes moving back and forth across the horizon. This suggests that the height and location of the peak in this diagram evolve with time. Simulations based on Equations (13.77), (13.79), (13.85), (13.86) and (13.87) confirm this, but also show that its shape does not change qualitatively. One may see these trends in Figure 13.8, which compares the matter power spectra calculated in the FLRW cosmology constrained by $R_\mathrm{h} = ct$ at three different epochs: today $(a = 1)$, in our past $(a = 0.001)$ and in our future $(a = 10)$. The peak shifts slowly with time towards higher k. Its location will have advanced from its current wavenumber by a factor of almost 2 when the Universe is ten times older than today. Thus, given this evolution, the consistency between the simulation and the data shown in Figure 13.7 is arguably just as remarkable as the confirmation of ΛCDM's prediction in Figure 13.6.

This begs the question concerning whether the different rates at which structure formed in ΛCDM versus $R_\mathrm{h} = ct$ could be discernible as a function of lookback time. After all, the comparison of $P(k, 0)$ with the data in Figures 13.6 and 13.7 merely tests for consistency at only one epoch. But what do the growth equations predict for large-scale structure across cosmic time? Today, we can answer this question rather well, and the prospects can only improve as large-scale surveys continue to greatly expand the available database. We shall address this topic next.

13.3 THE OBSERVED HALO MASS FUNCTION

In the standard picture, baryons accreted into the gravitational potential wells created by the dark matter condensations once they decoupled from the radiation, forming bound objects that would become stars, galaxies and clusters. This process involves many astrophysical inputs, some of which are not yet fully understood—and the data offer several anomalies to highlight this fact. There is better consensus concerning the (dark matter) halo evolution itself, however, traced by the growth equations we have developed in Section 13.2, and codified through the so-called halo mass function, $f(M, z)$ [421, 33, 422]. As we shall see, $f(M, z)$ is highly sensitive to the assumed parameters in ΛCDM, and is therefore critical to evaluating the formation of structure over cosmic time.

The halo mass function was introduced and derived analytically using several simplifying assumptions, including spherically symmetric collapse of the mass condensations and a Gaussian initial density field [383]. The distribution (known as Press-Schechter) calculated in this fashion is reasonable, though it underpredicts the number of high-mass halos and overpredicts the low-mass ones compared to detailed N-body simulations. An improvement has been seen with the more recent adoption of an ellipsoidal collapse model (called Sheth-Tormen), rather than spherical [421]. But until recently, these analytic and semi-analytic approaches were tested solely against numerical calculations, and even the Sheth-Tormen distribution overpredicts the number of halos at the high-mass end. Of course, the difficulty with testing these techniques using actual observations is that halos are not directly visible. The predicted $f(M, z)$ must be compared to the data indirectly, via the observation of galaxies and clusters, and therein lies the unavoidable hurdle because, as noted earlier, the theory of galaxy formation is far from being completely mapped out. We shall describe some of the significant tension between model predictions and the data below, but the reality is that some or all of it may simply be due to our incomplete understanding of how galaxies and halos co-evolved over cosmic time.

The fluctuation modes forming the decomposition in Equation (13.88) are present on all scales, so we do not actually see them individually. The perturbations $\delta_{\mathrm{dm},k}$ or $\delta_{\mathrm{de},k}$, as the case may be, are usually assumed to be a Gaussian random field, which means that the waves in this decomposition have random phases. The field may therefore be specified as a statistical average, entirely via its power spectrum,

$$P(\mathbf{k}, z) = \langle \delta_k \delta_k^* \rangle = \langle |\delta_k|^2 \rangle , \qquad (13.95)$$

where $\delta_k(z)$ is a generic representation of the fluctuation. For an isotropic distribution, the power spectrum averaged over all possible realizations must also be independent of direction:

$$P(k, z) = \frac{1}{4\pi} \oint \langle |\delta_k|^2 \rangle \, d\Omega . \qquad (13.96)$$

One may express this power spectrum as the Fourier transform of the autocorrelation function using the Wiener-Khinchin theorem:

$$\langle \delta^*(\mathbf{x}, z)\delta(\mathbf{x} + \mathbf{y}, z)\rangle = \left\langle \int \frac{d^3k'}{(2\pi)^3} \int \frac{d^3k}{(2\pi)^3} \delta_{k'}^*(z)\,\delta_k(z) \times \right.$$

$$\left. e^{i\mathbf{k}'\cdot\mathbf{x}}\, e^{-i\mathbf{k}\cdot(\mathbf{x}+\mathbf{y})} \right\rangle, \tag{13.97}$$

which may also be written

$$\langle \delta^*(\mathbf{x})\delta(\mathbf{x} + \mathbf{y})\rangle = \int \frac{d^3k}{(2\pi)^3} P(\mathbf{k}, z)e^{-i\mathbf{k}\cdot\mathbf{y}} . \tag{13.98}$$

Its inversion gives

$$P(\mathbf{k}, z) = \int d\mathbf{y} \, \langle \delta^*(\mathbf{x}, z)\delta(\mathbf{x} + \mathbf{y}, z)\rangle \, e^{i\mathbf{k}\cdot\mathbf{y}} . \tag{13.99}$$

And defining the vector $\mathbf{y}$ as the axis about which the poloidal and azimuthal angles are integrated, Equation (13.97) is reduced to a single integral over k:

$$\langle \delta^*(\mathbf{x}, z)\delta(\mathbf{x} + \mathbf{y}, z)\rangle = 4\pi \int \frac{k^2\,dk}{(2\pi)^3} P(k, z)\frac{\sin ky}{ky} , \tag{13.100}$$

where $y = |\mathbf{y}|$. Then, the blending of all the fluctuations together at a given point $\mathbf{x}$ may be characterized in terms of the *variance*, σ, of $\delta(\mathbf{x}, z)$, calculated from the autocorrelation function at $\mathbf{y} = 0$:

$$\sigma^2 = 4\pi \int \frac{k^2\,dk}{(2\pi)^3} P(k, z) . \tag{13.101}$$

Notice, however, that the fluctuations $\delta(\mathbf{x}, z)$ exist on all spatial scales, so the measure of σ as it stands is not very practical for comparing theory with observations. A more useful measure is the variance delimited to a specified volume, within which the measurement may be made. One typically introduces a window function $W_R(\mathbf{x})$, with a characteristic radius R, such that W_R is non-zero for $|\mathbf{x}| < R$ and decreases to zero for $|\mathbf{x}| \gg R$. The perturbation $\delta(\mathbf{x}, z)$ is then replaced by the convolution integral

$$\delta_R(\mathbf{x}, z) \equiv \int \delta(\mathbf{y}, z)W_R(|\mathbf{x} - \mathbf{y}|)\, d^3y . \tag{13.102}$$

The power spectrum $P(k, z)$ must then also be replaced with $P(k, z)\tilde{W}_R^2(k)$, where $\tilde{W}_R$ is the Fourier transform of $W_R(\mathbf{x})$. One may thus write the volume-delimited variance as

$$\sigma_R^2 = 4\pi \int \frac{k^2\,dk}{(2\pi)^3} P(k, z)\, \tilde{W}_R^2(k) . \tag{13.103}$$

Many workers adopt a conventional Gaussian window,

$$W_R(y) = \frac{1}{(2\pi)^{3/2}R^3}\, e^{-y^2/2R^2} , \tag{13.104}$$

for which the Fourier transform is simply

$$\tilde{W}_R(k) = e^{-(kR)^2/2} . \tag{13.105}$$

Putting all these relations together, one therefore arrives at the final expression for σ_R:

$$\sigma_R^2(z) = 4\pi \int \frac{k^2 \, dk}{(2\pi)^3} \langle |\delta_k(z)|^2 \rangle \, e^{-(kR)^2} . \tag{13.106}$$

In the local Universe ($z = 0$), this variance is typically calculated in spherical volumes with a radius of $8 \, h^{-1}$ Mpc, giving rise to the often used statistic (called σ_8) for the measurement of local structure:

$$\sigma_8^2(0) \equiv 4\pi \int \frac{k^2 \, dk}{(2\pi)^3} \langle |\delta_k(0)|^2 \rangle \, e^{-(8\bar{k})^2} , \tag{13.107}$$

where $\bar{k} \equiv k/(h \, \text{Mpc}^{-1})$ is the mode wavenumber in units of $h \, \text{Mpc}^{-1}$.

As noted earlier, the Press-Schechter halo mass function characterizes the fraction of matter bound in structures and their distribution, based on a spherical collapse model and the assumption of a Gaussian initial density field. The probability that this field has an overdensity $\delta(z)$ is given by

$$\Pi(\delta, R, z) = \left(\frac{1}{2\pi \, \sigma_R^2(z)} \right)^{1/2} \exp \left(-\frac{\delta^2}{2\sigma_R^2(z)} \right) , \tag{13.108}$$

where $\sigma_R^2(z)$ is the variance defined in Equation (13.106). In this simplified theory, a bound structure forms once δ exceeds a critical value δ_c, with an estimated mass

$$M_{\text{bound}} = \frac{4\pi R^3 \rho(z)}{3c^2} , \tag{13.109}$$

in terms of the background energy density $\rho(z)$ at the time of collapse.

But in addition to regions collapsing when $\delta > \delta_c$, there are also regions that appear to be underdense (i.e., $\delta < \delta_c$) when smoothed on a scale $R(M) = (3c^2 M/4\pi\rho)^{1/3}$, yet become overdense on scales bigger than R. There is therefore a non-zero probability that a region with $\delta < \delta_c$ on a scale R still becomes bound when the scale increases beyond R. The fraction of bound objects whose mass is greater than M must therefore be found from the expression

$$F(M, z) = \int_{\delta_c}^{\infty} \Pi(\delta, R, z) \, d\delta + \int_{-\infty}^{\delta_c} B(\delta_c, \delta) \, d\delta , \tag{13.110}$$

where B represents the probability of a region with $\delta < \delta_c$ on a scale R attaining an overdensity $\delta > \delta_c$ when smoothed over scales larger than R.

To understand how one might find B, let δ_i be the density field when smoothed over a scale of radius R_i, $i = 1...n$, with $R_{i+1} > R_i$. And let R_m (with $1 \leq m \leq n$) be the radius at which the condition $\delta > \delta_c$ is reached. The probability that a region remains underdense for all these filter scales is an integral over all the δ_m's of the joint probability distribution that the Gaussian variables δ_m simultaneously have a given set of values. And one minus this probability is then a representation of B, i.e., the probability that a point becomes overdense somewhere along the sequence R_i.

In fact, one does not actually calculate B directly, because the filtering process to find $\delta_m > \delta_c$ at some scale $R_m > R_i$, starting with δ_i at R_i, turns into a random-walk problem. In the continuum limit, this process becomes a diffusion equation [393],

$$\frac{\partial \Pi}{\partial \sigma_R^2} = \frac{1}{2}\frac{\partial^2 \Pi}{\partial \delta^2} , \tag{13.111}$$

whose solution, with a barrier of height δ_c, is

$$\Pi(\delta, \sigma_R^2) = \frac{1}{\sqrt{2\pi}\,\sigma_R}\left\{\exp\left[-\frac{\delta^2}{2\sigma_R^2}\right] - \exp\left[-\frac{(\delta - 2\delta_c)^2}{2\sigma_R^2}\right]\right\} . \tag{13.112}$$

Thus, to find the fraction, $F(M, z)$, of bound objects with mass greater than M, one must integrate Equation (13.112) from δ_c to ∞. The halo mass function $f(M, z)$ is defined to be the mass-derivative of $F(M, z)$:

$$f(M, z) \equiv \frac{\partial F(M, z)}{\partial M} . \tag{13.113}$$

The comoving number density of halos per unit mass may thus be found from the expression

$$N(M, z)\,dM = \frac{\rho(z)}{M} f(M, z) \equiv \frac{\rho(z)}{M}\frac{\partial F(M, z)}{\partial M}\,dM , \tag{13.114}$$

which may be further simplified with the use of the comoving number density

$$dn(M, z) \equiv N(M, z)\,dM , \tag{13.115}$$

which yields the number density of halos per unit comoving volume:

$$\frac{dn}{d\ln M} = M\frac{\rho(z)}{M^2} g_{PS}(\sigma_R)\left|\frac{d\ln\sigma_R}{d\ln M}\right| . \tag{13.116}$$

The quantity

$$g_{PS}(\sigma_R) \equiv \sqrt{\frac{2}{\pi}}\frac{\delta_c}{\sigma_R}\exp\left(-\frac{\delta_c^2}{2\sigma_R^2}\right) , \tag{13.117}$$

is known as the Press-Schechter mass function. The critical overdensity above which spherical collapse occurs is determined by local physics and is relatively independent of the background cosmology. Its value is estimated to be $\delta_c = 1.686$.

The Sheth-Tormen mass function, g_{ST}, analogous to g_{PS}, alleviates some of the inconsistencies seen in a comparison of the Press-Schechter formalism with numerical N-body calculations. Its derivation is somewhat more involved than the procedure outlined above, but is given as [421]

$$g_{ST}(\sigma_R) = A\sqrt{\frac{2a}{\pi}}\left[1 + \left(\frac{\sigma_R^2}{a\delta_c^2}\right)^p\right]\frac{\delta_c}{\sigma_R}\exp\left[-\frac{a\delta_c^2}{2\sigma_R^2}\right] , \tag{13.118}$$

where $A = 0.3222$ is a normalization factor, $a = 0.707$ and $p = 0.3$.

An important caveat to remember with the use of this method, however, is that while it improves upon Press-Schechter, it is not completely free of inconsistencies either. The Bolshoi simulations discussed in Section 1.2 have shown that, while discrepancies with the Sheth-Tormen mass function at $z \sim 0$ are $\lesssim 10\%$ for halo masses in the range $5 \times 10^9 - 5 \times 10^{14} \ M_\odot$, g_{ST} nevertheless over-predicts the halo density by $\sim 50\%$ at $z \sim 6$ for masses in the range $10^{11} - 10^{12} \ M_\odot$. The discrepancy worsens by an order of magnitude by $z \sim 10$. In principle, one could introduce correcting factors to improve upon the basic Sheth-Tormen formalism. As we shall see shortly, however, the disagreement between theory and observations appears to be so significant that even such corrections are of little help.

The predicted redshift-dependent halo-mass functions can now be compared with a growing database of halo number density measurements using three different techniques [423], including the clustering method [424, 425] based on the spatial distribution of galaxies to infer the halo masses. In this approach, one does not need to assume any physical properties of the galaxies themselves, but does need to have a working model for the dark-matter concentration. Other methods include template fitting [426], which assumes a correlation between the luminosity and stellar masses; and the abundance matching technique [427] based on the identification of certain features (such as the 'knee') in the galaxy's luminosity or mass function that relate to the halo mass distribution. If seen, these features may be used to match the galaxy and dark-matter densities and thereby to obtain the halo mass function. At higher redshifts ($z > 6$), halo masses may be derived from the UV luminosity function under the assumption that the mass-to-light ratio seen at lower redshifts persists to earlier times. This last method, however, more than the others, is still controversial because various arguments can be made concerning whether it—and its closely related halo to stellar-mass ratio—is in fact independent of redshift within the survey range. As we shall see shortly, the result of this cosmological test hinges on how this issue is resolved, and therefore remains open to revision as the quality of the observations continues to improve.

The measured halo-mass functions are shown in Figures 13.9 and 13.10, respectively, for $z = 5$ and 8. These were first calibrated using ΛCDM with the *Planck* optimized parameters as the background cosmology (Section 9.3), and are plotted in the lower panels of these figures. When comparing them to the FLRW cosmology constrained by $R_h = ct$ (upper panels), however, they must be recalibrated to take into account the differences in comoving volumes between these two models. The conversion factor is simply $[d_{com}^{\Lambda CDM} / d_{com}^{R_h = ct}]^3$, where $d_{com} \equiv d_L / (1 + z)$, and d_L is the luminosity distance in Equations (12.33) and (12.35). This is a modest effect, amounting to changes in volume by less than $\sim 10\%$ in the redshift range $0 \lesssim z \lesssim 10$, so a careful inspection of these figures would reveal only a slight shift in the number density between the panels. The theoretical curves in these figures are based on the g_{ST} mass function (Equation 13.118), calculated using the so-called HMFCalc code [428]. The model dependence in these simulations enters through the variance σ_R, which is calculated from Equation (13.106) using the solutions to the growth equations derived in Section 13.2.

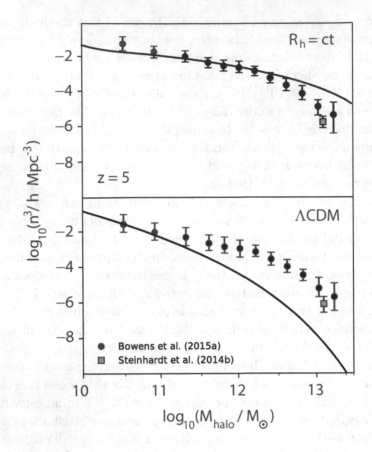

Figure 13.9 The halo mass function (Steinhardt et al. 2016) inferred from galaxy surveys at $z = 5$ compared with an FLRW cosmology constrained by $R_h = ct$ (top) and ΛCDM (bottom). (Adapted from ref. [411])

If these data are to be believed, they show quite emphatically that the halo distribution estimated from these surveys at $4 \lesssim z \lesssim 10$ is inconsistent with the evolution in halo formation predicted by ΛCDM. As noted in one of the earlier references, some authors have called this gross disparity "The Impossibly Early Galaxy Problem." At face value, the tension appears to be as large as 4 or 5 orders of magnitude, particularly at the high-mass end, and the disagreement worsens with redshift, suggesting that structure formation took place much earlier than predicted by the standard model. Several remedies have been proposed to reconcile these differences, including possible calibration errors based on observations at lower redshifts, but none have worked thus far. If anything, these subsequent studies have reinforced the view that the high-redshift galaxies used in estimating the halo mass function appear normal, fully consistent with what one would expect by extrapolating from lower redshifts.

The principal caveat with these results is that none of the data shown in these figures were obtained directly. All of these measurements are based on the use of relationships derived at lower redshifts and extrapolated into the survey range $4 \lesssim z \lesssim 10$. This approach is not universally accepted, however. In order to use galaxy evolution

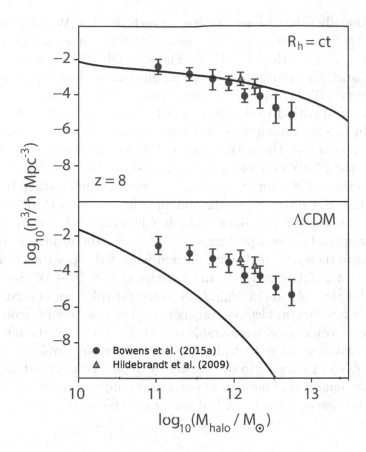

Figure 13.10 The halo mass function (Steinhardt et al. 2016) inferred from galaxy surveys at $z = 8$ compared with an FLRW cosmology constrained by $R_h = ct$ (top) and ΛCDM (bottom). (Adapted from ref. [411])

and their associated observational signatures, such their UV luminosity function or redshift-dependent clustering, one must have a viable model of dark-matter evolution [429], and realistic cosmological hydrodynamical simulations [430, 431], complemented by analytical and semi-analytical comparisons [432, 433]. Complications arise because the observed UV luminosity function—a key component in this analysis—depends strongly on redshift. One can change various inputs and assumptions and reach similar, degenerate results [434]. One therefore comes away with the impression that the degree of tension between theory and observations largely depends on one's point of view.

To bring this debate into clear focus, it is fair to say that the principal disagreement among the various proponents in this work is whether the mass-to-light ratio M_{halo}/L_{UV} and/or the halo mass to stellar-mass ratio were constant over the redshift range $z \lesssim 10$. Most investigators would agree that in order to significantly mitigate the disparity seen in Figures 13.9 and 13.10, an evolution in these ratios by about 0.8 dex would be required. That appears to be excessive given what is known today about galaxy formation and evolution during that epoch, but has not been ruled out observationally—the ultimate arbiter in such debates. Whether such significant evolution actually happened may be seen, e.g., through a careful scrutiny of the supernova rate as a function of redshift. These would be challenging observations to carry out at such high redshifts, but are planned to be done with the James Webb Space Telescope (JWST) over the coming decade. The outcome of this future work is nothing short of critical to the internal self-consistency of ΛCDM.

On the flip side, it is also quite clear from Figures 13.9 and 13.10 that the measured halo mass function over the redshift range $4 \lesssim z \lesssim 10$ is far more consistent with theory when the FLRW cosmology is constrained by $R_h = ct$. Given the debate we have just described, this comparison on its own would be interesting, though probably not compelling. But if one views this outcome in concert with other observational comparisons discussed in this book, notably Chapters 9, 11 and 12, there are good reasons to suspect that real problems do exist with the formation of structure in ΛCDM. Some of these have to do with the unusually early appearance of supermassive black holes at $z > 6$ (Section 9.3) and galaxies at $z \sim 10 - 12$ (Section 9.4). And although a drastic evolution in M_{halo}/L_{UV} is not yet ruled out observationally, fixing the halo mass function problem by forcing consistency between the observations and standard theory comes at a considerable cost [423]. To remove the tension, one must introduce—what appears to us today to be—implausible physics, such as the need to convert 100% of baryons into stars instantly upon halo virialization, which should otherwise take hundreds of millions of years to accomplish. Let us not forget that the age of the Universe at $z = 6$ in ΛCDM was only ~ 900 Myr.

CHAPTER 14

Future Prospects

T HE cosmic spacetime is no longer merely a highly symmetric solution to Einstein's equations, to be used prosaically for the purpose of optimizing the standard model's largely empirical parametrization. The lesson of Bishop James Ussher (Chapter 6), and other similar examples in history, have taught us that simply improving the precision of a measurement within an incomplete (perhaps even flawed) context does not necessarily foster a deeper understanding. We have seen in this book that each of its metric coefficients has profound physical meaning, whose careful consideration has provided us with a better appreciation for the origin of cosmological redshift and an apparent explanation for Einstein's famous formula, $E = mc^2$. This is just the beginning, for we are now entering a truly unprecedented era of cosmic discovery in the coming decades and century. In this final chapter, we shall attempt to convey the excitement of upcoming developments, for cosmology in general, and the cosmic spacetime in particular. And we shall highlight several major unresolved issues with the standard model that could benefit considerably from the introduction of new physics.

14.1 A DIRECT TEST OF THE COSMIC SPACETIME

We begin this survey with the cleanest, most direct test of the cosmic spacetime itself. We have seen an abundance of evidence in this book that the FLRW cosmology constrained by $R_\mathrm{h} = ct$ (i.e., the zero active mass condition) appears to be an attractor for the parametrization in the standard model. Time and again, when ΛCDM's parameters are optimized by fitting the data, the implied expansion profile and history are essentially as one would have expected in the former scenario, whose principal distinguishing feature compared to all the other cosmologies is its constant rate of expansion. One can therefore easily convince themselves that a measurement of the so-called *redshift drift* over a baseline of 5-20 years should provide the cleanest method for differentiating between $R_\mathrm{h} = ct$ and all other models (such as ΛCDM) that predict epochs of acceleration and deceleration. This kind of measurement will eventually produce a preference of one cosmology over the others at a confidence level approaching 3–5σ.

Other kinds of cosmological observation rely on at least some source properties, often dominated by redshift evolution, and we shall address several of these later in the chapter. The redshift drift of objects in the Hubble flow, however, is a direct non-geometric probe of the dynamics of the Universe that does not rely on how sources evolve, or issues of gravity and clustering [435]. This physical effect merely requires the validity of the Cosmological principle, specifically that the Universe is homogeneous and isotropic on large scales. In a Universe with a variable expansion rate, the redshift of a source at a fixed comoving distance changes with time, and it will be straightforward to see below how its first and second derivatives can distinguish between different models [436, 437, 438]. Suitable instruments for this kind of work include high-resolution spectrographs [439, 440], such as the high-resolution spectrograph on the Extremely Large Telescope (ELT-HIRES) [441], which will allow measurements over the redshift range $2 \lesssim z \lesssim 5$, and the Square Kilometre (Phase 2) Array (SKA) [442], which should be able to make analogous measurements below $z \sim 1$, and possibly also with 21-cm experiments, such as the Canadian Hydrogen Intensity Mapping Experiment (CHIME) [443].

From the expression for redshift (Equation 8.9) in terms of the expansion factor $a_e \equiv a(t_e)$ at the time (t_e) of emission and $a_0 \equiv a(t_0)$ at the time (t_0) of observation, one may write its first derivative in terms of t_0 as

$$\frac{dz}{dt_0} = [1 + z(t_0)]H(t_0) - \frac{a_0}{a_e^2}\frac{da_e}{dt_e}\frac{dt_e}{dt_0} \tag{14.1}$$

where, by definition,

$$H(t) = \frac{1}{a(t)}\frac{da(t)}{dt} \tag{14.2}$$

is the Hubble constant at time t. And since Equation (8.7) gives

$$dt_0 = [1 + z(t_0)]\, dt_e \, , \tag{14.3}$$

one also has

$$\frac{dz}{dt_0} = [1 + z]H_0 - H(z) \, . \tag{14.4}$$

In ΛCDM, one may write

$$H(z) = H_0\, E(z) \, , \tag{14.5}$$

for any arbitrary redshift z, where $H_0 \equiv H(t_0)$ is the Hubble constant today and $E(z)$ is given in Equation (9.4) in terms of the various density ratios.

During their monitoring campaign, the surveys will measure the spectroscopic velocity shift

$$\Delta v = \frac{c\Delta z}{1 + z} = \frac{c\Delta t}{1 + z}\frac{dz}{dt_0} \, , \tag{14.6}$$

associated with the redshift drift Δz over a time interval Δt. The quantity Δv is shown as a function of z for the *Planck* ΛCDM model [29] (solid black curve) in Figure 14.1, along with a slight variation (short-dash black curve) with $\Omega_m = 0.28$ to illustrate how the velocity shift will change with redshift and alternative values of the parameters.

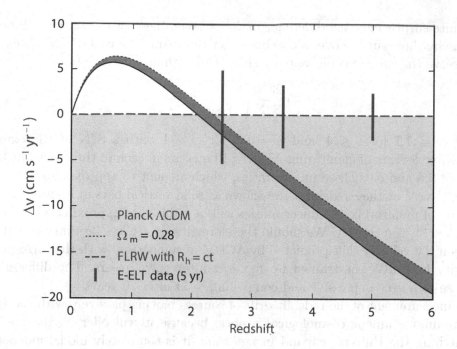

Figure 14.1 The spectroscopic velocity shift Δv produced by redshift drift in the *Planck* ΛCDM model (solid black curve). For comparison, the short-dash curve shows a slight variation with $\Omega_m = 0.28$. In the FLRW cosmology constrained by $R_h = ct$, $\Delta v = dz/dt_0 = 0$ at all redshifts (horizontal long-dash line). The vertical bars indicate the 1σ errors at $z = 2.5$, 3.5, and 5.0, as will be measured by the ELT-HIRES experiment after 5 years of monitoring, assuming the measured velocity shifts are zero. (Adapted from ref. [444])

By comparison, the redshift drift expected in an FLRW cosmology constrained by $R_h = ct$ is much simpler. Since $a(t) = t/t_0$ in this Universe, one has

$$H(t) \equiv \frac{\dot{a}}{a} = \frac{1}{t} , \tag{14.7}$$

and since

$$(1 + z) = \frac{t_0}{t_e} , \tag{14.8}$$

it is clear that $(1 + z) = H(t_e)/H(t_0)$, or

$$H(t_e) = H(t_0)[1 + z(t_0)] . \tag{14.9}$$

Equation (14.4) thus yields

$$\frac{dz}{dt_0} = 0 , \tag{14.10}$$

and therefore

$$\Delta v = 0 \tag{14.11}$$

at all redshifts, shown as a horizontal, long-dash line in Figure 14.1.

Quite surprisingly, the difference in Δv between these two scenarios will be measurable by these surveys over a baseline of 5 to 20 years. The ELT-HIRES is expected to observe the spectroscopic velocity shift [440] with an uncertainty of

$$\sigma_{\Delta v} = 1.35 \frac{2370}{\text{S/N}} \sqrt{\frac{30}{N_{\text{QSO}}}} \left(\frac{5}{1 + z_{\text{QSO}}}\right)^{\alpha}, \quad (14.12)$$

where $\alpha = 1.7$ for $z \leq 4$, and $\alpha = 0.9$ for $z > 4$, with a S/N of approximately 1,500 after 5 years of monitoring $N_{\text{QSO}} = 10$ quasars in each of three redshift bins at $z = 2.5$, 3.5 and 5.0. These uncertainties, which amount to approximately 12, 8 and 5 cm s^{-1} yr^{-1} at these redshifts, are shown as solid vertical bars in Figure 14.1. After 20 years of monitoring, the uncertainties will be reduced to approximately 6, 4 and 3 cm s^{-1} yr^{-1}, respectively. We should therefore already see a $\sim 3\sigma$ difference at $z = 5$ between the velocity shift predicted by ΛCDM in comparison with the corresponding quantity in FLRW constrained by $R_{\text{h}} = ct$ after only 5 years. The difference will increase to a very impressive and compelling $\sim 5\sigma$ after 20 years.

A measurement of the redshift drift of sources moving passively with the Hubble flow is unique among cosmological probes, because it will offer us the possibility of watching the Universe expand in *real time*. It is completely model independent, given that it does not require the use of integrated quantities, such as the luminosity distance. And its biggest impact will be felt on the cosmic spacetime itself, since it will clearly distinguish between the zero redshift drift at all z expected in an FLRW cosmology constrained by $R_{\text{h}} = ct$ from all other models, such as ΛCDM, in which the expansion rate is variable.

One should not underestimate the impact of this upcoming observational campaign because, while certain other observational signatures, such as the luminosity distance, are somewhat similar for ΛCDM and $R_{\text{h}} = ct$ at low redshifts (see, e.g., Figure 12.5), they diverge rather quickly in the early Universe. The difference is so large as one approaches the Big Bang that inflation is required to solve the horizon problem in ΛCDM, but not in an FLRW cosmology with the $R_{\text{h}} = ct$ constraint (Sections 11.3 and 11.4).

14.2 OBSERVATIONS OVER THE COMING DECADES

Of course, deepening our understanding of the cosmic spacetime cannot be the solitary goal of upcoming cosmological work. Almost certainly, extensions to the standard model of particle physics will be required to fully appreciate the nature of the constituents in the cosmic fluid, particularly dark matter and dark energy, about which we still know virtually nothing. For this reason, an extensive effort is underway to probe the dark Universe at unprecedented levels of sophistication and scope.

14.2.1 Dark Energy Surveys

The dark-energy equation-of-state will be studied with a broad range of instrumentation, from small ground-based facilities to major new satellites. Some will probe the expansion history of the Universe with unprecedented precision and accuracy (Section 12.3.2), others will measure how the distance between galaxies has changed over time, principally to ascertain how baryon acoustic oscillations (BAO) have evolved (Sections 6.4 and 11.1), and still more will trace the formation of large-scale structure (Chapter 13). All of these approaches will bear witness to the influence of dark energy over cosmic time, and thereby inform us if it is truly an unchanging cosmological constant, or whether its energy density has varied, implying that it is dynamic, evolving in concert with other particle species in the cosmic fluid. If the latter emerges unequivocally from the new data, dark energy will have to be incorporated into a new extended version of the standard model of particle physics (Chapter 9).

Receiving its first light in 2019, the Dark Energy Spectroscopic Instrument (DESI) [445] represents a major advance in capability for wide-field spectroscopy. Its multi-fiber architecture enables massively parallel measurements of galaxy redshifts, generating a 3-dimensional map of the Universe to allow us to measure the BAO and the cosmic expansion history. In addition, it will provide redshift space distortion measurements for us to trace the growth of structure. DESI will observe luminous red galaxies up to $z \sim 1$, and will target bright OII emission-line galaxies to extend the reach out to $z \sim 1.7$. Ly-α forest absorption features will be visible in the redshift range $2.1 \lesssim z \lesssim 3.5$, tracing the evolution and clumping of neutral hydrogen. By the end of its 5-year survey, DESI will have assembled a catalog of over 30 million galaxies and quasar redshifts.

Complementing the spectroscopic capability of DESI, the Large Synoptic Survey Telescope (LSST) [446] is a wide-field, ground-based telescope designed to image a large fraction of the sky in six optical bands every few nights. It is expected to observe for a decade following first light in 2020. The stacked images should provide sufficient S/N to detect galaxies beyond $z \sim 1$. LSST will carry out weak lensing studies, constraining cosmological parameters in the context of ΛCDM to percentage level accuracy, and will greatly enhance the precision with which the matter power spectrum is measured (Chapter 13). Its greatest contribution to the study of dark energy, however, is expected to come from its discovery of roughly 500 new Type Ia supernovae per year. With tens of thousands of well-measured supernova lightcurves spanning the region ($z \sim 0.7$) where the Universe is thought to have transitioned from deceleration to acceleration in the context of ΛCDM (Sections 12.2 and 12.3), we may finally see compelling evidence for—or against—this important event in cosmic history. Finally, LSST will also provide us with a sample of $\sim 2,600$ time-delay strong lenses (Section 12.5) with which to measure the cosmic distance scale, constituting a factor 100 improvement over the strong lens catalog available today.

The European Space Agency will be launching a new satellite mission called *Euclid* [447] in the early 2020's, with an expected mission lifetime of about 6 years. With a consortium size of over 1,200, this mission will involve a large fraction of Earth's observational cosmology community. Its 1.2 meter telescope will provide a wealth of data on weak gravitational lensing, BAO and redshift space distortions, that will be used to refine the dark-energy equation-of-state and to measure the redshift-dependent growth of structure. By the end of its survey period, *Euclid* will have accumulated a database of billions of galaxy shapes for the weak lensing analysis and star formation rates—critical to resolving the tension between the measured halo mass function and the predicted growth of structure in ΛCDM (Section 13.3)—out to $z \sim 2$, and tens of millions of redshifts measured with a precision of $\delta z < 0.001(1+z)$.

The last of the major upcoming missions is The Wide Field Infrared Survey Telescope (WFIRST) [448], a highly rated and anticipated space mission, whose cutting edge infrared detector technology will enable it to achieve dramatic advances across a broad range of astrophysical disciplines. One of its primary scientific goals will be to pin down the dark-energy equation-of-state. During its 1.9 years dedicated to cosmology, WFIRST will measure the weak-lensing shapes of about 500 million galaxies and the weak lensing mass profiles of 40,000 massive clusters. In addition, its spectroscopic survey will provide Hα emission-line redshifts of approximately 20 million galaxies in the redshift range $1 \lesssim z \lesssim 2$, and about 2 million OIII emission-line galaxies at $2 \lesssim z \lesssim 3$. Its supernova survey will produce a catalog of about 2,700 new Type Ia events over the redshift range $0.1 \lesssim z \lesssim 1.7$, almost doubling the redshift coverage of LSST. If the preparations continue to go well, WFIRST should be launched by the middle of the 2020's and have a mission lifetime of about 6 years.

14.2.2   B-mode Polarization in the CMB

With the release of its final CMB maps in 2018, the *Planck* mission brought to an end the current era of large-scale, space-based CMB-related instrumentation, at least for many years. Henceforth, cosmological observations of the microwave background will be conducted with smaller, more focused missions to study specific properties of the background radiation, particularly its polarization characteristics. The alignment of the fields carried by the microwaves can provide a signature of inflation, if detected and measured. There are two principal types of polarization associated with these waves: the first, called E mode, is due to Thomson scattering in an ionized medium near the LSS (Section 6.4). The fields in this kind of polarization are always perpendicular to the direction of propagation and therefore have zero curl. The second, called B mode, is also produced by Thomson scattering, but with an anisotropic background resulting from the passage of gravitational waves believed to have been seeded as tensor fluctuations during inflation (Equation 6.155). Gravitational waves produce a background anisotropy with a diagonal pattern relative to the direction of propagation, rather than perpendicular to it, and therefore generate polarization with a non-zero curl. E-mode polarization has been detected at the level of about 10%, consistent with the ionization fraction expected after recombination. But no confirmed B-mode polarization has been measured yet.

Given the lack of success with this endeavor thus far, the major agencies (such as The National Aeronautics and Space Administration in the USA and The European Space Agency in Europe) are very reluctant to assign significant funding for developing followup missions. The Background Imaging of Cosmic Extragalactic Polarization (BICEP2) experiment claimed in 2014 to have detected the tell-tale signature of polarization associated with the passage of gravitational waves created during inflation [449], but it turned out to be merely due to foreground dust emission in our own

Galaxy [450]. In the years since then, BICEP2 has coordinated its observations with *Planck* to refine the search, but nothing new has been discovered.

One of the most significant projects being planned for the next decade is called The Cosmic Microwave Background Stage-4 experiment (CMB-S4), comprising three 6-meter and 14 half-meter telescopes distributed across two sites in Antarctica and Chile, making it 100 times more sensitive than other current ground-based instrumentation. An alternative method is being used in the development of the Large Scale Polarization Explorer (LSPE), a stratospheric balloon mission dedicated to measuring the polarization of the CMB on large angular scales. The scientific motivation behind this approach is to achieve high sensitivity by exploiting circumpolar long duration flights over the Arctic night. Its frequency coverage should allow the removal of foreground contamination due to dust and synchrotron emission within our Galaxy, avoiding the pitfall encountered earlier by BICEP2.

Generally speaking, however, most of the efforts to measure B-mode polarization in the CMB are similar to The Simons Array CMB Polarization Experiment [451], which will consist of three cryogenic receivers each featuring multichroic bolometer arrays mounted onto separate 3.5m telescopes. The design calls for a combination of high sensitivity, multichroic frequency coverage and access to a large area of the sky from its planned mid-latitude Chilean observatory, allowing it to produce high fidelity polarization maps across a wide range of angular scales and to remove contaminating B-modes from the foreground. If the inflationary B-mode signal is present, the Simons Array should be able to detect it at a level $r > 0.01$ (Equation 6.164) with a significance exceeding 5σ. This should be compared with the current *Planck* limit, $r < 0.41$ (see Section 6.7), at a wavenumber $k = 0.002$ Mpc^{-1}.

One may come away from this description with the impression that measuring the B-mode polarization in the CMB does not have the same sense of urgency exhibited by the large survey missions seeking to pin down the dark-energy equation-of-state (Section 14.2.1), but the impact these observations will have on our view of the cosmos and its beginnings cannot be underestimated. B-mode polarization in the CMB is required at some level by inflation, though it is not unique to tensor fluctuations generated by the inflaton field (Sections 6.7 and 13.1). But if its detection continues to elude the ever-improving precision of these experiments, our confidence in the viability of inflation to fix the early Universe horizon problems will erode irretrievably (Chapter 11). And as we have been documenting throughout this book, the leading candidate to take the place of inflationary ΛCDM in that case would be a model without a period of decelerated expansion shortly after the Big Bang, such as an FLRW cosmology constrained by zero active mass (Sections 11.3 and 11.4).

14.2.3 Re-ionization History

The redshifted 21-cm signal from neutral hydrogen is a unique probe of the Epoch of Re-ionization (EoR) in the early Universe, spanning the redshift range $6 \lesssim z \lesssim 15$ (Figure 9.2), with the possibility of extending even deeper into the cosmic dawn (CD), up to $z \sim 30$. The challenge, however, is extracting the feeble 21-cm signal from a much brighter foreground contamination and numerous instrumental systematics.

Several experiments to statistically detect the 21-cm signal from the EoR are already underway, and have established useful upper limits. These include: (1) The Low-Frequency Array (LOFAR) [452], a large Murchison Widefield Array (MWA) telescope [453], another interferometric low-frequency radio telescope, located at the Murchison Radio-astronomy Observatory in Western Australia. One of MWA's critical roles is to function as a precursor instrument for the worldwide collaboration known as the Square Kilometre Array (SKA) project [454], currently under development. The SKA will be the world's largest radio telescope, providing unprecedented views of the early Universe over its expected 50-year lifetime. In the meantime, the MWA is carrying out a large survey of the sky in the Southern Hemisphere, with a primary goal of detecting intergalactic, neutral hydrogen surrounding early galaxies during the EoR. And (3), The Precision Array for Probing the Epoch of Re-ionization (PAPER) [455], a 64-antenna array in the Karoo reserve in South Africa and a 32-antenna array at the National Radio Astronomy Observatory near Green Bank, West Virginia. PAPER is a low-frequency radio interferometer designed to detect the birth of the first (Population III) stars and galaxies at $t \sim 500$ Myr. It maps the intensity of the 21-cm emission of neutral hydrogen in the redshift range $7 \lesssim z \lesssim 12$, and measures the power spectrum of fluctuations created by the first luminous sources in the Universe.

Along with SKA, the next generation of instruments designed to search for the 21-cm signal from the CD will include the Hydrogen Epoch of Re-ionization Array (HERA) [456], a large grid of 14-meter non-tracking telescopes packed into a hexagonal array 300 meters across. This large collecting area will ensure that HERA will have a sensitivity an order of magnitude greater than that of the first generation 21-cm instruments, enabling it to create the first images of large-scale HI structures.

Given the complexity of extracting the tiny 21-cm CD signal from the much more intense foreground field, there is no guarantee that any of these experiments will be completely successful, though simulations have raised expectations that a major breakthrough uncovering how the first stars and galaxies formed is on the horizon. Given the significant level of tension between the current measurements of the halo mass function (Chapter 13) and the predictions of ΛCDM, not to mention the very puzzling 'early' appearance of supermassive black holes (Section 9.3), any new discovery impacting the EoR and CD has a paradigm-shifting potential.

14.2.4   The Redshift-time Relation

So far-reaching will be the observational capability of the SKA that its mapping of the 21-cm line deep into the EoR and CD is but one highly anticipated outcome of this world-class facility. Phase 1 of the array, constituting about 10% of the overall design, should be constructed by 2025. Already, this completed portion will be more than sufficient for it to detect high-redshift, radio-loud quasars, well beyond the range ($z \sim 6$) where the uncomfortably early appearance of supermassive black holes has created significant tension with the standard model (Section 9.3). With its unprecedented capabilities, SKA is poised to probe the intermediate to high-redshift radio Universe with greater sensitivity than any other instrument before it.

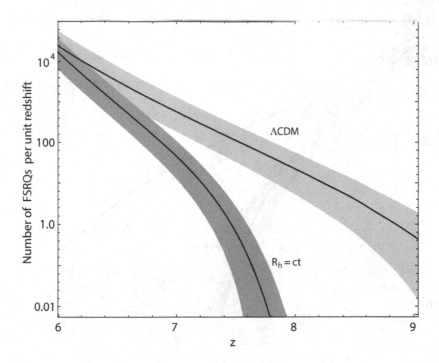

Figure 14.2 Estimated number of high-z FSRQs that will be detected per unit redshift in the SKA Wide survey, and an estimate of the errors represented by the shaded regions, for ΛCDM and an FLRW cosmology constrained by $R_{\rm h} = ct$. (Adapted from ref. [462])

The observation and detection of radio-loud quasars beyond $z \sim 6$ will produce a catalog of the number of such sources per unit comoving volume as a function of redshift, and critically test the redshift-time relation in various cosmologies (Section 9.2). Approximately half of all $z > 5.5$ radio-loud quasars known today are Flat Spectrum Radio Quasars (FSRQs), with the rest showing steeper spectra similar to Faranov-Riley II (FR-II) galaxies [457]. The commonly encountered components of radiative emission seen in the spectral energy distribution (SED) of such sources include (see also Section 12.3.1): a relativistic radio jet, pointing more or less towards the observer [458], and sometimes in X-rays and γ-rays via inverse Compton scattering [459]; a thermally-radiating accretion disk emitting predominantly in the UV [460]; a hot corona above the disk, emitting primarily in X-rays [224]; and an obscuring torus [461].

These FSRQs are members of a broader class of quasars beaming their emission towards us, known as *blazars*. Untangling the various processes contributing to the SED of a particular subgroup is often challenging, but the situation appears to be much more manageable from a phenomenological standpoint. In this regard, the so-called blazar sequence [463] created from *Fermi* high-energy observations [464] characterizes their average SED based on a sequence of γ-ray luminosity bins. An interpolation within this sequence may then be used to determine the emission profile of any FSRQ within a desired waveband, from radio to γ-rays [462].

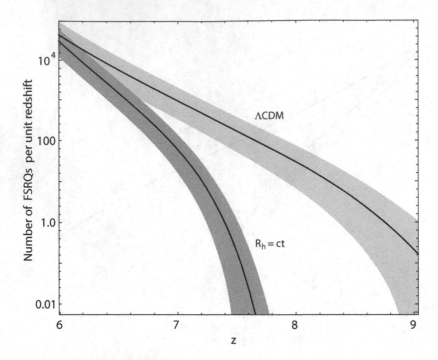

Figure 14.3 Estimated number of high-z FSRQs that will be detected per unit redshift in the SKA Medium-deep survey, and an estimate of the errors represented by the shaded regions, for ΛCDM and an FLRW cosmology constrained by $R_h = ct$. (Adapted from ref. [462])

The key constraint that will make these sources suitable for testing the redshift-time relation is the empirical evidence suggesting that most or all of the FSRQs in the high-redshift Universe ($z > 5.5$) are emitting at or near their Eddington luminosity (see Section 9.3, particularly Figure 9.4) [201]. In addition, circumstantial evidence, mostly based on the observation of superluminal motion, points to relativistic jets with a bulk Lorentz factor $\Gamma \sim 5 - 15$, reinforcing the underlying physical viability of the blazar sequence [463].

Simulations carried out in preparation for these upcoming observations, based on the blazar sequence and these empirical constraints, have predicted the number of FSRQs that should be detected by the SKA1-MID Wide Band I and Medium-deep Band 2 surveys in the context of ΛCDM and, for comparison, also for an FLRW cosmology constrained by $R_h = ct$. The known distribution of quasars at $z \sim 6$ normalizes the mass function, which is then devolved in lockstep towards higher redshifts, assuming a duty cycle close to 1. Assuming a fiducial efficiency of $\sim 10\%$, the expected e-folding time (i.e., the 'Salpeter time'; see Equation 9.11) is ~ 45 Myr. Thus, the overall growth rate of FSRQs—and therefore their detection rate as a function of redshift—is highly dependent on the background cosmology, given that the predicted timeline $t(z)$ changes considerably between models.

The predictions (summarized in Figures 14.2 and 14.3) reveal a striking result that will almost certainly be confirmed or rejected by SKA over the coming decade.

In a nutshell, ΛCDM predicts 40 times more blazar counts at $z > 7$ than the FLRW model constrained by $R_{\mathrm{h}} = ct$ in the Wide survey, and 80 times more at $z > 7.22$ in the Medium-deep survey. In terms of absolute numbers, the Wide survey should uncover 330^{+375}_{-221} FSRQs at $z > 7$ if the timeline in ΛCDM is correct, and only $8.1^{+12.4}_{-6.3}$ in the context of $R_{\mathrm{h}} = ct$. The disparity in the Medium-deep survey is even greater, with 96^{+120}_{-67} and $1.2^{+2.5}_{-1.1}$, respectively, above $z = 7.22$.

These predictions and test seem almost too good to be true, but there is already some evidence that the discovery rate of high-redshift quasars falls well below expectations. In the context of ΛCDM, about 20 AGNs should have been observed at $z > 5$ in the *Chandra* Deep Field South survey, while *none* were identified [465]— strongly favoring the count rate predicted by the FLRW cosmology constrained by $R_{\mathrm{h}} = ct$ [224]. Of course, the FSRQ catalog assembled by SKA will far exceed the quality and coverage of any analogous sample created before it, and produce results of paradigm-altering potential comparable to the other future prospects described in this chapter.

14.3 MAJOR UNSOLVED PROBLEMS IN ΛCDM

In spite of the general view that we have entered the era of 'precision' cosmology, in which all that remains is to measure the parameters of the standard model as accurately as possible, it would not be correct to believe that ΛCDM is complete. The physics underlying its parametrization is still not fully understood. Several inconsistencies and/or unknowns that have emerged over the years are yet to be resolved. Below we summarize the most prominent among them.

14.3.1 The Trans-Planckian Problem

If the scalar perturbations that seeded structure began as quantum fluctuations in an inflaton field, one must contend with their birth and evolution below the Planck scale (Sections 6.7 and 13.1). Certainly, to normalize them using canonical quantization in Minkowski space, one is obliged to evolve them in the so-called Bunch-Davies vacuum, corresponding to the infinite conformal past. The problem arises because the Compton wavelength λ_{C} grows below the Planck scale as the gravitational radius R_{h} shrinks, so quantum mechanics fails to provide any insight on how one should deal with modes whose wavelength is smaller than the Planck length. This difficulty may be mitigated using an alternative expansion history at early times, e.g., one based on scalar-field dynamics, though without inflation. Otherwise, the problem may simply be due to our overly naive attempts at marrying quantum mechanics with general relativity, pressing the need for the development of a truly viable theory of quantum gravity.

14.3.2   The Horizon Problems

With the discovery of the Higgs particle, we are now faced with two major horizon problems in cosmology: one due to the uniformity of the CMB temperature across the sky (Section 11.3), and the second arising from the electroweak phase transition that gave certain particles in the standard model their mass (Section 11.4). Inflation was invented to address the first of these, and has had a surprisingly large degree of success for such a complicated model (Sections 6.6 and 6.7). Nevertheless, as the quality and breadth of our cosmological measurements have continued to increase, basic slow-roll inflation looks less and less likely to be the correct paradigm (Sections 11.1 and 11.2). The principal difficulty facing us today is how to make inflation fix the temperature horizon problem, while simultaneously providing the necessary framework for generating perturbations consistent with the CMB anisotropies. The answer may lie with the introduction of new physics to create a viable foundation for the next generation of inflaton potentials, but the situation is far from being clear.

The electroweak horizon problem is a much more difficult task to address, and has thus far received scant attention. Even if a GUT-like transition did occur and create an inflationary expansion shortly after the Big Bang, it would not at all prevent the electroweak phase transition from giving rise to its own horizon inconsistencies. Yet the physics associated with the turning on of the Higgs field (at $\sim 10^{-11}$ seconds) is significantly better understood today than what may have transpired at $\sim 10^{-35}$ seconds, suggesting that we ought to be more concerned with the electroweak horizon problem than that associated with the CMB. It is fair to say, however, that both of these horizon problems persist as major stumbling blocks in our mission to acquire a fully comprehensive, self-consistent ΛCDM theory.

14.3.3   BBN and the ^{7}Li Anomaly

Big Bang nucleosynthesis (BBN) traces its origin to the seminal work of Gamow and his colleagues in the middle of the twentieth century, generally viewed as the beginning of 'physical' cosmology (Sections 1.1 and 6.3). Our working hypothesis regarding how and when the light elements were produced in the early Universe is at least roughly consistent with the physical conditions prevalent during the first 18 minutes of expansion following the Big Bang. Nevertheless, there are several basic problems with the standard picture that strongly suggest we are still missing some of the essential physics.

The first issue is not as well known as the second, though it is arguably the more serious inconsistency. In order for us to explain the primordial helium abundance, we must assume that the neutron to proton ratio at the beginning of nuclear burning was established by the Boltzmann factor in Equation (6.47), requiring the neutrons and protons to have been in thermal equilibrium with the cosmic fluid. If this were the case, however, one would also be obligated to accept the baryon to photon density ratio in Equation (6.45). The problem is that the latter misses the required ratio (Equation 6.48) by roughly seven orders of magnitude. In other words, for BBN to have worked as needed, baryogenesis could not possibly have been a thermal equilibrium process, at least not thermally coupled to the rest of the cosmic fluid, yet somehow the neutrons and protons had to remain in thermal equilibrium. How these two disparate conditions could have been satisfied simultaneously is a complete mystery today.

The second problem is that only the abundance of ^{7}Li predicted by BBN in the standard model is measurable and testable against the data once the observed helium abundance is used to fix the baryon-to-photon ratio in Equation (6.48). If we are to believe the accuracy of its measurement in Figure 6.5, however, this quantity disagrees with the observations at a level of confidence exceeding 10σ. This well-known *Lithium anomaly* has been studied for several decades, but the disparity between theory and observation seems to be getting worse rather than better as the precision of the measurements continues to improve.

14.3.4 Timeline for SMBHs and Early Galaxies

It is generally accepted that the early appearance of $\sim 10^9 \ M_\odot$ supermassive black holes at $z \sim 6-7$, barely 980 Myr after the Big Bang, is a major unsolved problem in standard cosmology. Either the timeline in ΛCDM is incorrect, or the astrophysics of black-hole formation and evolution needs major revision. Attempts at resolving this mystery generally fall into one of three categories: (1) black holes accreted at super-Eddington rates, (2) the black hole seeds were created with enormously large masses (several orders of magnitude larger than typically happens in a supernova explosion), and (3) black holes grew much faster than expected as a result of mergers in the early Universe. As discussed in Section 9.3, however, each of these mechanisms faces significant hurdles, either due to a lack of observational support, or because of astrophysical inconsistencies. This problem may become even worse than it is now when SKA will have completed its survey by the end of the 2020's, if the outcome shows that the number of quasars predicted at even larger redshifts in order to produce the $z \sim 6$ population is missing (see Figures 14.2 and 14.3).

The discovery of galaxies as far back as $z \sim 10-12$ may be even harder to explain than the supermassive black holes. Based on the astrophysics of galaxy formation we have today, there are rather strict limits on the time required for each of the steps along the way. In the end, however, we may simply be missing another important piece of the physics puzzle, in which case, observational clues may offer a way out. The bottom line is that in order to mitigate the implied time-compression problem in ΛCDM, these galaxies would need to have grown at a rate of $\sim 20 \ M_\odot \ \mathrm{yr}^{-1}$ instead of the $\sim 2 \ M_\odot \ \mathrm{yr}^{-1}$ currently inferred from the observations (Section 9.4). Fortunately, much improved measurements with upcoming missions—particularly the James Webb Space Telescope (Section 13.3)—should resolve this issue rather cleanly. Determining these early star-formation rates will be one of the principal goals of this observational campaign.

14.3.5 The Halo Mass Function

This is perhaps more of a significant controversy in cosmology rather than a straight out major problem, given that the principal issue is a well-publicized disagreement concerning how the galaxy and halo masses are related as a function of distance. If one naively extrapolates the observations at redshift $z \lesssim 4$ to the range $4 \lesssim z \lesssim 10$, our discussion in Chapter 13 would suggest that an unacceptably large tension exists between the predicted halo mass function and what is inferred from galaxies at high redshifts. That's the key to this whole debate, i.e., that the halos are not themselves directly visible; their characteristics must be inferred by assuming a specific mass-to-light ratio $M_{\text{halo}}/L_{\text{UV}}$, and/or the halo mass to stellar-mass ratio. The disparity can be as large as three to four orders of magnitude, but this arises only if the ratios are assumed to be more or less constant across the sample redshift range, as suggested by observational indicators at lower redshifts. The problem may be mitigated completely, however, if it turns out that the mass-to-light ratio varies by about 0.8 dex from $z \sim 10$ down to 4.

This problem can only be resolved observationally. Unfortunately, there is no completely established theory of galaxy formation. Instead, galaxy evolution is typically based on a hybrid approach in which observational constraints are necessary to flesh out the phenomenology, which makes the process very compliant to the model and one's subjective preferences [432, 433, 423]. Different approaches reach quite diverse results, such as one can see in Figures 13.10 and 13.11. Pinning down the evolution in $M_{\text{halo}}/L_{\text{UV}}$—if it does in fact change with z—will present a challenge to the observations at such high redshifts, but is being planned with the James Webb Space Telescope. One may therefore expect some sort of resolution to this debate in the coming decade.

14.3.6 The Hubble Constant

The newest entry into the pantheon of major problems with ΛCDM is a robust assessment that the value of the Hubble constant measured by *Planck* (i.e., $H_0 = 67.4\pm0.5$ km s^{-1} Mpc^{-1}) [29] is inconsistent with the value measured using the distance ladder associated with long-period Cepheids in the local Universe (now thought to be $H_0 = 74.03 \pm 1.42$ km s^{-1} Mpc^{-1}) [194].

Hubble Space Telescope observations of 70 long-period Cepheids in the Large Magellanic Cloud have resulted in a reduction of the overall uncertainty in the geometric calibration of the local distance ladder to an impressive 1.3%. The local value of H_0 was measured by combining this result with studies of detached eclipsing binaries [466], masers in NGC 4258 [467] and Milky Way parallaxes [468], yielding its unprecedentedly accurate value. Removing any one of these data sets changes H_0 by less than 0.7%. Given the accuracy with which these two values of H_0 have now been measured, the disparity between them is 6.6 ± 1.5 km s^{-1} Mpc^{-1}, or 4.4σ (corresponding to a Gaussian probability of 99.999%) in significance. Many now believe that such a discrepancy is beyond a reasonable level of chance, raising the odds that it is due to some unknown new physics beyond ΛCDM (see Chapter 9 and Section 12.3.2).

14.4   OPEN QUESTIONS IN FLRW WITH ZERO ACTIVE MASS

The evidence today suggests that the predictions of ΛCDM would be a better match to the data should the zero active mass condition (i.e., the equation-of-state $\rho + 3p = 0$ for the total energy density and pressure in the cosmic fluid) constrain its expansion dynamics. The FLRW cosmology guided by $R_h = ct$ has been successfully applied to a broad range of observations, from the local Universe to the very early quantum domain, from large-scale structure to the formation and evolution of supermassive black holes. Yet after a decade of development, this line of inquiry is starting to face significant hurdles going forward without the introduction of new physics. In this section, we shall briefly describe three of the most important questions that need to be resolved, presumably through extensions to the standard model of particle physics.

14.4.1   BBN

At first blush, the problem with BBN and the ^{7}Li anomaly in ΛCDM (Section 14.3.3) appears to become even worse in an FLRW cosmology constrained by $R_h = ct$. At least this has been the conclusion from several attempts at making BBN work with power-law cosmologies, even before the introduction of the zero active mass condition. Of course, neither model can be acceptable in the long run without a viable, physical explanation for the origin of the light elements.

The widespread view that a power-law evolution of the physical conditions during BBN is incompatible with the observations traces its roots back to the 1990's, when the earliest nucleosynthesis calculations were carried out for such a scenario. Indeed, the adjective "disaster" was invoked to describe the impact of power-law models on primordial nucleosynthesis at that time [469]. And a quick inspection of how these simulations were carried out offers clear insight into why the outcome was so negative. In ΛCDM, the deuterium bottleneck subsided when the temperature dropped to about 80 keV, corresponding to a time $t \sim 2.9$ minutes after the Big Bang (Section 6.3). Assuming the same kind of temperature dependence with redshift in power-law evolution, nucleosynthesis could not have started until ~ 45 years later if the conditions changed so slowly. But with a lifetime of only ~ 15 minutes, most of the neutrons would have decayed, mitigating any possibility of nuclear burning to occur. There are many differences that emerge with BBN in terms of whether or not the expansion was decelerated at the beginning, but the short neutron lifetime is without question the major hurdle.

The debate concerning whether BBN in power-law cosmologies can ever conform with the observations has broadened considerably since those early studies. For example, it has been pointed out that the correct ^{4}He abundance can be obtained with a constant expansion rate if the baryon density were much higher [470, 471] than the value ($\Omega_b = 0.0493 \pm 0.0002$ [67.4 km s^{-1} Mpc$^{-1}/H_0$]^2; see Section 6.1) optimized by *Planck* [29]. But in order to produce the correct ^{4}He yield with steady expansion, Ω_b would have to be at least ~ 15 times larger than this, which does not seem likely—

not to mention that the other light-element abundances would still disagree with the data.

Alternative approaches for circumventing the short neutron lifetime have used modifications to the basic premise of homogeneity or isotropy. For example, it has recently been suggested that a linear expansion with an inhomogeneous baryon density can effectively produce the observed abundances while reducing the required average baryon density to acceptable levels [472]. The reason for this is that neutrons can diffuse easily from high-density regions to lower-densities, rendering the former neutron-poor and proton-rich, while the low-density regions become neutron-rich and proton-poor. This effect can substantially alter the BBN abundance yields, even with the standard value of Ω_b. Ironically, this mechanism has the residual problem that the corresponding deuterium and lithium yields are very small, thus avoiding the lithium anomaly altogether, but then requiring a modification to the creation of the light elements.

Today, the most interesting and potentially fruitful approach to developing BBN in a constantly expanding cosmology was actually suggested quite early in this discussion [473], but elaborated upon only recently [474]. In power-law expansion, weak interactions remain in thermal equilibrium as the temperature continues to drop orders of magnitude below the BBN scale in ΛCDM, so the inverse β-decay of the proton can sustain the nuclear burning for much longer than the nominal neutron lifetime. In other words, while most of the light elemental abundances in ΛCDM need to be produced before all the neutrons decay, much of the BBN yield in a linearly expanding cosmos is actually created over a much longer period of time, with a neutron population sustained via weak interactions.

Simple simulations completed thus far have shown that the ^{4}He yield is comparable to that in standard BBN, though requiring a somewhat larger value of the primordial baryon-to-photon ratio (Equation 6.48): $\sim 10^{-8}$ instead of $\sim 6 \times 10^{-10}$. At the same time, the fact that the nuclear burning is more quiescent, means that the ^{7}Li yield is smaller, closer to its measured value, though still a factor ~ 3 too large.

But such results can only hint at what might happen in an FLRW cosmology constrained by the zero active mass condition. The outcome may be better, or it may be worse—possibly even a 'disaster,' echoing a description from long ago. There are several contributing factors one must take into account for a full resolution of this problem. First, the cosmic fluid has a very different makeup in $R_h = ct$ compared to ΛCDM. As we saw in Section 13.2.2, dark energy must have been present from the beginning in this cosmology (Equation 13.67), and the radiation energy density (Equation 13.68) would have evolved as $(1 + z)^2$ rather than the standard $(1 + z)^4$. It is not difficult to convince oneself that in this expansion scenario, the BBN could therefore have lasted over 100 Myr, rather than 18 minutes, or even 45 years, as previously hypothesized for power-law evolution.

To adequately handle this process, one must therefore include a much larger number of reactions (> 130) than is typically included in standard BBN (~ 30). Even then, one must contend with missing physics. Since dark energy is dynamic, one cannot ignore the transfer of energy from the dark sector to the standard model particles, so there are additional interaction equations that can only be written down once extensions to the standard model of particle physics are well understood. Tackling this difficult problem is all quite daunting, but necessary. Progress will be slow, but will undoubtedly be spurred on by continued refinements to the FLRW cosmology with the $R_h = ct$ condition if the observations eventually rule out the basic ΛCDM without its implied equation-of-state.

14.4.2   N-Body Simulations

The linear perturbation theory described in Chapter 13 is perfectly adequate to handle large-scale structure formation on scales exceeding ~ 100 Mpc. At smaller distances, however, this process becomes non-linear due to strong gravitational effects and the influence of pressure gradients, particularly in the context of bound systems that form galaxies and clusters. To study the evolution of perturbations on such scales, sophisticated methods based on N-body simulations have been developed, often with the inclusion of hydrodynamics as well.

One of the better known algorithms for carrying out simulations such as this is GADGET-3 [475], which integrates gas physics into cosmological simulations using the smoothed-particle hydrodynamics (SPH) method. With this approach, one assigns gas particles a density depending on an (adaptive) smoothing length, a smoothing kernel and a sum of the neighboring particle masses. A corresponding pressure is assigned from these densities and the information is fed through the equations of motion. A sophisticated code such as this can also account for viscosity by altering the dynamics equations when two or more particles are approaching each other. The N-body component is quite standard now. Much of the ongoing progress with these simulations emphasizes the gas hydrodynamics (sometimes even magnetohydrodynamics).

Though extensive work has been carried out using N-body codes in the context of ΛCDM, essentially nothing has been done so far for an FLRW cosmology constrained by $R_h = ct$. This constitutes a major gap in our ability to properly interpret the data acquired by upcoming surveys, e.g., in the area of weak lensing. Unlike strong lensing (Section 12.5), in which a single foreground galaxy redirects light rays from a background source (typically a quasar), weak lensing arises from the passage of light through an inhomogeneous medium. All three of the upcoming dark-energy surveys we highlighted in Section 14.2.1, i.e., the Large Synoptic Survey Telescope, *Euclid* and the Wide Field Infrared Survey Telescope, will feature in their observational program a dedicated acquisition of billions of galaxy shapes to incorporate into the subsequent analysis of weak lensing in the CMB. With the current absence of adequate model predictions, however, these observations will have very little impact on FLRW constrained by $R_h = ct$.

A related phenomenon is known as 'redshift-space distortions'—often viewed as a contaminating effect on the measurement of cosmological redshifts—arising from the internal motions and gravitational fields in bound structures, such as galaxy clusters. The matter power spectrum shown in Figures 13.6 and 13.7 is extracted primarily from CMB and Ly-α forest measurements, but there is another category of observations that can contribute significantly to these plots, based on galaxy correlations within clusters. The difficulty here, however, is that the true cosmological effects need to be disentangled from the aforementioned redshift space distortions, which are heavily model dependent. One must therefore simulate these effects individually for each assumed cosmology, and the method of choice is a hybrid N-body/hydrodynamics code, such as GADGET-3.

These cluster surveys are also used to extract the baryon acoustic oscillation (BAO) length scale, from which one may then 'measure' the redshift-dependent Hubble parameter $H(z)$. The analysis carried out in Section 12.2 was based entirely on measurements of this quantity using cosmic chronometers, which provide model-independent expansion rates. But the possibility of expanding the $H(z)$ database with the inclusion of BAO observations is enormous, especially with the breadth and coverage of upcoming surveys. Unlike the cosmic-chronometer measurements, however, these values of $H(z)$ extracted from the BAO measurements are heavily model-dependent due to the reshift space distortions.

The present hurdle in carrying out N-body simulations for large-scale structure formation on all scales in FLRW constrained by $R_{\rm h} = ct$ is the missing physics already introduced in Section 14.4.1. Returning to Figure 13.3 and Equations (13.65), (13.66), (13.67) and (13.68), the hindrance is evidently our lack of knowledge concerning the transfer of energy from the dark sector to radiation (first) and matter (second). The evolution described phenomenologically in Section 13.2.1 would be consistent with a partial decay of dark energy into standard model particles (during the transition era $t_{\rm r} < t < t_{\rm m}$ in Figure 13.3), or perhaps even the complete decay of the less dominant species of dark matter particle, should the dark sector be associated with multiple components. Again, without knowing how to extend the standard model of particle physics, one is left with too much ambiguity. Nevertheless, much can be learned by carrying out exploratory simulations based on empirical or phenomonological treatments, analogously to the formulation shown in Equation (13.71).

14.4.3   CMB Power Spectrum Below One Degree

One of the most striking confirmations of the ΛCDM cosmology has been the remarkable agreement between the measured CMB anisotropies and the predicted angular power spectrum shown in Figure 6.7—at least on sub-degree scales, corresponding to $\ell > 30$. The situation at angular scales $\gg 1°$ is not so clear, however, and there are several strong indications that the theory starts to break down in this so-called Sachs-Wolfe regime (see, e.g., Figure 11.2).

Our current understanding of how these anisotropies are produced holds that those with $\ell > 30$ are primarily due to metric perturbations related to scalar fluctuations in the matter field (the so-called Sachs-Wolfe effect), whereas the anisotropies on sub-degree scales (including the striking peaks in the power spectrum) are due to gravity-driven *acoustic* oscillations that propagate as sound waves until they freeze out at recombination. The physical mechanism for producing the CMB anisotropies is therefore quite different above and below the one-degree scale: above it, the perturbations are almost fully dependent on the cosmology; below it, they rely heavily on the astrophysics of the radiation-baryon fluid, especially its restoring forces that tightly constrain the sound speed. This dichotomy is the principal reason for viewing the angular correlation function of the CMB (Figure 11.2) as the better filter of cosmological models, while the power spectrum (Figure 6.7)—particularly at $\ell > 30$—is a powerful probe of the cosmic fluid's equation-of-state from the time the acoustic waves are produced to the era of recombination.

Calculating the angular correlation function in the context of FLRW constrained by $R_h = ct$ is thus straightforward, as we have seen in Section 11.1. This essentially amounts to permitting the power spectrum cutoff, k_{min}, to be non-zero. As such, results like those plotted in Figure 11.2 are easy to interpret: insofar as the Sachs-Wolfe anisotropies are concerned, slow-roll inflationary ΛCDM does not fare as well as the FLRW expansion with the zero active mass condition. But this leaves open the very crucial question of how the latter performs on sub-degree scales, and here too one must contend with the issue of missing physics.

The location of the peaks in the power spectrum is determined by the acoustic horizon at last scattering (Equation 6.69), the maximum comoving distance traversed by sound waves prior to recombination. This quantity, however, depends critically on how the sound speed, c_s, evolved with redshift prior to that time. In ΛCDM, the early Universe was dominated by radiation, but also contained some matter as well, and the strong coupling between photons, electrons and baryons typically leads to the estimate for c_s given in Equation (6.70). With only a tiny fraction of the energy density in baryonic form, however, the sound speed in the early ΛCDM Universe effectively reduced to that of a relativistic fluid, $c_s \to c/\sqrt{3}$, given that $\rho_b/\rho_r \to 0$.

But in an FLRW cosmology constrained by $R_h = ct$, the early Universe contained a significant concentration of dark energy, in addition to radiation and matter, which could not have been in the form of a cosmological constant. The sound speed could therefore not have been as straightforward as that indicated in Equation (6.70), with the simple relativistic limit $c_s \approx c/\sqrt{3}$. From Section 13.2.2, one expects that ρ_r/ρ_m was a decreasing function of t. In addition, ρ_r would have always been a small fraction of ρ, but in order for the constant equation-of-state $p = -\rho/3$ to be maintained, all three constituents—matter, radiation and dark energy—must have remained coupled during the acoustically important epoch, i.e., in the region $t \lesssim t_m$ in Figure 13.3.

One may thus write

$$c_s^2 = \left(+\frac{1}{3}\right)\frac{\partial \rho_r}{\partial \rho} + \frac{\partial p_{de}}{\partial \rho_{de}}\frac{\partial \rho_{de}}{\partial \rho} , \qquad (14.13)$$

under the reasonable assumption that $p_m \approx 0$ at all times. It is already clear that $\partial \rho_r/\partial \rho \leq 0.2$. Thus, the overall sound speed in the cosmic fluid is heavily influenced by the sound speed of dark energy, and c_s may or may not be much smaller than $c/\sqrt{3}$ in the early FLRW universe constrained by $R_h = ct$, depending on the particulate properties of the dark sector. The essential physics is, however, unfortunately still missing.

Perhaps a useful intermediate step may be to take the following phenomenological approach. One could estimate the sound speed by assuming for simplicity that

$$c_{\text{s}}(t) = c_{\text{s}}(t_*) \left(\frac{t_*}{t} \right)^{\beta} , \tag{14.14}$$

where t_* is the time at which the acoustic wave was produced and the index β is positive in order to reflect the decreasing importance of radiation with time. The acoustic horizon would then be given by the following modification to Equation (6.69):

$$r_{\text{s}}^{R_{\text{h}}=ct}(t_{\text{dec}}) = c_{\text{s}}(t_*) t_0 t_*^{\beta} \int_{t_*}^{t_{\text{dec}}} \frac{dt'}{(t')^{1+\beta}} . \tag{14.15}$$

Thus, as long as $t_{\text{dec}} \gg t_*$,

$$r_{\text{s}}^{R_{\text{h}}=ct} = \frac{c_{\text{s}}(t_*) t_0}{\beta} = \frac{R_{\text{h}}(t_0)}{\beta} \frac{c_{\text{s}}(t_*)}{c} , \tag{14.16}$$

and so

$$\frac{c_{\text{s}}(t_*)}{c} = \beta \frac{r_{\text{s}}^{R_{\text{h}}=ct}}{R_{\text{h}}(t_0)} . \tag{14.17}$$

Future work with this parametrized approach may yield several possible conclusions. For example, it could demonstrate, albeit qualitatively rather than quantitatively, whether a sound speed modified by the inclusion of dark energy (Equation 14.17) could nonetheless still account for the remarkably precise multi-peaked structure in the angular power spectrum shown in Figure 6.7. Fitting the CMB data may also constrain β and offer valuable clues concerning the unknown equation-of-state of this dynamical dark energy.

14.5 CONCLUDING REMARKS

Though our generation may not be the only one to justifiably anticipate a paradigm shift in cosmology, we may be entitled to harbor such lofty expectations for an evolution in our thinking over just a matter of years. We are poised to benefit considerably from an ambitious program of new ground-based and satellite facilities, offering state-of-the-art sensitivity and extensive monotoring capability. The Square Kilometre Array, for example, should deliver new data at a rate equivalent to the current world-wide web each and every day.

Those of us who strongly believe in the underlying correctness of Einstein's theory of gravity seek to understand the physical basis for the Friedmann-Lemaître-Robertson-Walker metric as fundamentally as possible. This has certainly been the principal goal of this book, and we have documented the breathtaking advances made in this endeavor over the past two or three decades. More importantly, we have also charted the path this work will take going forward.

But cosmologists also count among their ranks a meaningful minority of 'Einstein apostates,' those who firmly believe that our current version of the general theory of relativity applied to gravity is either incomplete, or simply wrong. A deeper under-standing into the nature of dark energy and matter would certainly help to alleviate existing discordance, and one cannot help but believe that the number of break-throughs resulting from such an explosive expansion in our information gathering technology will be enormous.

Though the upcoming measurement of real-time redshift drift may not be the only highly anticipated result of this work, for obvious reasons it will surely be its most impactful. This will essentially constitute a binary result—either zero or non-zero. A measurement of zero redshift drift will simultaneously solve myriad long-standing puzzles in cosmology and eliminate all the horizon problems that plague it—and inflation along with it. The stakes could not be higher.

We may then finally proceed with confidence to probe the Universe on sub-Planckian scales, quantitatively assessing various ideas for its origin and evolu-tion. Spatial curvature is a measure of the local expansion and gravitational energy densities—and flatness is an indicator that their sum is zero. Could the Big Bang therefore have 'simply' been a quantum event in vacuum, separating negative and positive energies in equal amounts and thereby initiating an expansion to recover the steady state? What could be more thoroughly pleasing than finally being able to answer such questions?

Bibliography

[1] R. Penrose. Gravitational Collapse and Spacetime Singularities. *Physical Review Letters*, 14(3):57–59, January 1965.

[2] S. W. Hawking and R. Penrose. *The Nature of Space and Time*. Princeton University Press, 1996.

[3] C. Blacker and M. Loewe. *Ancient Cosmologies*. George Allen and Unwin, 1975.

[4] V. Bourke. *The Essential Augustine* . Hackett Publishing Company, Indianapolis, 1964.

[5] Lucretius. *On the Nature of Things*. Prometheus Books, 1997.

[6] G. Freudenthal. Chemical Foundations for Cosmological Ideas: Ibn Sina on the Geology of an Eternal World. In Sabetai Unguru, editor, *Physics, Cosmology and Astronomy 1300-1700: Tensions and Accommodation*, pages 47–73, 1991.

[7] J. Barnes. *The Complete Works of Aristotle : the Revised Oxford Translation*. Princeton University Press, Princeton, N.J, 1984.

[8] B. Leftow. Why Didn't God Create the World Sooner? *Religious Studies*, 27(2):157–172, 1991.

[9] H. Bondi and T. Gold. The Steady-State Theory of the Expanding Universe. *Monthly Notices of the Royal Astronomical Society*, 108:252, January 1948.

[10] F. Hoyle. A New Model for the Expanding Universe. *Monthly Notices of the Royal Astronomical Society*, 108:372, January 1948.

[11] F. Hoyle. The Universe - Past and Present Reflections. *Annual Reviews of Astronomy and Astrophysics*, 20:1, January 1982.

[12] A. Einstein. Kosmologische Betrachtungen zur Allgemeinen Relativitätstheorie. *Sitzungsberichte der Königlich Preußischen Akademie der Wissenschaften (Berlin)*, pages 142–152, January 1917.

[13] H. Kragh. *Entropic Creation: Religious Contexts of Thermodynamics and Cosmology*. Ashgate Publishing, 2008.

[14] A. Friedmann. Über die Krümmung des Raumes. *Zeitschrift fur Physik*, 10:377–386, January 1922.

[15] G. Lemaître. The Beginning of the World from the Point of View of Quantum Theory. *Nature*, 127(3210):706, May 1931.

[16] G. Lemaître. L'Expansion de l'Espace. *Publications du Laboratoire d'Astronomie et de Geodesie de l'Universite de Louvain*, 8:101–120, January 1931.

[17] S. Chandrasekhar and Louis R. Henrich. An Attempt to Interpret the Relative Abundances of the Elements and Their Isotopes. *The Astrophysical Journal*, 95:288, March 1942.

[18] G. Gamow. *The Creation of the Universe*. Viking Press, 1952.

[19] R. A. Alpher, R. Herman, and G. A. Gamow. Thermonuclear Reactions in the Expanding Universe. *Physical Review*, 74(9):1198–1199, November 1948.

[20] P.J.E. Peebles. Discovery of the Hot Big Bang: What Happened in 1948. *European Physical Journal H*, 39(2):205–223, April 2014.

[21] R. A. Alpher and R. Herman. Evolution of the Universe. *Nature*, 162(4124):774–775, November 1948.

[22] A. A. Penzias and R. W. Wilson. A Measurement of Excess Antenna Temperature at 4080 Mc/s. *The Astrophysical Journal*, 142:419–421, July 1965.

[23] P. G. Roll and David T. Wilkinson. Cosmic Background Radiation at 3.2 cm—Support for Cosmic Blackbody Radiation. *Physical Review Letters*, 16(10):405–407, March 1966.

[24] P.J.E. Peebles. *Physical Cosmology*. Princeton University Press, 1971.

[25] R. H. Dicke. A Scientific Autobiography. Unpublished manuscript on file at the Membership office of the National Academy of Sciences, 1875.

[26] E. L. Wright, C. L. Bennett, K. Gorski, G. Hinshaw, and G. F. Smoot. Angular Power Spectrum of the Cosmic Microwave Background Anisotropy seen by the COBE DMR. *The Astrophysical Journal Letters*, 464:L21, June 1996.

[27] G. Hinshaw, A. J. Branday, C. L. Bennett, K. M. Gorski, A. Kogut, C. H. Lineweaver, G. F. Smoot, and E. L. Wright. Two-Point Correlations in the COBE DMR Four-year Anisotropy Maps. *The Astrophysical Journal Letters*, 464:L25, June 1996.

[28] D. N. Spergel, L. Verde, H. V. Peiris, E. Komatsu, M. R. Nolta, C. L. Bennett, M. Halpern, G. Hinshaw, N. Jarosik, A. Kogut, M. Limon, S. S. Meyer, L. Page, G. S. Tucker, J. L. Weiland, E. Wollack, and E. L. Wright. First-Year Wilkinson Microwave Anisotropy Probe (WMAP) Observations: Determination of Cosmological Parameters. *The Astrophysical Journal Supplement Series*, 148(1):175–194, September 2003.

[29] Planck Collaboration, N. Aghanim, Y. Akrami, M. Ashdown, J. Aumont, C. Baccigalupi, M. Ballardini, A. J. Banday, R. B. Barreiro, N. Bartolo, S. Basak, et al. Planck 2018 Results. VI. Cosmological Parameters. *arXiv e-prints*, page arXiv:1807.06209, July 2018.

[30] A. Kosowsky. The Atacama Cosmology Telescope. *New Astronomy Reviews*, 47(11-12):939–943, December 2003.

[31] J. E. Carlstrom, P.A.R. Ade, K. A. Aird, B. A. Benson, L. E. Bleem, S. Busetti, C. L. Chang, E. Chauvin, H. M. Cho, T. M. Crawford, et al. The 10-Meter South Pole Telescope. *Publications of the Astronomical Society of the Pacific*, 123(903):568, May 2011.

[32] A. A. Klypin, S. Trujillo-Gomez, and J. Primack. Dark Matter Halos in the Standard Cosmological Model: Results from the Bolshoi Simulation. *The Astrophysical Journal*, 740(2):102, October 2011.

[33] V. Springel, S.D.M. White, A. Jenkins, C. S. Frenk, N. Yoshida, L. Gao, J. Navarro, R. Thacker, D. Croton, J. Helly, et al. Simulations of the Formation, Evolution and Clustering of Galaxies and Quasars. *Nature*, 435(7042):629–636, June 2005.

[34] F. Zwicky. Die Rotverschiebung von Extragalaktischen Nebeln. *Helvetica Physica Acta*, 6:110–127, January 1933.

[35] F. Zwicky. Nebulae as Gravitational Lenses. *Physical Review*, 51(4):290–290, February 1937.

[36] SDSS Collaboration, P. Fischer, T. A. McKay, E. Sheldon, A. Connolly, A. Stebbins, J. A. Frieman, B. Jain, M. Joffre, D. Johnston, G. Bernstein, et al. Weak Lensing with Sloan Digital Sky Survey Commissioning Data: The Galaxy-Mass Correlation Function to 1 h^{-1} Mpc. *The Astronomical Journal*, 120(3):1198–1208, September 2000.

[37] D. Clowe, G. A. Luppino, N. Kaiser, and I. M. Gioia. Weak Lensing by High-redshift Clusters of Galaxies. I. Cluster Mass Reconstruction. *The Astrophysical Journal*, 539(2):540–560, August 2000.

[38] L. Van Waerbeke, Y. Mellier, M. Radovich, E. Bertin, M. Dantel-Fort, H. J. McCracken, O. Le Fèvre, S. Foucaud, J. C. Cuillandre, T. Erben, et al. Cosmic Shear Statistics and Cosmology. *Astronomy and Astrophysics*, 374:757–769, August 2001.

[39] A. G. Riess, A. V. Filippenko, P. Challis, A. Clocchiatti, A. Diercks, P. M. Garnavich, R. L. Gilliland, C. J. Hogan, S. Jha, R. P. Kirshner, et al. Observational Evidence from Supernovae for an Accelerating Universe and a Cosmological Constant. *The Astronomical Journal*, 116(3):1009–1038, September 1998.

[40] B. P. Schmidt, N. B. Suntzeff, M. M. Phillips, R. A. Schommer, A. Clocchiatti, R. P. Kirshner, P. Garnavich, P. Challis, B. Leibundgut, J. Spyromilio, et al. The High-z Supernova Search: Measuring Cosmic Deceleration and Global Curvature of the Universe Using Type Ia Supernovae. *The Astrophysical Journal*, 507(1):46–63, November 1998.

[41] The Supernova Cosmology Project, S. Perlmutter, G. Aldering, G. Goldhaber, R. A. Knop, P. Nugent, P. G. Castro, S. Deustua, S. Fabbro, A. Goobar, D. E. Groom, et al. Measurements of Ω and Λ from 42 High-redshift Supernovae. *The Astrophysical Journal*, 517(2):565–586, June 1999.

[42] S. Weinberg. The Cosmological Constant Problem. *Reviews of Modern Physics*, 61(1):1–23, January 1989.

[43] K. Rudnicki. *The Cosmological Principles*. Jagiellonian University, Krakow, 1995.

[44] J. J. Condon, W. D. Cotton, E. W. Greisen, Q. F. Yin, R. A. Perley, G. B. Taylor, and J. J. Broderick. The NRAO VLA Sky Survey. *The Astronomical Journal*, 115(5):1693–1716, May 1998.

[45] C.A.P. Bengaly, R. Maartens, N. Randriamiarinarivo, and A. Baloyi. Testing the Cosmological Principle in the Radio Sky. *Journal of Cosmology and Astroparticle Physics*, 2019(9):025, September 2019.

[46] A. G. Walker. Completely Symmetric Spaces. *Journal of the London Mathematical Society*, 19:219–226, October 1944.

[47] W.K.C. Guthrie. *A History of Greek Philosophy*. Cambridge University Press, 1981.

[48] B. Russell. *The Principles of Mathematics*. W. W. Norton and Company, 1996.

[49] B. Schutz. *A First Course in General Relativity*. Cambridge University Press, 2009.

[50] F. Melia. *Cracking the Einstein Code*. University of Chicago Press, 2009.

[51] K. Schwarzschild. Wikipedia, the free encyclopedia [online; accessed 27-april-2020] — https://commons.wikimedia.org/wiki/file:schwarzschild.jpg, 2015.

[52] R. P. Kerr. Gravitational Field of a Spinning Mass as an Example of Algebraically Special Metrics. *Physical Review Letters*, 11(5):237–238, September 1963.

[53] R. H. Boyer and R. W. Lindquist. Maximal Analytic Extension of the Kerr Metric. *Journal of Mathematical Physics*, 8(2):265–281, February 1967.

[54] H. Weyl. Zur Allgemeinen Relativitätstheorie. *General Relativity and Gravitation: Republication*, 41(7):1661–1666, July 2009.

[55] H. P. Robertson. Kinematics and World Structure. *The Astrophysical Journal*, 82:284, November 1935.

[56] A. G. Walker. On Milne's Theory of World Structure. *Proceedings of the London Mathematical Society*, 42:90–127, January 1937.

[57] A. Raychaudhuri. Relativistic Cosmology. I. *Physical Review*, 98(4):1123–1126, May 1955.

[58] A. Friedmann. Über die Möglichkeit einer Welt mit Konstanter Negativer Krümmung des Raumes. *Zeitschrift fur Physik*, 21(1):326–332, December 1924.

[59] E. Hubble. A Relation between Distance and Radial Velocity among Extragalactic Nebulae. *Proceedings of the National Academy of Science*, 15(3):168–173, March 1929.

[60] V. A. Fock. Special Issue: the Researches of A. A. Fridman on the Einstein Theory of Gravitation. *Soviet Physics Uspekhi*, 6(4):473–474, April 1964.

[61] E. A. Tropp, V. Y. Frenkel, and A. D. Chernin. *Friedmann: The Man who Made the Universe Expand*. Cambridge University Press, 1993.

[62] A. Einstein. Bemerkung zu der Arbeit von A. Friedmann Über die Krummung des Raumes. *Zeitschrift fur Physik*, 11(1):326–326, December 1922.

[63] N. S. Hetherington. The Delayed Response of Suggestions of an Expanding Universe. *Journal of the British Astronomical Association*, 84:22–28, January 1973.

[64] P. Kerzberg. *The Invented Universe: The Einstein-de Sitter Controversy (1916-1917) and the Rise of Relativistic Cosmology*. Clarendon Press, 1922.

[65] G. Lemaître. Un Univers Homogène de Masse Constante et de Rayon Croissant Rendant Compte de la Vitesse Radiale des Nébuleuses Extra-galactiques. *Annales de la Société Scientifique de Bruxelles*, 47:49–59, January 1927.

[66] S. Singh. *Big Bang*. Harper Perennial, 2005.

[67] A. Einstein. Zum Kosmologischen Problem der Allgemeinen Relativitätstheorie. *Sitzungsberichte der Königlich Preußischen Akademie der Wissenschaften (Berlin)*, pages 235–237, 1931.

[68] G. D. Birkhoff. *Relativity and Modern Physics*. Cambridge University Press, 1923.

[69] S. Weinberg. *Gravitation and Cosmology: Principles and Applications of the General Theory of Relativity*. Wiley, 1972.

[70] W. de Sitter. Einstein's Theory of Gravitation and its Astronomical Consequences. Third Paper. *Monthly Notices of the Royal Astronomical Society*, 78:3–28, November 1917.

[71] W. de Sitter. On the Relativity of Inertia. Remarks Concerning Einstein's Latest Hypothesis. *Koninklijke Nederlandse Akademie van Wetenschappen Proceedings Series B Physical Sciences*, 19:1217–1225, March 1917.

[72] E. A. Milne. Cosmological Theories. *The Astrophysical Journal*, 91:129, March 1940.

[73] A. Einstein and W. de Sitter. On the Relation between the Expansion and the Mean Density of the Universe. *Proceedings of the National Academy of Science*, 18(3):213–214, March 1932.

[74] P. S. Florides. The Robertson-Walker Metrics Expressible in Static Form. *General Relativity and Gravitation*, 12(7):563–574, July 1980.

[75] K. Lanczos. Über eine Stationäre Kosmologie im Sinne der Einsteinschen Gravitationstheorie. *Zeitschrift fur Physik*, 21(1):73–110, December 1924.

[76] J. P. Ostriker and Paul J. Steinhardt. The Observational Case for a Low-density Universe with a Non-zero Cosmological Constant. *Nature*, 377(6550):600–602, October 1995.

[77] J. Ussher. *Annales Veteris Testamenti, a Prima Mundi Origine Deducti, una cum Rerum Asiaticarum et Aegyptiacarum Chronico, a Temporis Historici Principio Usque ad Maccabaicorum Initia Producto.* EEBO Editions ProQuest, 2010.

[78] S. Borsanyi, Z. Fodor, J. Guenther, K. H. Kampert, S. D. Katz, T. Kawanai, T. G. Kovacs, S. W. Mages, A. Pasztor, F. Pittler, et al. Calculation of the Axion Mass Based on High-temperature Lattice Quantum Chromodynamics. *Nature*, 539(7627):69–71, November 2016.

[79] M. Laine and M. Meyer. Standard Model Thermodynamics Across the Electroweak Crossover. *Journal of Cosmology and Astroparticle Physics*, 2015(7):035, July 2015.

[80] G. Mangano, G. Miele, S. Pastor, and M. Peloso. A Precision Calculation of the Effective Number of Cosmological Neutrinos. *Physics Letters B*, 534(1-4):8–16, May 2002.

[81] Super-Kamiokande Collaboration, Y. Ashie, J. Hosaka, K. Ishihara, Y. Itow, J. Kameda, Y. Koshio, A. Minamino, C. Mitsuda, M. Miura, S. Moriyama, et al. Measurement of Atmospheric Neutrino Oscillation Parameters by Super-Kamiokande I. *Physical Review D*, 71(11):112005, June 2005.

[82] A. D. Sakharov. Violation of CP Invariance, C Asymmetry, and Baryon Asymmetry of the Universe. *ZhETF Pisma Redaktsiiu*, 5:32, January 1967.

[83] S. Weinberg. Cosmological Production of Baryons. *Physical Review Letters*, 42(13):850–853, March 1979.

[84] I. Affleck and M. Dine. A New Mechanism for Baryogenesis. *Nuclear Physics B*, 2:361–380, January 1985.

[85] J. M. Cline. Is Electroweak Baryogenesis Dead? *Philosophical Transactions of the Royal Society of London Series A*, 376(2114):20170116, January 2018.

[86] G. Servant. The Serendipity of Electroweak Baryogenesis. *Philosophical Transactions of the Royal Society of London Series A*, 376(2114):20170124, January 2018.

[87] Atlas Collaboration, G. Aad, T. Abajyan, B. Abbott, J. Abdallah, S. Abdel Khalek, A. A. Abdelalim, O. Abdinov, R. Aben, B. Abi, M. Abolins, et al. Observation of a New Particle in the Search for the Standard Model Higgs Boson with the ATLAS Detector at the LHC. *Physics Letters B*, 716(1):1–29, September 2012.

[88] P. Peter and J.-P. Uzan. *Primordial Cosmology*. Oxford University Press, 2010.

[89] K. A. Olive and Particle Data Group. Review of Particle Physics. *Chinese Physics C*, 38(9):090001, August 2014.

[90] J. Bernstein, L. S. Brown, and G. Feinberg. Cosmological Helium Production Simplified. *Reviews of Modern Physics*, 61(1):25–39, January 1989.

[91] A. Coc, S. Goriely, Y. Xu, M. Saimpert, and E. Vangioni. Standard Big Bang Nucleosynthesis up to CNO with an Improved Extended Nuclear Network. *The Astrophysical Journal*, 744(2):158, January 2012.

[92] A. Coc, J.-P. Uzan, and E. Vangioni. Standard Big Bang Nucleosynthesis and Primordial CNO Abundances after Planck. *Journal of Cosmology and Astroparticle Physics*, 2014(10):050, October 2014.

[93] Y. I. Izotov, T. X. Thuan, and N. G. Guseva. A New Determination of the Primordial He Abundance Using the He I $\lambda 10830$ Å Emission line: Cosmological Implications. *Monthly Notices of the Royal Astronomical Society*, 445(1):778–793, November 2014.

[94] E. Aver, K. A. Olive, R. L. Porter, and E. D. Skillman. The Primordial Helium Abundance from Updated Emissivities. *Journal of Cosmology and Astroparticle Physics*, 2013(11):017, November 2013.

[95] T. M. Bania, Robert T. Rood, and Dana S. Balser. The Cosmological Density of Baryons from Observations of $^3He^+$ in the Milky Way. *Nature*, 415(6867):54–57, January 2002.

[96] M. Pettini and R. Cooke. A New, Precise Measurement of the Primordial Abundance of Deuterium. *Monthly Notices of the Royal Astronomical Society*, 425(4):2477–2486, October 2012.

[97] R. J. Cooke, M. Pettini, R. A. Jorgenson, M. T. Murphy, and C. C. Steidel. Precision Measures of the Primordial Abundance of Deuterium. *The Astrophysical Journal*, 781(1):31, January 2014.

[98] E. Rollinde, E. Vangioni, and K. Olive. Cosmological Cosmic Rays and the Observed ^{6}Li Plateau in Metal-poor Halo Stars. *The Astrophysical Journal*, 627(2):666–673, July 2005.

[99] A. Heger, E. Kolbe, W. C. Haxton, K. Langanke, G. Martinez-Pinedo, and S. E. Woosley. Neutrino Nucleosynthesis. *Physics Letters B*, 606(3-4):258–264, January 2005.

[100] A.G.W. Cameron and W. A. Fowler. Lithium and the s-PROCESS in Red-Giant Stars. *The Astrophysical Journal*, 164:111, February 1971.

[101] F. Spite and M. Spite. Abundance of Lithium in Unevolved Stars and Old Disk Stars: Interpretation and Consequences. *Astronomy and Astrophysics*, 115:357–366, November 1982.

[102] L. Sbordone, P. Bonifacio, E. Caffau, H. G. Ludwig, N. T. Behara, J. I. González Hernández, M. Steffen, R. Cayrel, B. Freytag, et al. The Metal-poor End of the Spite Plateau. I. Stellar Parameters, Metallicities, and Lithium Abundances. *Astronomy and Astrophysics*, 522:A26, November 2010.

[103] F. Iocco. The Lithium Problem. A Phenomenologist's Perspective. *Memorie della Societa Astronomica Italiana Supplementi*, 22:19, January 2012.

[104] M. N. Saha. On a Physical Theory of Stellar Spectra. *Proceedings of the Royal Society of London Series A*, 99(697):135–153, May 1921.

[105] J. C. Mather, E. S. Cheng, D. A. Cottingham, Jr. Eplee, R. E., D. J. Fixsen, T. Hewagama, R. B. Isaacman, K. A. Jensen, S. S. Meyer, P. D. Noerdlinger, et al. Measurement of the Cosmic Microwave Background Spectrum by the COBE FIRAS Instrument. *The Astrophysical Journal*, 420:439, January 1994.

[106] C. L. Bennett, M. Halpern, G. Hinshaw, N. Jarosik, A. Kogut, M. Limon, S. S. Meyer, L. Page, D. N. Spergel, G. S. Tucker, et al. First-Year Wilkinson Microwave Anisotropy Probe (WMAP) Observations: Preliminary Maps and Basic Results. *The Astrophysical Journal Supplement Series*, 148(1):1–27, September 2003.

[107] S. Muller, A. Beelen, J. H. Black, S. J. Curran, C. Horellou, S. Aalto, F. Combes, M. Guélin, and C. Henkel. A Precise and Accurate Determination of the Cosmic Microwave Background Temperature at z = 0.89. *Astronomy and Astrophysics*, 551:A109, March 2013.

[108] P. Noterdaeme, P. Petitjean, R. Srianand, C. Ledoux, and S. López. The Evolution of the Cosmic Microwave Background Temperature. Measurements of T_{CMB} at High Redshift from Carbon Monoxide Excitation. *Astronomy and Astrophysics*, 526:L7, February 2011.

[109] G. Luzzi, M. Shimon, L. Lamagna, Y. Rephaeli, M. De Petris, A. Conte, S. De Gregori, and E. S. Battistelli. Redshift Dependence of the Cosmic Microwave Background Temperature from Sunyaev-Zeldovich Measurements. *The Astrophysical Journal*, 705(2):1122–1128, November 2009.

[110] J. Ge, J. Bechtold, and J. H. Black. A New Measurement of the Cosmic Microwave Background Radiation Temperature at z = 1.97. *The Astrophysical Journal*, 474(1):67–73, January 1997.

[111] D. S. Swetz, P. A. R. Ade, M. Amiri, J. W. Appel, E. S. Battistelli, B. Burger, J. Chervenak, M. J. Devlin, S. R. Dicker, W. B. Doriese, et al. Overview of the Atacama Cosmology Telescope: Receiver, Instrumentation, and Telescope Systems. *The Astrophysical Journal Supplement Series*, 194(2):41, June 2011.

[112] D. J. Fixsen, E. S. Cheng, D. A. Cottingham, Jr. Eplee, R. E., T. Hewagama, R. B. Isaacman, K. A. Jensen, J. C. Mather, D. L. Massa, S. S. Meyer, et al. Calibration of the COBE FIRAS Instrument. *The Astrophysical Journal*, 420:457, January 1994.

[113] A. Meiksin, M. White, and J. A. Peacock. Baryonic Signatures in Large-scale Structure. *Monthly Notices of the Royal Astronomical Society*, 304(4):851–864, April 1999.

[114] H.-J. Seo and D. J. Eisenstein. Baryonic Acoustic Oscillations in Simulated Galaxy Redshift Surveys. *The Astrophysical Journal*, 633(2):575–588, November 2005.

[115] D. Jeong and E. Komatsu. Perturbation Theory Reloaded: Analytical Calculation of Nonlinearity in Baryonic Oscillations in the Real-Space Matter Power Spectrum. *The Astrophysical Journal*, 651(2):619–626, November 2006.

[116] M. Crocce and R. Scoccimarro. Memory of Initial Conditions in Gravitational Clustering. *Physical Review D*, 73(6):063520, March 2006.

[117] N. Padmanabhan and M. White. Calibrating the Baryon Oscillation Ruler for Matter and Halos. *Physical Review D*, 80(6):063508, September 2009.

[118] R. K. Sachs and A. M. Wolfe. Perturbations of a Cosmological Model and Angular Variations of the Microwave Background. *The Astrophysical Journal*, 147:73, January 1967.

[119] M. White, D. Scott, and J. Silk. Anisotropies in the Cosmic Microwave Background. *Annual Reviews of Astronomy and Astrophysics*, 32:319–370, January 1994.

[120] M. Doran and M. Lilley. The Location of Cosmic Microwave Background Peaks in a Universe with Dark Energy. *Monthly Notices of the Royal Astronomical Society*, 330(4):965–970, March 2002.

[121] L. Page, M. R. Nolta, C. Barnes, C. L. Bennett, M. Halpern, G. Hinshaw, N. Jarosik, A. Kogut, M. Limon, S. S. Meyer, et al. First-Year Wilkinson Microwave Anisotropy Probe (WMAP) Observations: Interpretation of the TT and TE Angular Power Spectrum Peaks. *The Astrophysical Journal Supplement Series*, 148(1):233–241, September 2003.

[122] D. Scott, J. Silk, and M. White. From Microwave Anisotropies to Cosmology. *Science*, 268(5212):829–835, May 1995.

[123] A. A. Starobinsky. Spectrum of Relict Gravitational Radiation and the Early State of the Universe. *Soviet Journal of Experimental and Theoretical Physics Letters*, 30:682, December 1979.

[124] D. Kazanas. Dynamics of the Universe and Spontaneous Symmetry Breaking. *The Astrophysical Journal Letters*, 241:L59–L63, October 1980.

[125] A. H. Guth. Inflationary Universe: A Possible Solution to the Horizon and Flatness Problems. *Physical Review D*, 23(2):347–356, January 1981.

[126] A. D. Linde. A New Inflationary Universe Scenario: A Possible Solution of the Horizon, Flatness, Homogeneity, Isotropy and Primordial Monopole Problems. *Physics Letters B*, 108(6):389–393, February 1982.

[127] F. Melia. The $R_h = ct$ Universe Without Inflation. *Astronomy and Astrophysics*, 553:A76, May 2013.

[128] F. Englert and R. Brout. Broken Symmetry and the Mass of Gauge Vector Mesons. *Physical Review Letters*, 13(9):321–323, August 1964.

[129] P. W. Higgs. Broken Symmetries, Massless Particles and Gauge Fields. *Physics Letters*, 12(2):132–133, September 1964.

[130] LHCb Collaboration, R. Aaij, B. Adeva, M. Adinolfi, Z. Ajaltouni, S. Akar, J. Albrecht, F. Alessio, M. Alexander, S. Ali, G. Alkhazov, et al. Search for Long-lived Scalar Particles in $B^+ -> K^+ \chi (\mu^+ \mu^-)$ Decays. *Physical Review D*, 95(7):071101, April 2017.

[131] J. M. Bardeen. Gauge-invariant Cosmological Perturbations. *Physical Review D*, 22(8):1882–1905, October 1980.

[132] H. Kodama and M. Sasaki. Cosmological Perturbation Theory. *Progress of Theoretical Physics Supplement Series*, 78:1, January 1984.

[133] V. F. Mukhanov, H. A. Feldman, and R. H. Brandenberger. Theory of Cosmological Perturbations. *Physics Reports*, 215(5-6):203–333, June 1992.

[134] B. A. Bassett, S. Tsujikawa, and D. Wands. Inflation Dynamics and Reheating. *Reviews of Modern Physics*, 78(2):537–589, April 2006.

[135] M. Sasaki. Large Scale Quantum Fluctuations in the Inflationary Universe. *Progress of Theoretical Physics*, 76(5):1036–1046, November 1986.

[136] V. F. Mukhanov. The Quantum Theory of Gauge-invariant Cosmological Perturbations. *Zhurnal Eksperimentalnoi i Teoreticheskoi Fiziki*, 94:1–11, July 1988.

[137] T. S. Bunch and P.C.W. Davies. Quantum Field Theory in de Sitter Space - Renormalization by Point-splitting. *Proceedings of the Royal Society of London Series A*, 360(1700):117–134, March 1978.

[138] J. Martin and R. H. Brandenberger. Trans-Planckian Problem of Inflationary Cosmology. *Physical Review D*, 63(12):123501, June 2001.

[139] BICEP2 Collaboration, Keck Array Collaboration, P.A.R. Ade, Z. Ahmed, R. W. Aikin, K. D. Alexand er, D. Barkats, S. J. Benton, C. A. Bischoff, J. J. Bock, R. Bowens-Rubin, J. A. Brevik, et al. Improved Constraints on Cosmology and Foregrounds from BICEP2 and Keck Array Cosmic Microwave Background Data with Inclusion of 95 GHz Band. *Physical Review Letters*, 116(3):031302, January 2016.

[140] W. Rindler. Visual Horizons in World Models. *Monthly Notices of the Royal Astronomical Society*, 116:662, January 1956.

[141] G.F.R. Ellis and T. Rothman. Lost Horizons. *American Journal of Physics*, 61(10):883–893, October 1993.

[142] R. J. Nemiroff and B. Patla. Adventures in Friedmann Cosmology: A Detailed Expansion of the Cosmological Friedmann Equations. *American Journal of Physics*, 76(3):265–276, March 2008.

[143] V. Faraoni. *Cosmological and Black Hole Apparent Horizons*. Springer, 2015.

[144] R. Gautreau. Cosmological Schwarzschild Radii and Newtonian Gravitational Theory. *American Journal of Physics*, 64(12):1457–1467, December 1996.

[145] I. Ben-Dov. Outer Trapped Surfaces in Vaidya Spacetimes. *Physical Review D*, 75(6):064007, March 2007.

[146] V. Faraoni. Cosmological Apparent and Trapping Horizons. *Physical Review D*, 84(2):024003, July 2011.

[147] I. Bengtsson and J.M.M. Senovilla. Region with Trapped Surfaces in Spherical Symmetry, its Core, and their Boundaries. *Physical Review D*, 83(4):044012, February 2011.

[148] F. Melia. The Cosmic Horizon. *Monthly Notices of the Royal Astronomical Society*, 382(4):1917–1921, December 2007.

[149] T. M. Davis and C. H. Lineweaver. Expanding Confusion: Common Misconceptions of Cosmological Horizons and the Superluminal Expansion of the Universe. *Publications of the Astronomical Society of Australia*, 21(1):97–109, January 2004.

[150] P. van Oirschot, J. Kwan, and G. F. Lewis. Through the Looking Glass: Why the 'Cosmic Horizon' is not a Horizon. *Monthly Notices of the Royal Astronomical Society*, 404(4):1633–1638, June 2010.

[151] G. F. Lewis. Matter Matters: Unphysical Properties of the $R_h = ct$ Universe. *Monthly Notices of the Royal Astronomical Society*, 432(3):2324–2330, July 2013.

[152] D. Y. Kim, A. N. Lasenby, and M. P. Hobson. Spherically-symmetric Solutions in General Relativity Using a Tetrad-based Approach. *General Relativity and Gravitation*, 50(3):29, March 2018.

[153] O. Bikwa, F. Melia, and A. Shevchuk. Photon Geodesics in Friedmann-Robertson-Walker Cosmologies. *Monthly Notices of the Royal Astronomical Society*, 421(4):3356–3361, April 2012.

[154] F. Melia. The Gravitational Horizon for a Universe with Phantom Energy. *Journal of Cosmology and Astroparticle Physics*, 2012(9):029, September 2012.

[155] F. Melia. Proper Size of the Visible Universe in FRW Metrics with a Constant Spacetime Curvature. *Classical and Quantum Gravity*, 30(15):155007, August 2013.

[156] W. M. Stuckey. Can Galaxies Exist Within our Particle Horizon with Hubble Recessional Velocities Greater than c? *American Journal of Physics*, 60(2):142–146, February 1992.

[157] T. M. Davis and C. H. Lineweaver. Superluminal Recession Velocities. In Ruth Durrer, Juan Garcia-Bellido, and Mikhail Shaposhnikov, editors, *Cosmology and Particle Physics*, volume 555 of *American Institute of Physics Conference Series*, pages 348–351, February 2001.

[158] A. Kaya. Hubble's Law and Faster than Light Expansion Speeds. *American Journal of Physics*, 79(11):1151–1154, November 2011.

[159] F. Melia. *The Edge of Infinity–Supermassive Black Holes in the Universe*. Cambridge University Press, 2003.

[160] F. Melia. Physical Basis for the Symmetries in the Friedmann-Robertson-Walker Metric. *Frontiers of Physics*, 11(4):119801, August 2016.

[161] F. Melia. The Zero Active Mass Condition in Friedmann-Robertson-Walker Cosmologies. *Frontiers of Physics*, 12(1):129802, February 2017.

[162] F. Melia and M. Abdelqader. The Cosmological Spacetime. *International Journal of Modern Physics D*, 18(12):1889–1901, January 2009.

[163] F. Melia and A.S.H. Shevchuk. The R_h=ct Universe. *Monthly Notices of the Royal Astronomical Society*, 419(3):2579–2586, January 2012.

[164] E. R. Harrison. Mining Energy in an Expanding Universe. *The Astrophysical Journal*, 446:63, June 1995.

[165] M. J. Chodorowski. A Direct Consequence of the Expansion of Space? *Monthly Notices of the Royal Astronomical Society*, 378(1):239–244, June 2007.

[166] M. J. Chodorowski. The Kinematic Component of the Cosmological Redshift. *Monthly Notices of the Royal Astronomical Society*, 413(1):585–594, May 2011.

[167] Y. Baryshev. Problems of Practical Cosmology. In Yu Baryshev, Igor N. Taganov, and Pekka Teerikorpi, editors, *Problems of Practical Cosmology, Volume 1*, pages 60–67, Russian Geographical Society, St. Petersburg, January 2008.

[168] E. F. Bunn and D. W. Hogg. The Kinematic Origin of the Cosmological Redshift. *American Journal of Physics*, 77(8):688–694, August 2009.

[169] R. J. Cook and M. S. Burns. Interpretation of the Cosmological Metric. *American Journal of Physics*, 77(1):59–66, January 2009.

[170] Ø. Grøn and Ø. Elgarøy. Is Space Expanding in the Friedmann Universe Models? *American Journal of Physics*, 75(2):151–157, February 2007.

[171] Planck Collaboration, P.A.R. Ade, N. Aghanim, M. Arnaud, M. Ashdown, J. Aumont, C. Baccigalupi, A. J. Banday, R. B. Barreiro, J. G. Bartlett, N. Bartolo, et al. Planck 2015 Results. XIII. Cosmological Parameters. *Astronomy and Astrophysics*, 594:A13, September 2016.

[172] J. T. Jebsen. On the General Spherically Symmetric Solutions of Einstein's Gravitational Equations in Vacuo. *Arkiv for Matematik, Astronomi och Fysik*, 15(18):18, January 1921.

[173] C. W. Misner and D. H. Sharp. Relativistic Equations for Adiabatic, Spherically Symmetric Gravitational Collapse. *Physical Review*, 136(2B):571–576, October 1964.

[174] Jr. Hernandez, W. C. and C. W. Misner. Observer Time as a Coordinate in Relativistic Spherical Hydrodynamics. *The Astrophysical Journal*, 143:452, February 1966.

[175] A. Prain, V. Vitagliano, V. Faraoni, and M. Lapierre-Léonard. Hawking-Hayward Quasi-local Energy under Conformal Transformations. *Classical and Quantum Gravity*, 33(14):145008, July 2016.

[176] S. Chakraborty and N. Dadhich. Brown-York Quasilocal Energy in Lanczos-Lovelock Gravity and Black Hole Horizons. *Journal of High Energy Physics*, 2015:3, December 2015.

[177] Planck Collaboration, P.A.R. Ade, N. Aghanim, C. Armitage-Caplan, M. Arnaud, M. Ashdown, F. Atrio-Barand ela, J. Aumont, C. Baccigalupi, A. J. Banday, R. B. Barreiro, et al. Planck 2013 Results. XXIII. Isotropy and Statistics of the CMB. *Astronomy and Astrophysics*, 571:A23, November 2014.

[178] A. B. Nielsen and M. Visser. Production and Decay of Evolving Horizons. *Classical and Quantum Gravity*, 23(14):4637–4658, July 2006.

[179] G. Abreu and M. Visser. Kodama Time: Geometrically Preferred Foliations of Spherically Symmetric Spacetimes. *Physical Review D*, 82(4):044027, August 2010.

[180] J. R. Oppenheimer and G. M. Volkoff. On Massive Neutron Cores. *Physical Review*, 55(4):374–381, February 1939.

[181] F. Melia. The Apparent (Gravitational) Horizon in Cosmology. *American Journal of Physics*, 86(8):585–593, August 2018.

[182] G. F. Lewis and P. van Oirschot. How Does the Hubble Sphere Limit our View of the Universe? *Monthly Notices of the Royal Astronomical Society*, 423(1):L26–L29, June 2012.

[183] R. R. Caldwell. A Phantom Menace? Cosmological Consequences of a Dark Energy Component with Super-negative Equation of State. *Physics Letters B*, 545(1-2):23–29, October 2002.

[184] R. R. Caldwell, M. Kamionkowski, and N. N. Weinberg. Phantom Energy: Dark Energy with w<-1 Causes a Cosmic Doomsday. *Physical Review Letters*, 91(7):071301, August 2003.

[185] A. S. Eddington. *The Mathematical Theory of Relativity*. Cambridge University Press, 1924.

[186] C. Wirtz. Über die Bewegungen der Nebelflecke. *Astronomische Nachrichten*, 206(13):109, March 1918.

[187] C. Wirtz. Einiges zur Statistik der Radialbewegungen von Spiralnebeln und Kugelsternhaufen. *Astronomische Nachrichten*, 215:349, April 1922.

[188] K. Lundmark. The Determination of the Curvature of Space-time in de Sitter's World. *Monthly Notices of the Royal Astronomical Society*, 84:747–770, June 1924.

[189] W. Killing. Ueber die Grundlagen der Geometrie. *Journal fï die reine und angewandte Mathematik*, 1892(109):121–186, 1892.

[190] R. M. Wald. *General Relativity*. University of Chicago Press, 1984.

[191] S. Carrol. *An Introduction to General Relativity Spacetime and Geometry*. Addison Wesley, 2004.

[192] F. Melia. Cosmological Redshift in Friedmann-Robertson-Walker Metrics with Constant Space-time Curvature. *Monthly Notices of the Royal Astronomical Society*, 422(2):1418–1424, May 2012.

[193] M. Mizony and M. Lachièze-Rey. Cosmological Effects in the Local Static Frame. *Astronomy and Astrophysics*, 434(1):45–52, April 2005.

[194] A. G. Riess, S. Casertano, W. Yuan, L. M. Macri, and D. Scolnic. Large Magellanic Cloud Cepheid Standards Provide a 1% Foundation for the Determination of the Hubble Constant and Stronger Evidence for Physics beyond ΛCDM. *The Astrophysical Journal*, 876(1):85, May 2019.

[195] F. Melia. Tantalizing New Physics from the Cosmic Purview. *Modern Physics Letters A*, 34(26):1930004–30, August 2019.

[196] G. Kauffmann and M. Haehnelt. A Unified Model for the Evolution of Galaxies and Quasars. *Monthly Notices of the Royal Astronomical Society*, 311(3):576–588, January 2000.

[197] X. Fan. Evolution of High-redshift Quasars. *New Astronomy Reviews*, 50(9-10):665–671, November 2006.

[198] A. Wandel, B. M. Peterson, and M. A. Malkan. Central Masses and Broad-Line Region Sizes of Active Galactic Nuclei. I. Comparing the Photoionization and Reverberation Techniques. *The Astrophysical Journal*, 526(2):579–591, December 1999.

[199] M. Vestergaard and P. S. Osmer. Mass Functions of the Active Black Holes in Distant Quasars from the Large Bright Quasar Survey, the Bright Quasar Survey, and the Color-selected Sample of the SDSS Fall Equatorial Stripe. *The Astrophysical Journal*, 699(1):800–816, July 2009.

[200] D. P. Schneider, P. B. Hall, G. T. Richards, D. E. Vanden Berk, S. F. Anderson, X. Fan, S. Jester, C. Stoughton, M. A. Strauss, M. SubbaRao, et al. The Sloan Digital Sky Survey Quasar Catalog. III. Third Data Release. *The Astronomical Journal*, 130(2):367–380, August 2005.

[201] C. J. Willott, L. Albert, D. Arzoumanian, J. Bergeron, D. Crampton, P. Delorme, J. B. Hutchings, A. Omont, C. Reylé, and D. Schade. Eddington-limited Accretion and the Black Hole Mass Function at Redshift 6. *The Astronomical Journal*, 140(2):546–560, August 2010.

[202] Y. Shen, J. E. Greene, M. A. Strauss, G. T. Richards, and D. P. Schneider. Biases in Virial Black Hole Masses: An SDSS Perspective. *The Astrophysical Journal*, 680(1):169–190, June 2008.

[203] G. T. Richards, M. Lacy, L. J. Storrie-Lombardi, P. B. Hall, S. C. Gallagher, D. C. Hines, X. Fan, C. Papovich, D. E. Vanden Berk, G. B. Trammell, et al. Spectral Energy Distributions and Multiwavelength Selection of Type 1 Quasars. *The Astrophysical Journal Supplement Series*, 166(2):470–497, October 2006.

[204] F. Melia. *High-energy Astrophysics*. Princeton University Press, 2009.

[205] E. E. Salpeter. Accretion of Interstellar Matter by Massive Objects. *The Astrophysical Journal*, 140:796–800, August 1964.

[206] S. Zaroubi. The Epoch of Reionization. In Tommy Wiklind, Bahram Mobasher, and Volker Bromm, editors, *The First Galaxies*, volume 396 of *Astrophysics and Space Science Library*, page 45, 2013.

[207] V. Bromm and R. B. Larson. The First Stars. *Annual Reviews of Astronomy and Astrophysics*, 42(1):79–118, September 2004.

[208] N. Yoshida, V. Bromm, and L. Hernquist. The Era of Massive Population III Stars: Cosmological Implications and Self-termination. *The Astrophysical Journa*, 605(2):579–590, April 2004.

[209] J. L. Johnson, T. H. Greif, and V. Bromm. Local Radiative Feedback in the Formation of the First Protogalaxies. *The Astrophysical Journal*, 665(1):85–95, August 2007.

[210] E. Bañados, B. P. Venemans, C. Mazzucchelli, E. P. Farina, F. Walter, F. Wang, R. Decarli, D. Stern, X. Fan, F. B. Davies, et al. An 800-million-solar-mass Black Hole in a Significantly Neutral Universe at a Redshift of 7.5. *Nature*, 553(7689):473–476, January 2018.

[211] F. Melia. J1342+0928 Supports the Timeline in the $R_h = ct$ Cosmology. *Astronomy and Astrophysics*, 615:A113, July 2018.

[212] F. Melia and T. M. McClintock. Supermassive Black Holes in the Early Universe. *Proceedings of the Royal Society of London Series A*, 471(2184):20150449, December 2015.

[213] X.-B. Wu, F. Wang, X. Fan, W. Yi, W. Zuo, F. Bian, L. Jiang, I. D. McGreer, R. Wang, J. Yang, et al. An Ultraluminous Quasar with a Twelve-billion-solar-mass Black Hole at Redshift 6.30. *Nature*, 518(7540):512–515, February 2015.

[214] M. Volonteri and M. J. Rees. Rapid Growth of High-redshift Black Holes. *The Astrophysical Journal*, 633(2):624–629, November 2005.

[215] F. Pacucci, M. Volonteri, and A. Ferrara. The Growth Efficiency of High-redshift Black Holes. *Monthly Notices of the Royal Astronomical Society*, 452(2):1922–1933, September 2015.

[216] K. Inayoshi, Z. Haiman, and J. P. Ostriker. Hyper-Eddington Accretion Flows on to Massive Black Holes. *Monthly Notices of the Royal Astronomical Society*, 459(4):3738–3755, July 2016.

[217] J. Yoo and J. Miralda-Escudé. Formation of the Black Holes in the Highest Redshift Quasars. *The Astrophysical Journal Letters*, 614(1):L25–L28, October 2004.

[218] M. A. Latif, D.R.G. Schleicher, W. Schmidt, and J. Niemeyer. Black Hole Formation in the Early Universe. *Monthly Notices of the Royal Astronomical Society*, 433(2):1607–1618, August 2013.

[219] T. Tanaka and Z. Haiman. The Assembly of Supermassive Black Holes at High Redshifts. *The Astrophysical Journal*, 696(2):1798–1822, May 2009.

[220] Z. Lippai, Z. Frei, and Z. Haiman. On the Occupation Fraction of Seed Black Holes in High-redshift Dark Matter Halos. *The Astrophysical Journal*, 701(1):360–368, August 2009.

[221] M. Hirschmann, S. Khochfar, A. Burkert, T. Naab, S. Genel, and R. S. Somerville. On the Evolution of the Intrinsic Scatter in Black Hole Versus Galaxy Mass Relations. *Monthly Notices of the Royal Astronomical Society*, 407(2):1016–1032, September 2010.

[222] D. J. Mortlock, S. J. Warren, B. P. Venemans, M. Patel, P. C. Hewett, R. G. McMahon, C. Simpson, T. Theuns, E. A. Gonzáles-Solares, A. Adamson, et al. A Luminous Quasar at a Redshift of z = 7.085. *Nature*, 474(7353):616–619, June 2011.

[223] G. De Rosa, R. Decarli, F. Walter, X. Fan, L. Jiang, J. Kurk, A. Pasquali, and H. W. Rix. Evidence for Non-evolving Fe II/Mg II Ratios in Rapidly Accreting z~6 QSOs. *The Astrophysical Journal*, 739(2):56, October 2011.

[224] M. Fatuzzo and F. Melia. Unseen Progenitors of Luminous High-z Quasars in the R_h=ct Universe. *The Astrophysical Journal*, 846(2):129, September 2017.

[225] B. M. Peterson, M. C. Bentz, Louis-Benoit Desroches, Alexei V. Filippenko, Luis C. Ho, Shai Kaspi, Ari Laor, Dan Maoz, Edward C. Moran, Richard W. Pogge, et al. Multiwavelength Monitoring of the Dwarf Seyfert 1 Galaxy NGC 4395. I. A Reverberation-based Measurement of the Black Hole Mass. *The Astrophysical Journal*, 632(2):799–808, October 2005.

[226] T. J. Maccarone, A. Kundu, S. E. Zepf, and K. L. Rhode. A Black Hole in a Globular Cluster. *Nature*, 445(7124):183–185, January 2007.

[227] H. Baumgardt, J. Makino, P. Hut, S. McMillan, and S. Portegies Zwart. A Dynamical Model for the Globular Cluster G1. *The Astrophysical Journal Letters*, 589(1):L25–L28, May 2003.

[228] LIGO Scientific Collaboration, Virgo Collaboration, B. P. Abbott, R. Abbott, T. D. Abbott, F. Acernese, K. Ackley, C. Adams, T. Adams, P. Addesso, R. X. Adhikari, V. B. Adya, et al. GW170814: A Three-Detector Observation of Gravitational Waves from a Binary Black Hole Coalescence. *Physical Review Letters*, 119(14):141101, October 2017.

[229] J. L. Johnson, C. Dalla Vecchia, and S. Khochfar. The First Billion Years Project: the Impact of Stellar Radiation on the Co-evolution of Populations II and III. *Monthly Notices of the Royal Astronomical Society*, 428(3):1857–1872, January 2013.

[230] R. J. Bouwens, P. A. Oesch, G. D. Illingworth, I. Labbé, P. G. van Dokkum, G. Brammer, D. Magee, L. R. Spitler, M. Franx, R. Smit, et al. Photometric Constraints on the Redshift of z~10 Candidate UDFj-39546284 from Deeper WFC3/IR+ACS+IRAC Observations over the HUDF. *The Astrophysical Journal Letters*, 765(1):L16, March 2013.

[231] R. S. Ellis, R. J. McLure, J. S. Dunlop, B. E. Robertson, Y. Ono, M. A. Schenker, A. Koekemoer, R.A.A. Bowler, M. Ouchi, A. B. Rogers, et al. The Abundance of Star-forming Galaxies in the Redshift Range 8.5 − 12: New Results from the 2012 Hubble Ultra Deep Field Campaign. *The Astrophysical Journal Letters*, 763(1):L7, January 2013.

[232] D. Coe, A. Zitrin, M. Carrasco, X. Shu, W. Zheng, M. Postman, L. Bradley, A. Koekemoer, R. Bouwens, T. Broadhurst, et al. CLASH: Three Strongly Lensed Images of a Candidate z ≈ 11 Galaxy. *The Astrophysical Journal*, 762(1):32, January 2013.

[233] P. A. Oesch, R. J. Bouwens, G. D. Illingworth, I. Labbé, M. Franx, P. G. van Dokkum, M. Trenti, M. Stiavelli, V. Gonzalez, and D. Magee. Probing the Dawn of Galaxies at z~9-12: New Constraints from HUDF12/XDF and CANDELS data. *The Astrophysical Journal*, 773(1):75, August 2013.

[234] J. H. Wise and T. Abel. Resolving the Formation of Protogalaxies. I. Virialization. *The Astrophysical Journal*, 665(2):899–910, August 2007.

[235] J. Jaacks, K. Nagamine, and J. H. Choi. Duty Cycle and the Increasing Star Formation History of z ≥ 6 Galaxies. *Monthly Notices of the Royal Astronomical Society*, 427(1):403–414, November 2012.

[236] M. Postman, D. Coe, H. Ford, A. Riess, W. Zheng, M. Donahue, L. Moustakas, and CLASH Team. CLASH: Cluster Lensing and Supernova survey with Hubble. In *American Astronomical Society Meeting Abstracts #217*, volume 217 of *American Astronomical Society Meeting Abstracts*, page 227.06, January 2011.

[237] W. Zheng, M. Postman, A. Zitrin, J. Moustakas, X. Shu, S. Jouvel, O. Høst, A. Molino, L. Bradley, D. Coe, et al. A Magnified Young Galaxy from About 500 Million Years After the Big Bang. *Nature*, 489(7416):406–408, September 2012.

[238] F. Melia. The Premature Formation of High-redshift Galaxies. *The Astronomical Journal*, 147(5):120, May 2014.

[239] G. C. McVittie. Gravitational Collapse to a Small Volume. *The Astrophysical Journal*, 140:401, August 1964.

[240] I. H. Thompson and G. J. Whitrow. Time-dependent Internal Solutions for Spherically Symmetrical Bodies in General Relativity. I, Adiabatic Collapse. *Monthly Notices of the Royal Astronomical Society*, 136:207, January 1967.

[241] B. C. Nolan. A Point Mass in an Isotropic Universe: Existence, Uniqueness, and Basic Properties. *Physical Review D*, 58(6):064006, September 1998.

[242] B. C. Nolan. A Point Mass in an Isotropic Universe: III. The Region R ≤ 2M. *Classical and Quantum Gravity*, 16(10):3183–3191, October 1999.

[243] N. Kaloper, M. Kleban, and D. Martin. McVittie's Legacy: Black Holes in an Expanding Universe. *Physical Review D*, 81(10):104044, May 2010.

[244] R. Moradi, C. Stahl, J. T. Firouzjaee, and S.-S. Xue. Charged Cosmological Black Hole. *Physical Review D*, 96(10):104007, November 2017.

[245] K. Lake and M. Abdelqader. More on McVittie's Legacy: A Schwarzschild-de Sitter Black and White Hole Embedded in an Asymptotically ΛCDM Cosmology. *Physical Review D*, 84(4):044045, August 2011.

[246] C. L. Bennett, D. Larson, J. L. Weiland, N. Jarosik, G. Hinshaw, N. Odegard, K. M. Smith, R. S. Hill, B. Gold, M. Halpern, et al. Nine-year Wilkinson Microwave Anisotropy Probe (WMAP) Observations: Final Maps and Results. *The Astrophysical Journal Supplement Series*, 208(2):20, October 2013.

[247] C. J. Copi, D. Huterer, D. J. Schwarz, and G. D. Starkman. Lack of Large-angle TT Correlations Persists in WMAP and Planck. *Monthly Notices of the Royal Astronomical Society*, 451(3):2978–2985, August 2015.

[248] J. Kim and P. Naselsky. Lack of Angular Correlation and Odd-parity Preference in Cosmic Microwave Background Data. *The Astrophysical Journal*, 739(2):79, October 2011.

[249] F. Melia. Angular Correlation of the Cosmic Microwave Background in the R_h=ct Universe. *Astronomy and Astrophysics*, 561:A80, January 2014.

[250] A. Gruppuso, N. Kitazawa, N. Mandolesi, P. Natoli, and A. Sagnotti. Pre-inflationary Relics in the CMB? *Physics of the Dark Universe*, 11:68–73, March 2016.

[251] C. L. Bennett, R. S. Hill, G. Hinshaw, D. Larson, K. M. Smith, J. Dunkley, B. Gold, M. Halpern, N. Jarosik, A. Kogut, et al. Seven-year Wilkinson Microwave Anisotropy Probe (WMAP) Observations: Are There Cosmic Microwave Background Anomalies? *The Astrophysical Journal Supplement Series*, 192(2):17, February 2011.

[252] G. Efstathiou. A Maximum Likelihood Analysis of the Low Cosmic Microwave Background Multipoles from the Wilkinson Microwave Anisotropy Probe. *Monthly Notices of the Royal Astronomical Society*, 348(3):885–896, March 2004.

[253] C. R. Contaldi, M. Peloso, L. Kofman, and A. Linde. Suppressing the Lower Multipoles in the CMB Anisotropies. *Journal of Cosmology and Astroparticle Physics*, 2003(7):002, July 2003.

[254] C. Destri, H. J. de Vega, and N. G. Sanchez. Preinflationary and Inflationary Fast-roll Eras and their Signatures in the Low CMB Multipoles. *Physical Review D*, 81(6):063520, March 2010.

[255] A. Niarchou, A. H. Jaffe, and L. Pogosian. Large-scale Power in the CMB and New Physics: An Analysis Using Bayesian Model Comparison. *Physical Review D*, 69(6):063515, March 2004.

[256] F. Melia and M. López-Corredoira. Evidence of a Truncated Spectrum in the Angular Correlation Function of the Cosmic Microwave Background. *Astronomy and Astrophysics*, 610:A87, March 2018.

[257] J. R. Bond and G. Efstathiou. Cosmic Background Radiation Anisotropies in Universes Dominated by Nonbaryonic Dark Matter. *The Astrophysical Journal Letters*, 285:L45–L48, October 1984.

[258] W. Hu and N. Sugiyama. Anisotropies in the Cosmic Microwave Background: an Analytic Approach. *The Astrophysical Journal*, 444:489, May 1995.

[259] J. Liu and F. Melia. Viability of Slow-roll Inflation in the Light of the Non-zero k_{min} Measured in the CMB Power Spectrum. *Proceedings of the Royal Society A*, in press (e-print: arXiv:2006.02510), 2020.

[260] J. L. Cook and L. M. Krauss. Large Slow Roll Parameters in Single Field Inflation. *Journal of Cosmology and Astroparticle Physics*, 2016(3):028, March 2016.

[261] E. Ramirez. Low Power on Large Scales in Just-enough Inflation Models. *Physical Review D*, 85(10):103517, May 2012.

[262] W. J. Handley, S. D. Brechet, A. N. Lasenby, and M. P. Hobson. Kinetic Initial Conditions for Inflation. *Physical Review D*, 89(6):063505, March 2014.

[263] A. Scacco and A. Albrecht. Transients in Finite Inflation. *Physical Review D*, 92(8):083506, October 2015.

[264] M. A. Santos, M. Benetti, J. S. Alcaniz, F. A. Brito, and R. Silva. CMB Constraints on β-exponential Inflationary Models. *Journal of Cosmology and Astroparticle Physics*, 2018(3):023, March 2018.

[265] P. Fileviez Pérez, H. H. Patel, M. J. Ramsey-Musolf, and K. Wang. Triplet Scalars and Dark Matter at the LHC. *Physical Review D*, 79(5):055024, March 2009.

[266] Ya. B. Zel'Dovich, I. Yu. Kobzarev, and L. B. Okun'. Cosmological Consequences of a Spontaneous Breakdown of a Discrete Symmetry. *Soviet Journal of Experimental and Theoretical Physics*, 40:1, July 1975.

[267] T.W.B. Kibble. Topology of Cosmic Domains and Strings. *Journal of Physics A Mathematical General*, 9(8):1387–1398, August 1976.

[268] A. Vilenkin and E.P.S. Shellard. *Cosmic Strings and Other Topological Defects.* Cambridge University Press, 1994.

[269] A. Lazanu, C.J.A.P. Martins, and E.P.S. Shellard. Contribution of Domain Wall Networks to the CMB Power Spectrum. *Physics Letters B*, 747:426–432, July 2015.

[270] L. Sousa and P. P. Avelino. Cosmic Microwave Background Anisotropies Generated by Domain Wall Networks. *Physical Review D*, 92(8):083520, October 2015.

[271] L. Randall and S. Thomas. Solving the Cosmological Moduli Problem with Weak Scale Inflation. *Nuclear Physics B*, 449:229–247, February 1995.

[272] D. H. Lyth and E. D. Stewart. Thermal Inflation and the Moduli Problem. *Physical Review D*, 53(4):1784–1798, February 1996.

[273] L. Randall, M. Soljacic, and A. H. Guth. Supernatural Inflation: Inflation from Supersymmetry with No (Very) Small Parameters. *Nuclear Physics B*, 472(1-2):377–405, February 1996.

[274] G. Germán, G. Ross, and S. Sarkar. Low-scale Inflation. *Nuclear Physics B*, 608(1-2):423–450, August 2001.

[275] H. Davoudiasl, D. Hooper, and S. D. McDermott. Inflatable Dark Matter. *Physical Review Letters*, 116(3):031303, January 2016.

[276] F. Melia. A Solution to the Electroweak Horizon Problem in the R_h=ct Universe. *European Physical Journal C*, 78(9):739, September 2018.

[277] J. F. Donoghue, K. Dutta, A. Ross, and M. Tegmark. Likely Values of the Higgs Vacuum Expectation Value. *Physical Review D*, 81(7):073003, April 2010.

[278] V. Agrawal, S. M. Barr, John F. Donoghue, and D. Seckel. Viable Range of the Mass Scale of the Standard Model. *Physical Review D*, 57(9):5480–5492, May 1998.

[279] C. J. Hogan. Why the Universe is Just So. *Reviews of Modern Physics*, 72(4):1149–1161, October 2000.

[280] T. Damour and J. F. Donoghue. Constraints on the Variability of Quark Masses from Nuclear Binding. *Physical Review D*, 78(1):014014, July 2008.

[281] N. Arkani-Hamed, S. Dimopoulos, and G. Dvali. The Hierarchy Problem and New Dimensions at a Millimeter. *Physics Letters B*, 429(3-4):263–272, June 1998.

[282] T. T. Takeuchi. Application of the Information Criterion to the Estimation of Galaxy Luminosity Function. *Astrophysics and Space Science*, 271:213, January 2000.

[283] A. R. Liddle. Information Criteria for Astrophysical Model Selection. *Monthly Notices of the Royal Astronomical Society*, 377(1):L74–L78, May 2007.

[284] M.Y.J. Tan and R. Biswas. The Reliability of the Akaike Information Criterion Method in Cosmological Model Selection. *Monthly Notices of the Royal Astronomical Society*, 419(4):3292–3303, February 2012.

[285] H. Akaike. A New Look at the Statistical Model Identification. *IEEE Transactions on Automatic Control*, 19(6):716–723, 1974.

[286] K. P. Burnham and D. R. Anderson. *Model Selection and Multimodel Inference.* Springer-Verlag, 2002.

[287] H. Yanagihara and C. Ohmoto. On Distribution of AIC in Linear Regression Models. *Journal of Statistical Planning and Inference*, 133:417–433, 2005.

[288] J. E. Cavanaugh. A Large-sample Model Selection Criterion Based on Kullback's Symmetric Divergence. *Statistics and Probability Letters*, 42(4):333–343, 1999.

[289] G. Schwarz. Estimating the Dimension of a Model. *Annals of Statistics*, 6(2):461–464, July 1978.

[290] A. R. Liddle. How Many Cosmological Parameters? *Monthly Notices of the Royal Astronomical Society*, 351(3):L49–L53, July 2004.

[291] K. Shi, Y. F. Huang, and T. Lu. A Comprehensive Comparison of Cosmological Models from the Latest Observational Data. *Monthly Notices of the Royal Astronomical Society*, 426(3):2452–2462, November 2012.

[292] J. Kuha. AIC and BIC: Comparisons of Assumptions and Performance. *Sociological Methods and Research*, 33(2):188–229, 2004.

[293] III Gott, J. R., M. S. Vogeley, S. Podariu, and B. Ratra. Median Statistics, H_0, and the Accelerating Universe. *The Astrophysical Journal*, 549(1):1–17, March 2001.

[294] K. Leaf and F. Melia. Analysing H(z) Data Using Two-point Diagnostics. *Monthly Notices of the Royal Astronomical Society*, 470(2):2320–2327, September 2017.

[295] R. Jimenez and A. Loeb. Constraining Cosmological Parameters Based on Relative Galaxy Ages. *The Astrophysical Journal*, 573(1):37–42, July 2002.

[296] J. Dunlop, J. Peacock, H. Spinrad, A. Dey, R. Jimenez, D. Stern, and R. Windhorst. A 3.5-Gyr-old Galaxy at Redshift 1.55. *Nature*, 381(6583):581–584, June 1996.

[297] H. Spinrad, A. Dey, D. Stern, J. Dunlop, J. Peacock, R. Jimenez, and R. Windhorst. LBDS 53W091: An Old, Red Galaxy at z = 1.552. *The Astrophysical Journal*, 484(2):581–601, July 1997.

[298] B. Panter, R. Jimenez, A. F. Heavens, and S. Charlot. The Star Formation Histories of Galaxies in the Sloan Digital Sky Survey. *Monthly Notices of the Royal Astronomical Society*, 378(4):1550–1564, July 2007.

[299] D. Thomas, C. Maraston, R. Bender, and C. Mendes de Oliveira. The Epochs of Early-Type Galaxy Formation as a Function of Environment. *The Astrophysical Journal*, 621(2):673–694, March 2005.

[300] T. Treu, R. S. Ellis, T. X. Liao, P. G. van Dokkum, P. Tozzi, A. Coil, J. Newman, M. C. Cooper, and M. Davis. The Assembly History of Field Spheroidals: Evolution of Mass-to-Light Ratios and Signatures of Recent Star Formation. *The Astrophysical Journal*, 633(1):174–197, November 2005.

[301] D. Stern, R. Jimenez, L. Verde, S. A. Stanford, and M. Kamionkowski. Cosmic Chronometers: Constraining the Equation of State of Dark Energy. II. A Spectroscopic Catalog of Red Galaxies in Galaxy Clusters. *The Astrophysical Journal Supplement Series*, 188(1):280–289, May 2010.

[302] M. Moresco, A. Cimatti, R. Jimenez, L. Pozzetti, G. Zamorani, M. Bolzonella, J. Dunlop, F. Lamareille, M. Mignoli, H. Pearce, et al. Improved Constraints on the Expansion Rate of the Universe up to z~1.1 from the Spectroscopic Evolution of Cosmic Chronometers. *Journal of Cosmology and Astroparticle Physics*, 2012(8):006, August 2012.

[303] M. Moresco. Raising the Bar: New Constraints on the Hubble Parameter with Cosmic Chronometers at z~2. *Monthly Notices of the Royal Astronomical Society Letters*, 450:L16–L20, June 2015.

[304] F. Melia and T. M. McClintock. A Test of Cosmological Models Using High-z Measurements of H(z). *The Astronomical Journal*, 150(4):119, October 2015.

[305] C. Blake, S. Brough, M. Colless, C. Contreras, W. Couch, S. Croom, D. Croton, T. M. Davis, M. J. Drinkwater, K. Forster, et al. The WiggleZ Dark Energy Survey: Joint Measurements of the Expansion and Growth History at z<1. *Monthly Notices of the Royal Astronomical Society*, 425(1):405–414, September 2012.

[306] D. O. Jones, S. A. Rodney, A. G. Riess, B. Mobasher, T. Dahlen, C. McCully, T. F. Frederiksen, S. Casertano, J. Hjorth, C. R. Keeton, et al. The Discovery of the Most Distant Known Type Ia Supernova at Redshift 1.914. *The Astrophysical Journal*, 768(2):166, May 2013.

[307] G. Risaliti and E. Lusso. Cosmological Constraints from the Hubble Diagram of Quasars at High Redshifts. *Nature Astronomy*, 3:272–277, January 2019.

[308] Y. Avni and H. Tananbaum. X-Ray Properties of Optically Selected QSOs. *The Astrophysical Journal*, 305:83, June 1986.

[309] G. Risaliti and E. Lusso. A Hubble Diagram for Quasars. *The Astrophysical Journal*, 815(1):33, December 2015.

[310] S. R. Rosen, N. A. Webb, M. G. Watson, J. Ballet, D. Barret, V. Braito, F. J. Carrera, M. T. Ceballos, M. Coriat, R. Della Ceca, et al. The XMM-Newton Serendipitous Survey. VII. The Third XMM-Newton Serendipitous Source Catalogue. *Astronomy and Astrophysics*, 590:A1, May 2016.

[311] Y. Shen, G. T. Richards, M. A. Strauss, P. B. Hall, D. P. Schneider, S. Snedden, D. Bizyaev, H. Brewington, V. Malanushenko, E. Malanushenko, et al. A Catalog of Quasar Properties from Sloan Digital Sky Survey Data Release 7. *The Astrophysics Journal Supplement Series*, 194(2):45, June 2011.

[312] I. Pâris, P. Petitjean, N. P. Ross, A. D. Myers, É. Aubourg, A. Streblyanska, S. Bailey, É. Armengaud, N. Palanque-Delabrouille, C. Yèche, et al. The Sloan Digital Sky Survey Quasar Catalog: Twelfth Data Release. *Astronomy and Astrophysics*, 597:A79, January 2017.

[313] D. W. Just, W. N. Brandt, O. Shemmer, A. T. Steffen, D. P. Schneider, G. Chartas, and G. P. Garmire. The X-Ray Properties of the Most Luminous Quasars from the Sloan Digital Sky Survey. *The Astrophysical Journal*, 665(2):1004–1022, August 2007.

[314] M. Young, M. Elvis, and G. Risaliti. The X-ray Energy Dependence of the Relation Between Optical and X-ray Emission in Quasars. *The Astrophysical Journal*, 708(2):1388–1397, January 2010.

[315] F. Melia. Cosmological Test Using the Hubble Diagram of High-z Quasars. *Monthly Notcies of the Royal Astronomical Society*, 489(1):517–523, October 2019.

[316] F. Melia. The Cosmic Equation of State. *Astrophysics and Space Science*, 356(2):393–398, April 2015.

[317] D. Branch and J. C. Wheeler. *Supernova Explosions*. Springer-Verlag, 2017.

[318] K. Maguire. Type Ia Supernovae. In A. Alsabti and P. Murdin, editors, *Handbook of Supernovae*, pages 1–24, October 2016.

[319] S. Taubenberger. The Extremes of Thermonuclear Supernovae. In Athem W. Alsabti and Paul Murdin, editors, *Handbook of Supernovae*, pages 1–24, 2017.

[320] M. M. Phillips. The Absolute Magnitudes of Type Ia Supernovae. *The Astrophysical Journal Letters*, 413:L105, August 1993.

[321] S. Dhawan, S. W. Jha, and B. Leibundgut. Measuring the Hubble Constant with Type Ia Supernovae as Near-infrared Standard Candles. *Astronomy and Astrophysics*, 609:A72, January 2018.

[322] P. E. Nugent, M. Sullivan, S. B. Cenko, R. C. Thomas, D. Kasen, D. A. Howell, D. Bersier, J. S. Bloom, S. R. Kulkarni, M. T. Kandrashoff, et al. Supernova SN 2011fe from an Exploding Carbon-oxygen White Dwarf Star. *Nature*, 480(7377):344–347, December 2011.

[323] S. W. Jha. Type Iax Supernovae. In Athem W. Alsabti and Paul Murdin, editors, *Handbook of Supernovae*, page 375, 2017.

[324] M. Betoule, R. Kessler, J. Guy, J. Mosher, D. Hardin, R. Biswas, P. Astier, P. El-Hage, M. Konig, S. Kuhlmann, et al. Improved Cosmological Constraints from a Joint Analysis of the SDSS-II and SNLS Supernova Samples. *Astronomy and Astrophysics*, 568:A22, August 2014.

[325] DES Collaboration, T.M.C. Abbott, S. Allam, P. Andersen, C. Angus, J. Asorey, A. Avelino, S. Avila, B. A. Bassett, K. Bechtol, G. M. Bernstein, et al. First Cosmology Results using Type Ia Supernovae from the Dark Energy Survey: Constraints on Cosmological Parameters. *The Astrophysical Journal Letters*, 872(2):L30, February 2019.

[326] J. Guy, M. Sullivan, A. Conley, N. Regnault, P. Astier, C. Balland, S. Basa, R. G. Carlberg, D. Fouchez, D. Hardin, et al. The Supernova Legacy Survey 3-year Sample: Type Ia Supernovae Photometric Distances and Cosmological Constraints. *Astronomy and Astrophysics*, 523:A7, November 2010.

[327] A. Conley, M. Sullivan, E. Y. Hsiao, J. Guy, P. Astier, D. Balam, C. Balland, S. Basa, R. G. Carlberg, D. Fouchez, et al. SiFTO: An Empirical Method for Fitting SN Ia Light Curves. *The Astrophysical Journal*, 681(1):482–498, July 2008.

[328] M. Kowalski, D. Rubin, G. Aldering, R. J. Agostinho, A. Amadon, R. Amanullah, C. Balland , K. Barbary, G. Blanc, P. J. Challis, et al. Improved Cosmological Constraints from New, Old, and Combined Supernova Data Sets. *The Astrophysical Journal*, 686(2):749–778, October 2008.

[329] The Supernova Cosmology Project, N. Suzuki, D. Rubin, C. Lidman, G. Aldering, R. Amanullah, K. Barbary, L. F. Barrientos, J. Botyanszki, M. Brodwin, N. Connolly, et al. The Hubble Space Telescope Cluster Supernova Survey. V.

Improving the Dark-energy Constraints above z > 1 and Building an Early-type-hosted Supernova Sample. *The Astrophysical Journal*, 746(1):85, February 2012.

[330] F. Melia. Fitting the Union2.1 Supernova Sample with the R_h=ct Universe. *The Astronomical Journal*, 144(4):110, October 2012.

[331] A. G. Kim. Type Ia Supernova Intrinsic Magnitude Dispersion and the Fitting of Cosmological Parameters. *Publications of the Astronomical Society of the Pacific*, 123(900):230, February 2011.

[332] J.-J. Wei, X.-F. Wu, F. Melia, and R. S. Maier. A Comparative Analysis of the Supernova Legacy Survey Sample With ΛCDM and the R_h=ct Universe. *The Astronomical Journal*, 149(3):102, March 2015.

[333] G. D'Agostini. Fits, and Especially Linear Fits, with Errors on Both Axes, Extra Variance of the Data Points and Other Complications. *arXiv e-prints*, page physics/0511182, November 2005.

[334] M. Sako, B. Bassett, A. C. Becker, P. J. Brown, H. Campbell, R. Wolf, D. Cinabro, C. B. D'Andrea, K. S. Dawson, F. DeJongh, et al. The Data Release of the Sloan Digital Sky Survey-II Supernova Survey. *Publications of the Astronomical Society of the Pacific*, 130(988):064002, June 2018.

[335] J. Guy, P. Astier, S. Baumont, D. Hardin, R. Pain, N. Regnault, S. Basa, R. G. Carlberg, A. Conley, S. Fabbro, et al. SALT2: Using Distant Supernovae to Improve the use of Type Ia Supernovae as Distance Indicators. *Astronomy and Astrophysics*, 466(1):11–21, April 2007.

[336] F. Melia, J. J. Wei, R. S. Maier, and X. F. Wu. Cosmological Tests with the Joint Lightcurve Analysis. *EPL (Europhysics Letters)*, 123(5):59002, September 2018.

[337] I.M.H. Etherington. Republication of: LX. On the Definition of Distance in General Relativity. *General Relativity and Gravitation*, 39(7):1055–1067, July 2007.

[338] B. A. Bassett and M. Kunz. Cosmic Distance-duality as a Probe of Exotic Physics and Acceleration. *Physical Review D*, 69(10):101305, May 2004.

[339] S. L. Adler. Photon Splitting and Photon Dispersion in a Strong Magnetic Field. *Annals of Physics*, 67:599–647, January 1971.

[340] J.-P. Uzan, N. Aghanim, and Y. Mellier. Distance Duality Relation from X-ray and Sunyaev-Zel'dovich Observations of Clusters. *Physical Review D*, 70(8):083533, October 2004.

[341] G.F.R. Ellis, R. Poltis, J.-P. Uzan, and A. Weltman. Blackness of the Cosmic Microwave Background Spectrum as a Probe of the Distance-duality Delation. *Physical Review D*, 87(10):103530, May 2013.

[342] R.F.L. Holanda, J.A.S. Lima, and M. B. Ribeiro. Testing the Distance-Duality Relation with Galaxy Clusters and Type Ia Supernovae. *The Astrophysical Journal Letters*, 722(2):L233–L237, October 2010.

[343] F. Melia. A Comparison of the R_h = ct and ΛCDM Cosmologies Using the Cosmic Distance Duality Relation. *Monthly Notices of the Royal Astronomical Society*, 481(4):4855–4862, December 2018.

[344] R. D. Blandford and A. Königl. Relativistic Jets as Compact Radio Sources. *The Astrophysical Journal*, 232:34–48, August 1979.

[345] L. I. Gurvits, K. I. Kellermann, and S. Frey. The "Angular Size - Redshift" Relation for Compact Radio Structures in Quasars and Radio Galaxies. *Astronomy and Astrophysics*, 342:378–388, February 1999.

[346] S. Cao, M. Biesiada, X. Zheng, and Z.-H. Zhu. Exploring the Properties of Milliarcsecond Radio Sources. *The Astrophysical Journal*, 806(1):66, June 2015.

[347] S. Cao, X. Zheng, M. Biesiada, J. Qi, Y. Chen, and Z.-H. Zhu. Ultra-compact Structure in Intermediate-luminosity Radio Quasars: Building a Sample of Standard Cosmological Rulers and Improving the Dark Energy Constraints up to z~3. *Astronomy and Astrophysics*, 606:A15, September 2017.

[348] F. Melia and M. K. Yennapureddy. The Maximum Angular-diameter Distance in Cosmology. *Monthly Notices of the Royal Astronomical Society*, 480(2):2144–2152, October 2018.

[349] F. Hoyle. The Relation of Radio Astronomy to Cosmology. In Ronald N. Bracewell, editor, *URSI Symp. 1: Paris Symposium on Radio Astronomy*, volume 9 of *IAU Symposium*, page 529, January 1959.

[350] K. Nilsson, M. J. Valtonen, J. Kotilainen, and T. Jaakkola. On the Redshift–Apparent Size Diagram of Double Radio Sources. *The Astrophysical Journal*, 413:453, August 1993.

[351] A. Buchalter, D. J. Helfand, R. H. Becker, and R. L. White. Constraining Ω_0 with the Angular Size-Redshift Relation of Double-lobed Quasars in the FIRST Survey. *The Astrophysical Journal*, 494(2):503–522, February 1998.

[352] K. I. Kellermann and I.I.K. Pauliny-Toth. Compact Radio Sources. *Annual Reviews of Astronomy and Astrophysics*, 19:373–410, January 1981.

[353] J. C. Jackson. Is There a Standard Measuring Rod in the Universe? *Monthly Notices of the Royal Astronomical Society*, 390(1):L1–L5, October 2008.

[354] F. Melia. Model Selection Based on the Angular-diameter Distance to the Compact Structure in Radio Quasars. *EPL (Europhysics Letters)*, 123(3):39001, August 2018.

[355] M. Seikel, C. Clarkson, and M. Smith. Reconstruction of Dark Energy and Expansion Dynamics Using Gaussian Processes. *Journal of Cosmology and Astroparticle Physics*, 2012(6):036, June 2012.

[356] C. Rasmussen and C. Williams. *Gaussian Processes for Machine Learning*. MIT Press, 2006.

[357] T. Holsclaw, U. Alam, B. Sansó, H. Lee, K. Heitmann, S. Habib, and D. Higdon. Nonparametric Dark Energy Reconstruction from Supernova Data. *Physical Review Letters*, 105(24):241302, December 2010.

[358] ALMA Partnership, C. Vlahakis, T. R. Hunter, J. A. Hodge, L. M. Pérez, P. Andreani, C. L. Brogan, P. Cox, S. Martin, M. Zwaan, S. Matsushita, et al. The 2014 ALMA Long Baseline Campaign: Observations of the Strongly Lensed Submillimeter Galaxy HATLAS J090311.6+003906 at z = 3.042. *The Astrophysical Journal Letters*, 808(1):L4, July 2015.

[359] C. S. Kochanek, E. E. Falco, C. D. Impey, J. Lehár, B. A. McLeod, H. W. Rix, C. R. Keeton, J. A. Muñoz, and C. Y. Peng. The Fundamental Plane of Gravitational Lens Galaxies and The Evolution of Early-Type Galaxies in Low-Density Environments. *The Astrophysical Journal*, 543(1):131–148, November 2000.

[360] R. Emden. *Gaskugeln*. Teubner, 1907.

[361] A. Einstein. Lens-Like Action of a Star by the Deviation of Light in the Gravitational Field. *Science*, 84(2188):506–507, December 1936.

[362] T. Treu and L.V.E. Koopmans. Massive Dark Matter Halos and Evolution of Early-Type Galaxies to z~1. *The Astrophysical Journal*, 611(2):739–760, August 2004.

[363] M. W. Auger, T. Treu, A. S. Bolton, R. Gavazzi, L.V.E. Koopmans, P. J. Marshall, K. Bundy, and L. A. Moustakas. The Sloan Lens ACS Survey. IX. Colors, Lensing, and Stellar Masses of Early-Type Galaxies. *The Astrophysical Journal*, 705(2):1099–1115, November 2009.

[364] Y. Shu, J. R. Brownstein, A. S. Bolton, L.V.E. Koopmans, T. Treu, A. D. Montero-Dorta, M. W. Auger, O. Czoske, R. Gavazzi, P. J. Marshall, et al. The Sloan Lens ACS Survey. XIII. Discovery of 40 New Galaxy-scale Strong Lenses. *The Astrophysical Journal*, 851(1):48, December 2017.

[365] DES Collaboration, H. T. Diehl, E. J. Buckley-Geer, K. A. Lindgren, B. Nord, H. Gaitsch, S. Gaitsch, H. Lin, S. Allam, T. E. Collett, C. Furlanetto, et al. The DES Bright Arcs Survey: Hundreds of Candidate Strongly Lensed Galaxy Systems from the Dark Energy Survey Science Verification and Year 1 Observations. *The Astrophysical Journal Supplement Series*, 232(1):15, September 2017.

[366] I. Jorgensen, M. Franx, and P. Kjaergaard. Spectroscopy for E and S0 Galaxies in Nine Clusters. *Monthly Notices of the Royal Astronomical Society*, 276(4):1341–1364, October 1995.

[367] S. Cao, M. Biesiada, R. Gavazzi, A. Piórkowska, and Z.-H. Zhu. Cosmology with Strong-lensing Systems. *The Astrophysical Journal*, 806(2):185, June 2015.

[368] K. Leaf and F. Melia. Model Selection with Strong-lensing Systems. *Monthly Notices of the Royal Astronomical Society*, 478(4):5104–5111, August 2018.

[369] C. Grillo, M. Lombardi, and G. Bertin. Cosmological Parameters from Strong Gravitational Lensing and Stellar Dynamics in Elliptical Galaxies. *Astronomy and Astrophysics*, 477(2):397–406, January 2008.

[370] A. Kosowsky, M. Milosavljevic, and R. Jimenez. Efficient Cosmological Parameter Estimation from Microwave Background Anisotropies. *Physical Review D*, 66(6):063007, September 2002.

[371] V. F. Mukhanov and G. V. Chibisov. Quantum Fluctuations and a Nonsingular Universe. *Soviet Journal of Experimental and Theoretical Physics Letters*, 33:532, May 1981.

[372] A. H. Guth and S. Y. Pi. Fluctuations in the New Inflationary Universe. *Physical Review Letters*, 49(15):1110–1113, October 1982.

[373] S. W. Hawking. The Development of Irregularities in a Single Bubble Inflationary Universe. *Physics Letters B*, 115(4):295–297, September 1982.

[374] A. A. Starobinsky. Dynamics of Phase Transition in the New Inflationary Universe Scenario and Generation of Perturbations. *Physics Letters B*, 117(3-4):175–178, November 1982.

[375] SDSS Collaboration, M. Tegmark, M. R. Blanton, M. A. Strauss, F. Hoyle, D. Schlegel, R. Scoccimarro, M. S. Vogeley, D. H. Weinberg, I. Zehavi, A. Berlind, et al. The Three-Dimensional Power Spectrum of Galaxies from the Sloan Digital Sky Survey. *The Astrophysical Journal*, 606(2):702–740, May 2004.

[376] Y. Sofue, Y. Tutui, M. Honma, A. Tomita, T. Takamiya, J. Koda, and Y. Takeda. Central Rotation Curves of Spiral Galaxies. *The Astrophysical Journal*, 523(1):136–146, September 1999.

[377] F. Combes. Properties of Dark Matter Haloes. *New Astronomy Reviews*, 46(12):755–766, November 2002.

[378] J. P. Kneib, R. S. Ellis, I. Smail, W. J. Couch, and R. M. Sharples. Hubble Space Telescope Observations of the Lensing Cluster Abell 2218. *The Astrophysical Journal*, 471:643, November 1996.

[379] F. Hammer, I. M. Gioia, E. J. Shaya, P. Teyssand ier, O. Le Fèvre, and G. A. Luppino. Detailed Lensing Properties of the MS 2137-2353 Core and Reconstruction of Sources from Hubble Space Telescope Imagery. *The Astrophysical Journal*, 491(2):477–482, December 1997.

[380] I. M. Gioia, E. J. Shaya, O. Le Fèvre, E. E. Falco, G. A. Luppino, and F. Hammer. The Lensing Cluster MS 0440+0204 Seen by HST, ROSAT, and ASCA. I. Cluster Properties. *The Astrophysical Journal*, 497(2):573–586, April 1998.

[381] J. A. Tyson, G. P. Kochanski, and I. P. Dell'Antonio. Detailed Mass Map of CL 0024+1654 from Strong Lensing. *The Astrophysical Journal Letters*, 498(2):L107–L110, May 1998.

[382] K. Freese. Status of Dark Matter in the Universe. *International Journal of Modern Physics D*, 26(6):1730012–223, January 2017.

[383] W. H. Press and P. Schechter. Formation of Galaxies and Clusters of Galaxies by Self-Similar Gravitational Condensation. *The Astrophysical Journal*, 187:425–438, February 1974.

[384] G. R. Bengochea, P. Cañate, and D. Sudarsky. Inhomogeneities from Quantum Collapse Scheme Without Inflation. *Physics Letters B*, 743:484–491, April 2015.

[385] S. Hollands and R. M. Wald. Essay: An Alternative to Inflation. *General Relativity and Gravitation*, 34(12):2043–2055, December 2002.

[386] A. Perez, H. Sahlmann, and D. Sudarsky. On the Quantum Origin of the Seeds of Cosmic Structure. *Classical and Quantum Gravity*, 23(7):2317–2354, April 2006.

[387] J. Khoury, B. A. Ovrut, P. J. Steinhardt, and N. Turok. Ekpyrotic Universe: Colliding Branes and the Origin of the Hot Big Bang. *Physical Review D*, 64(12):123522, December 2001.

[388] P. J. Steinhardt and N. Turok. A Cyclic Model of the Universe. *Science*, 296(5572):1436–1439, May 2002.

[389] L. D. Landau and E. M. Lifshitz. *The Classical Theory of Fields*. Pergamon Press, 1975.

[390] P.J.E. Peebles. *The Large-scale Structure of the Universe*. Princeton University Press, 1980.

[391] W. H. Press and E. T. Vishniac. Tenacious Myths About Cosmological Perturbations Larger than the Horizon Size. *The Astrophysical Journal*, 239:1–11, July 1980.

[392] E. W. Kolb and M. Turner. *The Early Universe*. Addison-Wesley, 1990.

[393] T. Padmanabhan. *Structure Formation in the Universe*. Cambridge University Press, 1993.

[394] P.J.E. Peebles. *Principles of Physical Cosmology*. Princeton University Press, 1993.

[395] P. Coles and F. Lucchin. *Cosmology. The Origin and Evolution of Cosmic Structure*. Wiley, 1995.

[396] J. A. Peacock. *Cosmological Physics*. Cambridge University Press, 1999.

[397] A. R. Liddle and K. A. Lyth. *Cosmological Inflation and Large-scale Structure*. Cambridge University Press, 2000.

[398] C. G. Tsagas, A. Challinor, and R. Maartens. Relativistic Cosmology and Large-scale Structure. *Physics Reports*, 465(2-3):61–147, August 2008.

[399] L. F. Abbott and Mark B. Wise. Constraints on Generalized Inflationary Cosmologies. *Nuclear Physics B*, 244(2):541–548, October 1984.

[400] F. Lucchin and S. Matarrese. Power-law Inflation. *Physical Review D*, 32(6):1316–1322, September 1985.

[401] J. D. Barrow. Cosmic No-hair Theorems and Inflation. *Physics Letters B*, 187(1-2):12–16, March 1987.

[402] A. R. Liddle. Power-law Inflation with Exponential Potentials. *Physics Letters B*, 220(4):502–508, April 1989.

[403] R. Penrose. *The Road to Reality*. Vintage Books, 2004.

[404] V. F. Mukhanov. *Physical Foundations of Cosmology*. Cambridge University Press, 2005.

[405] S. Weinberg. *Cosmology*. Oxford University Press, 2008.

[406] D. H. Lyth and A. R. Liddle. *The Primordial Density Perturbations*. Cambridge University Press, 2009.

[407] N. Brouzakis, J. Rizos, and N. Tetradis. On the Dynamics of Classicalization. *Physics Letters B*, 708(1-2):170–173, February 2012.

[408] G. Dvali and C. Gomez. Ultra-high Energy Probes of Classicalization. *Journal of Cosmology and Astroparticle Physics*, 2012(7):015, July 2012.

[409] S. Dodelson. Coherent Phase Argument for Inflation. In José F. Nieves and Raymond R. Volkas, editors, *Neutrinos, Flavor Physics, and Precision Cosmology*, volume 689 of *American Institute of Physics Conference Series*, pages 184–196, October 2003.

[410] C.-P. Ma and E. Bertschinger. Cosmological Perturbation Theory in the Synchronous and Conformal Newtonian Gauges. *The Astrophysical Journal*, 455:7, December 1995.

[411] M. K. Yennapureddy and F. Melia. Structure Formation and the Matter Power-spectrum in the R_h=ct Universe. *Physics of the Dark Universe*, submitted, January 2020.

[412] E. Bertschinger. COSMICS: Cosmological Initial Conditions and Microwave Anisotropy Codes. *arXiv e-prints*, pages astro–ph/9506070, June 1995.

[413] F. Melia and M. Fatuzzo. The Epoch of Reionization in the R_h=ct Universe. *Monthly Notices of the Royal Astronomical Society*, 456(4):3422–3431, March 2016.

[414] M. Tegmark and M. Zaldarriaga. Separating the Early Universe from the Late Universe: Cosmological Parameter Estimation Beyond the Black Box. *Physical Review D*, 66(10):103508, November 2002.

[415] R.A.C. Croft, D. H. Weinberg, M. Pettini, L. Hernquist, and N. Katz. The Power Spectrum of Mass Fluctuations Measured from the Lyα Forest at Redshift z = 2.5. *The Astrophysical Journal*, 520(1):1–23, July 1999.

[416] V. K. Narayanan, D. N. Spergel, R. Davé, and C.-P. Ma. Constraints on the Mass of Warm Dark Matter Particles and the Shape of the Linear Power Spectrum from the Lyα Forest. *The Astrophysical Journal Letters*, 543(2):L103–L106, November 2000.

[417] M. White and R.A.C. Croft. Suppressing Linear Power on Dwarf Galaxy Halo Scales. *The Astrophysical Journal*, 539(2):497–504, August 2000.

[418] P. McDonald, J. Miralda-Escudé, M. Rauch, W.L.W. Sargent, T. A. Barlow, R. Cen, and J. P. Ostriker. The Observed Probability Distribution Function, Power Spectrum, and Correlation Function of the Transmitted Flux in the Lyα Forest. *The Astrophysical Journal*, 543(1):1–23, November 2000.

[419] M. Zaldarriaga, L. Hui, and M. Tegmark. Constraints from the Lyα Forest Power Spectrum. *The Astrophysical Journal*, 557(2):519–526, August 2001.

[420] N. Katz, D. H. Weinberg, and L. Hernquist. Cosmological Simulations with TreeSPH. *The Astrophysical Journal Supplement Series*, 105:19, July 1996.

[421] R. K. Sheth, H. J. Mo, and G. Tormen. Ellipsoidal Collapse and an Improved Model for the Number and Spatial Distribution of Dark Matter Haloes. *Monthly Notices of the Royal Astronomical Society*, 323(1):1–12, May 2001.

[422] M. Vogelsberger, S. Genel, V. Springel, P. Torrey, D. Sijacki, D. Xu, G. Snyder, D. Nelson, and L. Hernquist. Introducing the Illustris Project: Simulating the Coevolution of Dark and Visible Matter in the Universe. *Monthly Notices of the Royal Astronomical Society*, 444(2):1518–1547, October 2014.

[423] C. L. Steinhardt, P. Capak, D. Masters, and J. S. Speagle. The Impossibly Early Galaxy Problem. *The Astrophysical Journal*, 824(1):21, June 2016.

[424] H. Hildebrandt, J. Pielorz, T. Erben, L. van Waerbeke, P. Simon, and P. Capak. CARS: the CFHTLS-Archive-Research Survey. II. Weighing Dark Matter Halos of Lyman-break Galaxies at z = 3-5. *Astronomy and Astrophysics*, 498(3):725–736, May 2009.

[425] K.-S. Lee, H. C. Ferguson, T. Wiklind, T. Dahlen, M. E. Dickinson, M. Giavalisco, N. Grogin, C. Papovich, H. Messias, Y. Guo, and L. Lin. How Do Star-forming Galaxies at z>3 Assemble Their Masses? *The Astrophysical Journal*, 752(1):66, June 2012.

[426] O. Ilbert, H. J. McCracken, O. Le Fèvre, P. Capak, J. Dunlop, A. Karim, M. A. Renzini, K. Caputi, S. Boissier, S. Arnouts, et al. Mass Assembly in Quiescent and Star-forming Galaxies Since z~4 from UltraVISTA. *Astronomy and Astrophysics*, 556:A55, August 2013.

[427] S. L. Finkelstein, M. Song, P. Behroozi, R. S. Somerville, C. Papovich, M. Milosavljević, A. Dekel, D. Narayanan, M.L.N. Ashby, et al. An Increasing Stellar Baryon Fraction in Bright Galaxies at High Redshift. *The Astrophysical Journal*, 814(2):95, December 2015.

[428] S. G. Murray, C. Power, and A.S.G. Robotham. HMFcalc: An Online Tool for Calculating Dark Matter Halo Mass Functions. *Astronomy and Computing*, 3:23–34, November 2013.

[429] C. G. Lacey, C. M. Baugh, C. S. Frenk, and A. J. Benson. The Evolution of Lyman-break Galaxies in the Cold Dark Matter Model. *Monthly Notices of the Royal Astronomical Society*, 412(3):1828–1852, April 2011.

[430] K. Finlator, B. D. Oppenheimer, and R. Davé. Smoothly Rising Star Formation Histories During the Reionization Epoch. *Monthly Notices of the Royal Astronomical Society*, 410(3):1703–1724, January 2011.

[431] D. Waters, S. M. Wilkins, T. Di Matteo, Y. Feng, R. Croft, and D. Nagai. Monsters in the Dark: Predictions for Luminous Galaxies in the Early Universe from the BLUETIDES Simulation. *Monthly Notices of the Royal Astronomical Society*, 461(1):L51–L55, September 2016.

[432] P. S. Behroozi and J. Silk. A Simple Technique for Predicting High-redshift Galaxy Evolution. *The Astrophysical Journal*, 799(1):32, January 2015.

[433] N. Mashian, P. A. Oesch, and A. Loeb. An Empirical Model for the Galaxy Luminosity and Star Formation Rate Function at High Redshift. *Monthly Notices of the Royal Astronomical Society*, 455(2):2101–2109, January 2016.

[434] R. J. Bouwens, G. D. Illingworth, P. A. Oesch, M. Trenti, I. Labbé, L. Bradley, M. Carollo, P. G. van Dokkum, V. Gonzalez, B. Holwerda, et al. UV Luminosity

Functions at Redshifts z∼4 to z∼10: 10,000 Galaxies from HST Legacy Fields. *The Astrophysical Journal*, 803(1):34, April 2015.

[435] A. Sandage. The Change of Redshift and Apparent Luminosity of Galaxies due to the Deceleration of Selected Expanding Universes. *The Astrophysical Journal*, 136:319, September 1962.

[436] P.-S. Corasaniti, D. Huterer, and A. Melchiorri. Exploring the Dark Energy Redshift Desert with the Sandage-Loeb Test. *Physical Review D*, 75(6):062001, March 2007.

[437] C. Quercellini, L. Amendola, A. Balbi, P. Cabella, and M. Quartin. Real-time Cosmology. *Physics Reports*, 521(3):95–134, December 2012.

[438] C.J.A.P. Martins, M. Martinelli, E. Calabrese, and M. P. L. P. Ramos. Real-time Cosmography with Redshift Derivatives. *Physical Review D*, 94(4):043001, August 2016.

[439] A. Loeb. Direct Measurement of Cosmological Parameters from the Cosmic Deceleration of Extragalactic Objects. *The Astrophysical Journal Letters*, 499(2):L111–L114, June 1998.

[440] J. Liske, A. Grazian, E. Vanzella, M. Dessauges, M. Viel, L. Pasquini, M. Haehnelt, S. Cristiani, F. Pepe, G. Avila, et al. Cosmic Dynamics in the Era of Extremely Large Telescopes. *Monthly Notices of the Royal Astronomical Society*, 386(3):1192–1218, May 2008.

[441] J. Liske. Status of the European Extremely Large Telescope. In *Thirty Meter Telescope Science Forum*, page 52, July 2014. Available online at http://conference.ipac.caltech.edu/tmtsf2014/, id. 52

[442] H. R. Kloeckner, D. Obreschkow, C. Martins, A. Raccanelli, D. Champion, A. L. Roy, A. Lobanov, J. Wagner, and R. Keller. Real Time Cosmology - A Direct Measure of the Expansion Rate of the Universe with the SKA. In *Advancing Astrophysics with the Square Kilometre Array (AASKA14)*, page 27, April 2015. Available online at http://pos.sissa.it/cgi-bin/reader/conf.cgi?confid=215, id.27

[443] H.-R. Yu, T.-J. Zhang, and U.-L. Pen. Method for Direct Measurement of Cosmic Acceleration by 21-cm Absorption Systems. *Physical Review Letters*, 113(4):041303, July 2014.

[444] F. Melia. Definitive Test of the $R_h = ct$ Universe Using Redshift Drift. *Monthly Notices of the Royal Astronomical Society Letters*, 463(1):L61–L63, November 2016.

[445] DESI Collaboration, A. Aghamousa, J. Aguilar, S. Ahlen, S. Alam, L. E. Allen, C. Allende Prieto, J. Annis, S. Bailey, C. Balland, O. Ballester, et al. The DESI Experiment Part I: Science,Targeting, and Survey Design. *arXiv e-prints*, page arXiv:1611.00036, October 2016.

[446] J. A. Tyson. Large Synoptic Survey Telescope: Overview. In J. Anthony Tyson and Sidney Wolff, editors, *Proceedings of the International Society for Optical Engineering*, volume 4836 of *Society of Photo-Optical Instrumentation Engineers (SPIE) Conference Series*, pages 10–20, 2002.

[447] Euclid Collaboration, R. Laureijs, J. Amiaux, S. Arduini, J. L. Auguères, J. Brinchmann, R. Cole, M. Cropper, C. Dabin, L. Duvet, A. Ealet, et al. Euclid Definition Study Report. *arXiv e-prints*, page arXiv:1110.3193, October 2011.

[448] J. Green, P. Schechter, C. Baltay, R. Bean, D. Bennett, R. Brown, C. Conselice, M. Donahue, X. Fan, B. S. Gaudi, et al. Wide-Field InfraRed Survey Telescope (WFIRST) Final Report. *arXiv e-prints*, page arXiv:1208.4012, August 2012.

[449] BICEP2 Collaboration, P.A.R. Ade, R. W. Aikin, D. Barkats, S. J. Benton, C. A. Bischoff, J. J. Bock, J. A. Brevik, I. Buder, E. Bullock, C. D. Dowell, et al. Detection of B-Mode Polarization at Degree Angular Scales by BICEP2. *Physical Review Letters*, 112(24):241101, June 2014.

[450] R. Cowan. Gravitational Waves Discovery Now Officially Dead. *Nature News*, January 2015.

[451] N. Stebor, P. Ade, Y. Akiba, C. Aleman, K. Arnold, C. Baccigalupi, B. Barch, D. Barron, S. Beckman, A. Bender, et al. The Simons Array CMB Polarization Experiment. In *Proceedings of the International Society for Optical Engineering*, volume 9914 of *Society of Photo-Optical Instrumentation Engineers (SPIE) Conference Series*, page 99141H, 2016.

[452] M. P. van Haarlem, M. W. Wise, A. W. Gunst, G. Heald, J. P. McKean, J.W.T. Hessels, A. G. de Bruyn, R. Nijboer, J. Swinbank, R. Fallows, et al. LOFAR: The LOw-Frequency ARray. *Astronomy and Astrophysics*, 556:A2, August 2013.

[453] R. B. Wayth, S. J. Tingay, C. M. Trott, D. Emrich, M. Johnston-Hollitt, B. McKinley, B. M. Gaensler, A. P. Beardsley, T. Booler, B. Crosse, et al. The Phase II Murchison Widefield Array: Design Overview. *Publications of the Astronomical Society of Australia*, 35:33, November 2018.

[454] C. A. Blake, F. B. Abdalla, S. L. Bridle, and S. Rawlings. Cosmology with the SKA. *New Astronomy Reviews*, 48(11-12):1063–1077, December 2004.

[455] A. R. Parsons, A. Liu, J. E. Aguirre, Z. S. Ali, R. F. Bradley, C. L. Carilli, D. R. DeBoer, M. R. Dexter, N. E. Gugliucci, D. C. Jacobs, et al. New Limits on 21 cm Epoch of Reionization from PAPER-32 Consistent with an X-Ray Heated Intergalactic Medium at z = 7.7. *The Astrophysical Journal*, 788(2):106, June 2014.

[456] D. R. DeBoer, A. R. Parsons, J. E. Aguirre, P. Alexander, Z. S. Ali, A. P. Beardsley, G. Bernardi, J. D. Bowman, R. F. Bradley, C. L. Carilli, et al. Hydrogen Epoch of Reionization Array (HERA). *Publications of the Astronomical Society of the Pacific*, 129(974):045001, April 2017.

[457] R. Coppejans, S. Frey, D. Cseh, C. Müller, Z. Paragi, H. Falcke, K. É. Gabányi, L. I. Gurvits, T. An, and O. Titov. On the Nature of Bright Compact Radio Sources at z > 4.5. *Monthly Notices of the Royal Astronomical Society*, 463(3):3260–3275, December 2016.

[458] C. M. Urry and P. Padovani. Unified Schemes for Radio-Loud Active Galactic Nuclei. *Publications of the Astronomical Society of the Pacific*, 107:803, September 1995.

[459] F. Melia and A. Königl. The Radiative Deceleration of Relativistic Jets in Active Galactic Nuclei. *The Astrophysical Journal*, 340:162, May 1989.

[460] N. I. Shakura and R. A. Sunyaev. Black Holes in Binary Systems. Observational Appearance. *Astronomy and Astrophysics*, 500:33–51, June 1973.

[461] J. H. Krolik and M. C. Begelman. An X-ray Heated Wind in NGC 1068. *The Astrophysical Journal Letters*, 308:L55–L58, September 1986.

[462] K. Leaf and F. Melia. Cosmological Test Using the High-redshift Detection Rate of FSRQs with the Square Kilometre Array. *Monthly Notices of the Royal Astronomical Society*, 487(2):2030–2037, August 2019.

[463] G. Ghisellini and T. Sbarrato. Dark Bubbles Around High-redshift Radio-loud Active Galactic Nucleus. *Monthly Notices of the Royal Astronomical Society*, 461(1):L21–L25, September 2016.

[464] M. Ackermann, M. Ajello, A. Allafort, W. B. Atwood, L. Baldini, J. Ballet, G. Barbiellini, D. Bastieri, K. Bechtol, A. Belfiore, et al. The First Fermi-LAT Catalog of Sources above 10 GeV. *The Astrophysical Journal Supplement Series*, 209(2):34, December 2013.

[465] A. K. Weigel, K. Schawinski, E. Treister, C. M. Urry, M. Koss, and B. Trakhtenbrot. The Systematic Search for z ≳ 5 Active Galactic Nuclei in the Chandra Deep Field South. *Monthly Notices of the Royal Astronomical Society*, 448(4):3167–3195, April 2015.

[466] G. Pietrzyński, D. Graczyk, W. Gieren, I. B. Thompson, B. Pilecki, A. Udalski, I. Soszyński, S. Kozlowski, P. Konorski, K. Suchomska, et al. An Eclipsing-binary Distance to the Large Magellanic Cloud Accurate to Two Percent. *Nature*, 495(7439):76–79, March 2013.

[467] E.M.L. Humphreys, M. J. Reid, J. M. Moran, L. J. Greenhill, and A. L. Argon. Toward a New Geometric Distance to the Active Galaxy NGC 4258. III. Final Results and the Hubble Constant. *The Astrophysical Journal*, 775(1):13, September 2013.

[468] G. F. Benedict, B. E. McArthur, M. W. Feast, T. G. Barnes, T. E. Harrison, R. J. Patterson, J. W. Menzies, J. L. Bean, and W. L. Freedman. Hubble Space Telescope Fine Guidance Sensor Parallaxes of Galactic Cepheid Variable Stars: Period-Luminosity Relations. *The Astronomical Journal*, 133(4):1810–1827, April 2007.

[469] M. Kaplinghat, G. Steigman, I. Tkachev, and T. P. Walker. Observational Constraints on Power-law Cosmologies. *Physical Review D*, 59(4):043514, February 1999.

[470] M. Sethi, A. Batra, and D. Lohiya. Comment on "Observational Constraints on Power-law Cosmologies". *Physical Review D*, 60(10):108301, November 1999.

[471] M. Kaplinghat, G. Steigman, and T. P. Walker. Nucleosynthesis in Power-law Cosmologies. *Physical Review D*, 61(10):103507, May 2000.

[472] G. Singh and D. Lohiya. Inhomogeneous Nucleosynthesis in Linearly Coasting Cosmology. *Monthly Notices of the Royal Astronomical Society*, 473(1):14–19, January 2018.

[473] A. Batra, D. Lohiya, S. Mahajan, A. Mukherjee, and A. Ashtekar. Nucleosynthesis in a Universe with a Linearly Evolving Scale Factor. *International Journal of Modern Physics D*, 9(6):757–773, January 2000.

[474] A. Benoit-Lévy and G. Chardin. Introducing the Dirac-Milne Universe. *Astronomy and Astrophysics*, 537:A78, January 2012.

[475] V. Springel. E pur si Muove: Galilean-invariant Cosmological Hydrodynamical Simulations on a Moving Mesh. *Monthly Notices of the Royal Astronomical Society*, 401(2):791–851, January 2010.

Index

Printed in the United States